Insect Management
for Food Storage
and Processing

Dr. Dennis S. Hill
'Haydn House'
20 Saxby Avenue
Skegness
Lincs.

Dr Dennis S. Hill
Associate Professor
I.B.E.C.
Universiti Malaysia Sarawak
94300 Kota Samarahan
SARAWAK

Insect Management
for Food Storage
and Processing

Edited by

Fred J. Baur
(retired)

The Procter & Gamble Company
Cincinnati, Ohio

Dr. Dennis S. Hill
'Haydn House'
20 Saxby Avenue
Skegness
Lincs.

Published by the
American Association of Cereal Chemists
St. Paul, Minnesota

Cover photograph courtesy of Fred J. Baur

This book has been reproduced directly from typewritten copy
submitted in final form to the American Association of Cereal Chemists
by the editor of the volume. No editing or proofreading has been
done by the Association.

Library of Congress Catalog Card Number: 84-72814
International Standard Book Number: 0-913250-38-4

©1984 by the American Association of Cereal Chemists
 Second printing, 1985

Printed in the United States of America

The American Association of Cereal Chemists
3340 Pilot Knob Road
St. Paul, Minnesota 55121, USA

FOREWORD

The number of species of living things on earth is estimated at 4-10 million and higher, with insect species constituting by far the majority. The estimate of insect species runs as high as 6 million, with the number of described beetles alone exceeding 500,000. Science and history have yet to note the disappearance of a single insect species, which attests to their aptitude for survival.

It has been said by some entomologists that flour beetles are the "number one" insect pest to the food industry in the United States of America. The key basis for this belief is that population explosions are possible within a short period of time. If one assumes each female can lay 400 eggs and that 60% of these eggs live to maturity, then in a four-month period under favorable environmental conditions, a population of almost 2 million beetles can result. The German cockroach is capable of comparable reproductive efforts. It has been estimated that the insect population of the earth is one billion billion or 1,000,000,000,000,000,000. If the average weight of an insect is 2.5 mg, less than 0.0001 oz, the weight of the insect population exceeds man's by a factor of about ten.

The potentially large numbers of any given species and the large numbers of species with their attendant variability in behavior are but two reasons why insect control is difficult. A third factor is the size of insects. The insects of greatest risk to industry, the stored-products or infesting types, generally are less than one-seventh of an inch in length with some as small as one-tenth of an inch. This makes detection difficult, particularly in the ever present cracks and crevices of buildings and equipment.

This book results from the desire of the American Association of Cereal Chemists (AACC) and the editor to make available to consumer products industries a text that will provide readily available information on how to do a better job of avoiding insect problems and eliminating such problems when they arise. Insects are the largest visible pest load; industry can and should do a better job of control; and the literature lacks a publication which attempts to treat completely the need for information on avoidance, detection, and elimination of insects.

The AACC has long been a concerned, professional, scientific society. In 1950, it formed a committee to deal with sanitation, including cleanup and associated benefits such as insect control. In 1960 AACC assumed sponsorship of a check sample program which has been extremely important in training and certifying microanalytical entomologists for the running of F&E (foreign and extraneous) analyses, most of which are insect related. It is not surprising that AACC has this interest. The raw materials used by the food industry that are most attractive to insects are those produced by segments of the food industry active in the AACC. Also, many of the finished food products marketed by AACC corporate members are among the most attractive to insects. A partial listing of the primary market coverage of AACC membership will serve to illustrate these two points: flour and other grain mill products; dog, cat, and other pet food; prepared feed and feed ingredients; bread and other bakery products; malt; macaroni, spaghetti, vermicelli, and noodles; and other food preparations such as prepared baking mixes.

A more recent demonstration of AACC's interest in pest control has been the sponsorship at its annual meetings of symposia on insects. Eight of the papers presented in 1982 are included in this publication.

The editor brings to this effort a background of over 20 years of experience in sanitation, product protection, and quality assurance/Good Manufacturing Practices. This experience was with a company that had few problems, with products and raw materials highly attractive to insects and other pests. The reason for the good track record was that prevention – the key word – was practiced. However, because insects are ubiquitous, problems occasionally arose. As the editor's knowledge grew on how to handle these problems, his awareness grew about how little he understood insects and how difficult it was to find in the literature information that was of practical value for a given need. Consultation with "experts" continuously disclosed that some "facts" were available but had to be dug out.

The aim of this book, therefore, is to present in one volume information for the consumer products industries, with emphasis on foods, on how to avoid, control, and eliminate insects.

F. J. Baur
Cincinnati, Ohio
December 1983

ACKNOWLEDGMENTS

The initial expression of appreciation goes to Robert Davis, one of the contributors to this book. Bob, who heads the USDA Insect Laboratory in Savannah, freely gave of his scarce time for consulting, editing, and assisting in locating last minute replacements of authors.

I am grateful to all the experts who contributed to the book, taking time from their always busy schedules.

Thanks go to my previous and main career employer, Procter & Gamble Company. It provided me with the opportunity to learn and to become concerned about pest control, permitted me to accept this responsibility well prior to retirement, and supported the effort with people and facilities.

An additional expression of appreciation goes to the American Association of Cereal Chemists for its interest in this book, support indicative of its on-going dedication to helping industry and consumers.

Thanks are due to many others who have contributed to varying degrees, including Dick Patterson of USDA Gainesville, Fran O'Melia of Dow Chemical, the late Al Yerington of USDA Fresno, Bob Yeager of Rose Exterminator, Bobby Corrigan of Purdue University, and Bill Breidster, a previous colleague at Procter & Gamble.

Last, but far from least, affectionate appreciation goes to my wife, Elaine, an immaculate housekeeper, who patiently tolerated the mounds of manuscripts in our residence.

F. J. Baur
Cincinnati, Ohio
December 1983

CONTENTS

Insect Management for Food Storage and Processing

INTRODUCTORY BACKGROUND INFORMATION

Fred J. Baur

The Procter & Gamble Company
6090 Center Hill Road
Cincinnati, Ohio 45224[1]

This introduction consists of two parts; the first part mentions briefly the main reasons industry* is concerned with improved insect control, and the second gives some editorial comments relative to the contents of this book and the control of insects.

STIMULI FOR IMPROVED INSECT CONTROL

Economic

There are two interrelated reasons for food industry's concerns with insects. First and foremost is economic. In response to a recent question to the United States Department of Agriculture requesting information on dollar losses to finished food products from insects, the writer was told no data are available but losses to all pests in this country run 30 billion dollars annually. In the report of the Council of Environmental Quality on Integrated Pest Management by Dale R. Bottrell of December 1979, the preharvest losses of agricultural commodities to pests in the U.S. were said to run about 33% with postharvest loss another 6%. The 1983 ARS (Agriculture Research Service) Commodity and Quarantine Research Plan for Insects states that most of the US farm-produced foods during the marketing process (food assembled, graded, stored, processed, packaged, wholesaled, and retailed) "is subject to damage by insects and the loss is estimated to be about 9%, or 36 million tons of lost food each year." The General Accounting Office report of September 16, 1977 entitled "Food Waste: An Opportunity to Improve Resource Use" said that "about one-fifth of all food produced for human consumption is lost annually in the United States." The report lists losses for six categories—harvest, storage, transportation, processing, wholesale/retail, and institution/consumer. In three of these categories—storage, transportation, and institution/consumer—insects were listed as the greatest loss factors. Industry also bears the costs of dealing with consumer complaints, loss of sales, and the costs associated with recalls, seizures, prosecutions, and injunctions. In the absence of direct data the annual losses to the food industry from insects are judged to run in the low billions of dollars.** Further, the total sales of insecticides in 1980 have been stated (personal communication from the Environmental Protection

---

[1]Present address: 1545 Larry Avenue, Cincinnati, Ohio 45224
* NOTE: The need to control insects is not limited to the food industry.
** NOTE: Benefits of insects to agriculture also run in the low billions of dollars per year.

Agency) to be 5.8 billion dollars. A significant portion of this cost was borne by the food industry which, under proper integrated (holistic) pest management, could be reduced appreciably.

Regulatory Concern

The second stimulus for improved control which is closely entwined with economics is concern with regulatory compliance and possible regulatory action, principally by the Food and Drug Administration. Chapter 27 describes FDA's concerns.

The presence of insects or insect residues in a food raw material or product makes the product adulterated or illegal.*** This adulteration is described as filth, as listed under Section 402(a)3 of the Food and Drug Act. The presence of insects or their parts in food can be harmful to health, as is discussed in chapter 21, but, FDA has never to the writer's knowledge cited the appropriate poisonous or deleterious Section 402(a)2 in an insect enforcement action. And FDA says (personal communication) such a possibility is "unlikely." Another section of the Act which is frequently cited with insect problems is 402(a)4. This section says that a food is adulterated if it has been prepared, packed, or held under conditions whereby it may have become contaminated, or to paraphrase, where the infestation of the premises is such that product contamination is likely to occur unless corrective measures are taken promptly. The FDA investigator on noting excessive evidence of insects in a plant and/or a warehouse will seek to find products contaminated with insects, but such a finding is not required for regulatory action.

Regulatory Actions

One can keep track of the regulatory actions by FDA through perusal of the Weekly Enforcement Reports. They contain information on prosecution, seizures, injunctions, and recalls. The actions in the reports are explained as follows:

PROSECUTION: A criminal action filed by FDA against a company or individual charging violation of the law. Prosecutions listed have been filed with a court but not yet tried or concluded.

SEIZURE: An action taken to remove a product from commerce because it is in violation of the law. FDA initiates a seizure by filing a complaint with the U.S. District Court where the goods are located. A U.S. marshal is then directed by the court to take possession of the goods until the matter is resolved. The date listed is the date a seizure request is filed, not the date of seizure.

INJUNCTION: A civil action filed by FDA against an individual or company seeking, in most cases, to stop a company from continuing to manufacture or distribute products that are in violation of the law. Injunctions listed have been filed with the court but not concluded.

RECALL: Voluntary removal by a firm of a defective product from the market. Some recalls begin when the firm finds a problem, others are conducted at FDA's request. Recalls may involve the physical removal of products from the market or correction of the problem where the product is located.

The listings in the weekly reports are not considered "finalized" legal actions by FDA but are "recommended actions." The numbers, therefore, in the upcoming tables may not be accurate. The quarterly FDA reports, which are available from the Freedom of Information service, have accurate figures on seizures, recalls, etc., but no breakdown on contaminants or violations.

Data from 1981 and 1982 on foods, for all causes not just insects, are given in Table 1.

*** NOTE: Chapter 25 speaks to "tolerances" providing Good Manufacturing Practices are met.

Table 1. FDA Actions on Foods
Weekly Enforcement Reports, 1/7/81 through 12/29/82

	Number of Actions[1]	% of Visible Pests
For Insects	67	62
For Rodents	40	37
For Birds	1	1
For Microbes[2]	73	
For Leaking, Swollen, Defective Cans	56	
For Other Reasons[3]	181	
Total[4]	397	

[1]Recalls; Seizures; Injunctions; Prosecutions.
[2]Aflatoxins, molds, yeasts, decomposition.
[3]Filth, held under insanitary conditions, metal, glass, wood, plastic, foreign material, labeling, misbranded, mis- or over-formulation, product mix-up, undeclared nuts, chemicals − such as chemicals, ammonia, arsanilic not ascorbic acid, cleaning substances, colors, histamine (an excess), hydrogen sulfide, lead, mercury, penicillin, phosphates, saccharin, sanitizers and sodium hydroxide, sodium nitrite (in excess), xylene, unsafe food additives, unprescribed or unsafe color additive, not approved food additive or gras (generally recognized as safe), deficient in nutrients, short fill, short weight, component left out, no apple juice, too much water, poisonous or deleterious substance, unidentified deleterious substance, medicated feed mixed with nonmedicated, lubricating oil, not sterile, diseased animals, toxic berry, off odor, machine oil smell, chip or crack at top of jar.
[4]A few actions involve more than one pest problem (most commonly rodents and insects); hence, total is less than sum of specific problems.

This information shows that insects were the number one visible pest risk, as reported by FDA, in 1981 and 1982.* In the years immediately previous to 1981, rodents were responsible for as many if not more actions than insects. Why has this change occurred? Have the control needs changed? Has FDA shifted, intentionally or otherwise, its attention? Has the insect load increased? Or is it a combination of these and possibly other factors? The answer is immaterial. Insects are the largest visible pest risk and are likely to remain such. Industry needs to recognize this fact and react in a proactive manner. (The information on contaminants other than visible pests was included for general interest.)

It was noted above that there has been a recent trend toward insects as the number one pest on which FDA has taken regulatory action. A breakdown of the data into the two separate years, 1981 vs. 1982, is given in Table 2.

These data show that there was an overall decrease in total actions, actions resulting from pests, both visible and invisible, and actions from "other" reasons, but an increase in problems with canned goods resulting from processing difficulties.** The percentage of insects as causes of the total visible pest problems remained the same, 62%, with the ratio of insects to rodents actions remaining about the same or 1.6 to 1.7:1.0.

 * NOTE: Insects were the number one visible pest for the past three and one-half calendar years. In the Weekly Enforcement Reports for 1983 there were 26 instances involving insects and 22 involving rodents. In the first one-half of 1984 the numbers were 10 and 6, respectively.
** NOTE: Total actions for 1983 were 190; for the first one-half of 1984, 122. Microbes continued to be the number one reported problem.

Table 2. FDA Actions on Foods
Weekly Enforcement Reports, Comparison of 1981 vs. 1982

Number of Actions	1981	1982
Total	207	190
For Insects	38 (62)*	29 (62)*
For Rodents	23 (38)	17 (36)
For Birds	0	1 (2)
For Microbes	39	34
For Leaking, Swollen, Defective Cans	19	37
For Other Reasons	103	78

*Numbers in parentheses show percentage of visible pests. For 1983, the percentages were 54 and 46, and for the first half of 1984, 62 and 38, for insects and rodents, respectively.

Table 3 summarizes the description of insect involvement as cited in the enforcement actions. Note that the wording is variable. No significance should be placed on wording differences such as insect infested or insect contamination insofar as the magnitude of the problem or contamination is concerned. The main reason for those differences is that the wording in the enforcement actions, and their handling, will vary between FDA districts.

Table 3 also includes mention of the five actions, out of 67, in which specific species comments were made.

Table 3. Insect Actions
FDA Weekly Enforcement Reports, 1981-82, Description of Insect Involvement

Cited More Than One Time
    Contains insects, insect larvae
    Live insects
    Insects
    Insect infestation
    Presence of insects
    Insect cast skins, contamination, excreta, filth, fragments, infested parts
    Held under insanitary conditions
    Processed under insanitary conditions

Cited One Time
    Active insect infestation in raw materials
    Insect activity, adulteration, damage, damaged seed, infestation in plant,
        infested almonds, larval cast skins, particles, pellets
    Live insect infestation
    Manufactured from insect infested ingredients
    May be adulterated with insect filth
    Nesting material
    Presence of insect eggs
    Worm infestation

Comments on Species
    Indianmeal larvae (two distinct actions)
    Contains mites [not insects]
    Fruit fly (insect filth)
    Weevil equivalent and fragments

Table 4. Products/Commodities Involved in Insect Actions,
1/7/81 through 12/29/82

Recalls
    Flour   3
    Brandy sauce   2
    Bread   2
    Mole in pasta (a condiment)   2
    Peanut Butter   2
    Sesame seed   2
    Others--almonds, cake mix, cheese, cheese flavored
        snack food, chilies, cookies, date-walnut-raisin
        ingredient, diet bars, fruit cake, hot sauce,
        maraschino cherries, mixed nuts, noodles, rice,
        root beer, shelled walnuts, snack bars, yogurt.

Prosecutions
    Stored food   2
    Holding food   1

Table 5. Products/Commodities Involved in Insect Actions,
1/7/81 through 12/29/82

Seizures
    Stored foods, various foods   5
    Mixes   4
    Flour   3
    Macaroni   3
    Mole in pasta   3
    Rice   3
    Cocoa powder   2
    Texturized vegetable protein   2
    Others--blackeyes, coffee whitener, corn meal, dried
        garlic bulbs, durum wheat, maraschino cherries,
        raw peanuts, sunflower seeds, yogurt

Complaint for Injunction
    Candy   1

Tables 4 and 5 show what commodities were involved in the actions taken by FDA in 1981 and 1982.

Measurement of Insect Fragments in Foods

The Foreword mentioned briefly AACC's sponsorship of the check sample series on analytical methods relative to the microscopic detection and measurement of insect fragments in foods. FDA is involved in this effort via the Association of Official Analytical Chemists. The methods investigation and validation take place under the category of "Extraneous Materials in Food & Drugs." Of the 25 subtopics or specific assignments (associate refereeships for 1983-84), 20 deal with insects. Examples are "Insect Excreta in Flour," "Mites in Stored Foods," and "Soluble Insect and Other Animal Filth." The project leaders of these and 12 of the other insect topics and the overall supervisor (general referee) are FDA and HPB (Health Protection Branch - Canada) scientists, thirteen and two, respectively.

A Third Stimulus

The two main stimuli for industry's concern with insects are economic and regulatory compliance needs. It would be improper, however, if a third reason for concern was not mentioned. It is that maintenance of an insect-free plant or warehouse on a sustained basis is, for practical purposes, impossible. Insects will enter any and every plant. Also, once an infestation has occurred, complete eradication is most difficult given the usual building design, construction, and maintenance. Great reliance, therefore, must be placed on prevention or avoidance of problems.

INTRODUCTORY COMMENTS BY THE EDITOR

This book gives information on how to better handle insect control problems and needs. The book covers in varying depths essentially all aspects of insects and their control including regulatory concerns and requirements. Some of the chapters are appropriately review efforts recognizing the lack of such coverage in the literature.

Five Key Steps for Control

There are probably five key steps in dealing with insects in a plant or warehouse. In chronological order these are:

1. Have an inspection or surveillance program or system for the facility which will yield prompt awareness of a possible problem;
2. Determine the extent and nature of the possible problem--what species, how many, and where;
3. Devise a plan for control (or elimination) of the problem. Make use of your own basic knowledge and information, this book and/or other literature, a quality pest control operator (PCO), consultant experts, etc., whatever combination is required;
4. Implement the devised plan being willing to modify it as indicated; and
5. Monitor the results of the effort.

Key Aspects of Insect Control

The above steps are covered in Table 6. This table gives the reader a quick and easy reference to many of the key chapters and what the pertinent pages are. As expected, the next or second chapter on "Insect Pest Management and Control" dominates. Also as an aid to the reader, the authors were encouraged to refer to other chapters in their texts.

Two items in Table 6 deserve separate mention here as they are not discussed elsewhere in the book.

The first deals with the use of checklists. Mr. Gentry, in the third chapter, on inspections, gives the most comprehensive list of "where to looks" the writer has ever seen. You are strongly encouraged to use this list and to develop others as aids to avoid missing an important aspect and to build a firmer base for your sanitation or quality assurance/good manufacturing practices efforts. As an example of another type of checklist, Table 7 is offered.

The two stimuli for the development of Table 7 were: 1) most of the insect problems encountered by the writer had their origins with raw materials, particularly packaging materials, and 2) these problems too often disappeared as

Table 6. Key Aspects of Insect Control

Prevention/Avoidance	Chapter(s)	Page(s)
1. Keep facility clean ("sanitation")	2, 8, 9, 10	18, 90, 92, 123, 141
2. Don't bring the insects into the plant		
--check conveyances	2	18, 22
--check raw materials including packaging supplies	2, 9, 13, 14, 23	18, 21, 22, 122, 173, 181-192, 312
--maintain building tightness	2, 8, 9	18, 19, 95, 123
3. Have clearly defined inspection program for early detection	2, 4, 9	22, 34-42, 128
--use checklist(s) of where to look	4	40-42
--use traps as monitoring devices	7, 8, 16	70-86 (particularly 80-82), 91, 213, 214
Presence Noted - Assessment of "Problem"		
1. Species definitely identified	2, 5, 8	21, 22, 44-50, 97
2. Determine behavior/habits - think like an insect	5, 6, 9	44, 45, 47, 52-67, 123, 124, 128
3. Determine source(s) or origin(s) of "infestation" (see also "Don't bring the insects into the plant," above)	7, 8	80, 81, 97
Plan for Control ("Elimination")		
1. Define magnitude of "problem" - species, numbers, locations	7, 8	80, 81, 97
2. Select control treatment(s) based on	assorted chapters	
--finished product(s)	10, 20, 26	137, 270, 271, 337, 338, 342, 343
--potential for contamination or product stream risk	10, 11, 18, 26	137, 147, 150, 251, 337, 338, 342, 343
--physical facilities	11, 12, 13	152-154, 162-170, 173-178
--risk to employees	10, 11, 12, 13, 14, 22	137, 158, 168, 169, 175, 176, 190, 295-307
--costs (control measures, possible loss of production, etc.)	18, 20	250, 273

a matter of time as opposed to being eliminated by an expended effort. In the continuing war with insects it is important that attempts to learn be made simultaneously with the control or elimination of any problems. The last chapter, "Illustrations of Insect Problems and Solutions - Case Histories," gives examples of such learning.

The second item in Table 6 that deserves special mention here deals with "think like an insect." The need to think like a pest to properly control it is true for any of the visible pests. Behavior of pests is so atypical of humans yet so vital for proper control that one needs to put oneself in the pest's place. In the bird control area this is sometimes referred to as becoming a "bird brain."

Table 7. Insect Contamination "Checklist"--Raw Material
Factors to Consider

Environment - Geography
Time of Year (temperatures/humidities)
Plant/Warehouse - Control Program
    Tightness
    Clean-up
    Other storage/attractants
    Insecticides
    Electrocutors
Handling of Reshippers (if used)
    Age
    Cleanliness
    Humidity/temperature of storage
    Adhesives used
Shipment (particularly rail)
    Duration of shipment
    Car condition
Outside Warehouse (if involved)
    Environment
    Time of year
    Length of storage
    Control program
For any problems: document codes, times,
    carriers, species, numbers, live, dead,
    where found, stages, webbing, etc.

Explanatory/Background Comments

Some background comments for explanatory and reinforcement purposes need to be made. They are:

1. The title of the book speaks of food storage and processing. Should industries other than foods be concerned? The answer is a resounding YES! Insects are found and cause problems in non-food plants and non-food products. Just one example is paper products. The possibility of termites and other cellulose digesting insects infesting paper products should be well known. Other "environmental" type insects such as psocids (book lice) contaminate paper products. Even flour beetles have been found in paper products.

    It has been this writer's observation that non-food plants downplay concern for pest control. It is true that visible pests are generally less important, but they are still important. All plants are at some risk, particularly to insects. Some non-food plants use edible raw materials like dextrin as an adhesive. All plants have employees some of whom may even be permitted to eat on the job in processing areas. Since the appetites of insects are not large, most plants have ample edible materials on hand to sustain a significant population. Even airborne dust usually contains a high percentage of edible materials. Apart from food, another basic need of insects is housing or a sanctuary. Sanctuary is provided in every plant by darkness furnished by cracks/crevices/voids and packaging materials. In addition to darkness, most insects like warmth and moisture, both of which are available in plants.

Every plant, therefore, is susceptible to plant if not product contamination by insects.

2. Many of the authors in their chapters speak only about stored-product insects. Are these the only insects of concern? The answer is a resounding NO! Environmental insects (USDA generally refers to as household) particularly will get into containers such as bottles and cans. A single house fly in a processing area for the manufacture of a sterile drug may be one too many.

   Even though an author may not speak of environmental insects, keep them in mind and be aware that the messages and the information presented are of value in the control of environmental insects. Their biology is sufficiently similar to the stored-product types that are discussed.

3. What is FDA's attitude relative to stored-product or infesting insects compared to the environmental types such as fleas, roaches, termites, psocids, etc.? It varies with the species, product(s), and the plant. As one would guess, in the food area, FDA's concern is greater for the stored-product types, because they are infesting and because their presence can indicate insanitary conditions (402(a)4). "Contamination" from environmental species is generally considered as incidental in nature, except for insects of strong filth connotation such as flies and cockroaches. Keep in mind that any insect species in a product is filth; all can carry microorganisms, and stored-product types can be in the environment should there be an appropriate crop or an industry, like a grain elevator, nearby.

4. How many insects are too many? This question is sometimes asked, particularly in reference to an internal portion of a plant. The answer can only be made on an individual case-by-case basis. FDA and any reputable consultant will say that it varies with the species but more with the situation - particularly with the plant, its products, and its operation. No two situations, no two problems are ever identical.

5. Clean-up. Nothing should really need to be said about the importance of clean-up but it is so basic, so important. This importance has been strengthened by the Environmental Protection Agency's (EPA) recent actions against ethylene dibromide (EDB) as a permitted material.

6. The use of insect traps needs mentioning. Pest control programs and concerned industry have tended to ignore the usefulness of the food attractant, sticky or adhesive, and now pheromone traps for monitoring species and levels of insects and to even exert some control. This has been unwise. The past tense is appropriate since I understand from Dr. Wendell Burkholder that the use of traps was catching on in 1983, better late than never. The writer has also learned that traps are being used in the boots of elevators, another step to combat the loss of EDB.

7. Detection of halide also merits a comment. You may note that the authors do not agree on the sensitivity of the halide leak detector. This is understandable since the measurement is not exact. One's eyesight, the available light, and the gradations between blue and green are the main variables. Dow Chemical (personal communication) stated that the detection level is in the range of 25-50 ppm. Perhaps exceptional eyesight, at night, and with good color discernment one could detect 15 ppm. I suggest you use the 25-50 ppm range.

8.   Where can a "sanitarian" go for help? This has already been touched
     on. Go to the literature, or to your favorite PCO, or consultant, or to
     the National Pest Control Association (as Mr. Robert Yeager mentions).
     To these suggestions are added: a) send staff to meetings - Purdue Pest
     Control Conferences, Lauhoff Grain GMPs Seminars, seminars by pest
     control companies, seminars by FDA, seminars by the Environmental
     Management Association (EMA), and to AACC meetings - the most recent one
     being an "Integrated Pest Management Workshop" in cooperation with the
     EPA, FDA, and USDA in early 1984. Also consider membership in the
     Environmental Management Association. This is the only professional
     society devoted solely to sanitation. It has a Food Sanitation
     Institute. The key points, however, are that the association membership
     is small aiding one-on-one interchange, and the style of the association
     and its members is to help one another on problems of mutual concern,
     interest, or knowledge. Proper sanitation is one. Lastly, do not
     forget about the USDA entomologists. The USDA insect laboratories (see
     chapter 16) contain many of the expert entomologists in the United
     States, and they are very public service oriented professionals.

9.   Have checklists. This recommendation has already been made. A
     cautionary note - make them as complete as is practical but do not "etch
     them in stone or you could get burned"--always be thinking and flexible.

10. Try to document problems. This is also repetitious but importantly so.
     Document and be willing to share as some have done in this book.
     Insects are our biggest visible pest enemies; we need to combine forces
     to combat them.

11. One last piece of advice - undergirding any efforts to control insects
     or pests - is "prevention." The adage, an ounce of prevention is worth
     a pound of cure, was never truer. Prevention needs to be effected by
     --keeping the insects out of the plant or warehouse to whatever extent
        possible,
     --keeping the facility clean since some insects will get in, and
     --keeping good records of the on-going inspection and the control
        program in dealing with the probable presence of insects.

12. You will note that Indianmeal moth when mentioned in the text of this
     book is spelled with two words as opposed to the prevalent three (Indian
     meal moth) as seen in the literature references. This change recognizes
     the adoption of the former spelling by the Entomological Society of
     America as listed in its most recent edition, 1982, of "Common Names of
     Insects and Related Organisms."

13. Be aware that some chapters were completed prior to EPA's recent actions
     against the use of ethylene dibromide (EDB). See chapter 13, "Spot
     Fumigation," for an up-to-date report on the status of EDB.

14. Immediately preceding each chapter are brief comments by the editor on
     the chapter varying from abstracting the contents to value judgments.

SELECTED[1] GLOSSARY

aggregation   both sexes attracted, males and females group together in
                            large numbers

---

[1] Selected, reflecting words used in this book and to identify insect fragments
frequently reported as part of F&E.

antenna	a pair of head appendages, varies greatly in form (shape) and number of segments
ca.	circa, about
caterpillar	the larva of moths, butterflies, and other species
caudal	pertaining to the tail or posterior end
cervix	the neck, region between head and prothorax
chitin	a nitrogeneous polysaccharide in outer layer of body wall
commensal	a living together of 2 or more species (can include man), ex. commensal rodents
coxa	basal segment of leg
crochets	hooked spines at apex of proleg of moth and butterfly larva
cuticula	non-cellular layer of the chitinous body wall
dentate	toothed, or with tooth-like projections
diapause	a period of arrested development or suspended animation
diurnal	active during daytime
elytron (pl. elytra)	a thickened, leathery, or horny front wing
F&E	"Foreign and Extraneous" - frequently refers to animal residues such as insect fragments (IF), rodent hairs (RH), and feather barbules (FB) which FDA calls filth. In a broader sense also includes evidence of decomposition (mold) and material such as sticks, stones, sand, glass, metal, fibers, etc. Or, any foreign, unintended, atypical, non-chemical, and non-bacterial matter. Dirt and hairs from animals other than rodents are also filth. Sample preparation usually involves physical separation, frequently with the aid of solvents, and then examination of the particulate matter under magnification (microscopic). Some analysts speak of light (hairs, fibers, insect fragments) and heavy (sticks, sand, etc.) extraneous or filth.
femur	third leg segment
feral	wild, as feral pigeon
filament	a slender thread-like structure
frass	excrement usually mixed with remains from feeding
frons	front part of head
grub	larva (of weevils)--head and thoracic legs, no abdominal prolegs
host	the organism in or on which a parasite feeds, the material on which an insect feeds
instar	insects between successive molts, from egg to adult
larva	immature stage between egg and pupa
lepidopterous	pertaining to moths (and butterflies)
maggot	a legless fly larva without well-developed head capsule
mandibles	jaws
maxillae	segmented mouth parts
mites	small arachnids (class of arthropods or invertebrate animals) whose adults have 4 pairs of legs (as do spiders); insects have 3 pairs
molt	process of shedding skin
nymph	immature stage of insect lacking pupal stage
obligate	cannot live away from its host
ocelli	simple eyes
order	a subdivision of a class containing a group of families
patella	a segment of legs of arachnida (mites, spiders, ticks, etc.)
predator	an animal that attacks and feeds on other animals
proleg	fleshy abdominal legs of certain larvae
pubescent	hairy (soft down)

pupa	non-feeding, usually inactive stage between larva and adult
serrate	toothed along an edge like a saw
setae	hairs or hairlike bristles
spinneret	a structure with which silk is spun, usually fingerlike in shape
spiracle	a breathing pore
tarsus (pl. tarsi)	part of a leg
thorax	region between head and abdomen; comprised of prothorax, mesothorax, and metathorax
tibia	fourth segment of leg
tubercle	extrusion outward of body wall wing
veins	wing venation

STRUCTURE AND PARTS OF INSECTS

Figures 1 and 2 illustrate a few of the above terms.

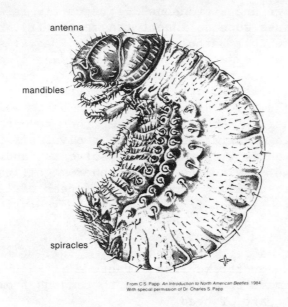

From C.S. Papp. *An Introduction to North American Beetles.* 1984
With special permission of Dr. Charles S. Papp.

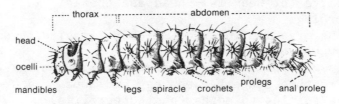

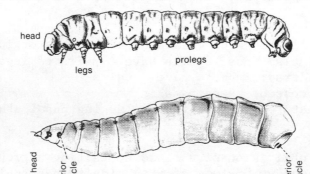

From C.S. Papp: *An Encyclopedia of North American Insects.*
With special permission of Dr. Charles S. Papp.

Fig. 1. Copyright Charles S. Papp. Used by permission.

12

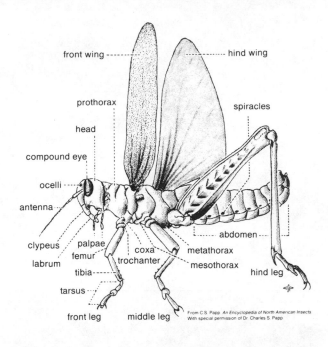

From C.S. Papp *An Encyclopedia of North American Insects*
With special permission of Dr. Charles S. Papp

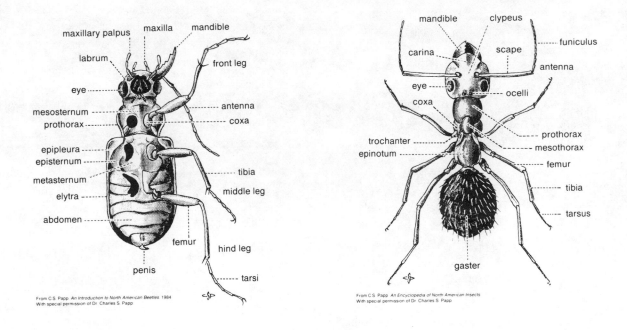

From C.S. Papp *An Introduction to North American Beetles* 1984
With special permission of Dr. Charles S. Papp

From C.S. Papp *An Encyclopedia of North American Insects*
With special permission of Dr. Charles S. Papp

Fig. 2. Copyright Charles S. Papp. Used by permission.

INSECT PEST MANAGEMENT
AND CONTROL

John V. Osmun

Editor's Comments

This chapter gives an overview of the systems
management of insect control. It is must reading. Synonyms
for Dr. Osmun's "Pest Management and Control" as he mentions
are IPM (Integrated Pest Management) and UPM (Urban Pest
Management). All speak to the consideration of all possible
approaches with continuing but lessened use of chemicals.
Dr. Osmun in his presentation has importantly expanded
on the tools which should be considered.
The summary chart "Insect Pest Management" on the last
page will be of interest.

                                        F. J. Baur

INSECT PEST MANAGEMENT AND CONTROL

John V. Osmun

Department of Entomology
Purdue University
West Lafayette, IN 47907

CONCEPTS OF PEST MANAGEMENT

In order to appreciate the application of pest management in the food industry, we need to review the development of it in agriculture. During the decade of the 1960's, attention turned from "spray and kill" to insect control based on devising systems involving a more holistic approach than had been previously practiced. Not only was more reliance placed on biological, cultural, and mechanical controls, but on integrating them with judicious and timely use of pesticides. Key to the success of these multi-faceted programs has been reliance on predictive approaches such as scouting to determine pest species and their population levels while monitoring the environmental conditions that favor or discourage pest development. The underlying principle is that of assessing the ecology of a pest and the natural factors regulating pest populations, then managing these populations in a combination of ways that prevent a pest from causing economically significant damage. This is in sharp contrast to instituting single factor pest control such as simply spraying an insecticide. Pest management (often referred to as Integrated Pest Management or IPM) has proven highly successful in some phases of agricultural production.

Before we turn our attention to management of insect pests in the food industry, there are other aspects that need discussion. The use of some elements of management presupposes that eradication is not always the goal. For example, if the strategy is using parasites and predators, then there must always exist low levels of pest populations which support the continuing existence of the beneficial organisms. Another is the notion that the use of pesticides can be greatly reduced and, in some cases, eliminated. Frequently, this is not the case because pesticides are useful and dependable tools of technology that can quickly reduce pest level when it is essential to do so.

Pesticides and related chemicals will always be part of most pest management systems. Further, there are new insect control chemicals that provide an expanded range of approaches: pheromones for attraction; growth regulators which modify physiological processes in such ways as to disrupt normal development. These will be discussed later in this book.

An effective pest management system therefore employs all suitable techniques and methods in a compatible and environmentally acceptable manner (Glass, 1975) to reduce pests to tolerable levels. The approach is possible today because we have become systems oriented, even to using computer-generated models, and we are more inclined to bring together the expertise of specialists from a number of disciplines in seeking solutions to problems. These ingredients make the concept of pest management different today from those previously practiced, even when several approaches may have been used in controlling a particular pest.

PEST MANAGEMENT IN THE FOOD INDUSTRY

Many aspects of pest management are applicable to problems in the food industry. There is frequently a great range of options that may be incorporated into a system. For example, Pederson and Mills (1978) noted that in no other type of pest control has a greater variety of techniques been used than in control of stored-product pests. This is true of many food industry situations. One initial advantage, as compared to pest management in production agriculture, is the widely understood and accepted practice of good sanitation in the food industries. For most situations, this is a key element of the system. Sanitation not only reduces many types of infestations, it facilitates inspection and diagnosis.

The food processing industry and food service establishments have changed rather dramatically in the last generation. As Schoenherr (1983) points out, food product ingredients are now more frequently shipped long distances or stored for longer periods. Likewise, both storage and distant shipping of finished products occur after products are processed. All of these elements of the production-marketing chain increase the chances for insect infestation and necessitate greater reliance on preventive measures.

Ecological Variants

Situations favoring insect development in foods commence on the farm. Even the harvesting equipment bringing raw commodities in from the field may be involved. A common example is that of previous season insect-infested grain remaining in small quantities in trucks used in the harvest of new grain. Grain may be stored with too high moisture content that favors insect development. Grain bins and other food storage areas may be neglected and seldom cleaned. The same may be true of rail cars and trucks used to transport food to elevators, warehouses, and processing plants.

The dried food industry is especially vulnerable to pest problems at intermediate points such as elevators. Here there may exist conditions of warmth and isolation that remain undisturbed for extended periods. It is not surprising that grain often reaches the mills infested with a variety of insects.

The ecology of food processing plants is varied and often optimum for insect development. Because of types of materials and design of building construction, especially those in older buildings, combined with the particular procedures of plant operation, many species of insects are provided with havens for development. All the ingredients are there: food ideally prepared, moisture, warmth, protected niches ranging from processing machinery to electrical conduits, and prolonged static conditions that favor development. In-plant conditions are further complicated by the presence of areas such as receiving docks, repackaging areas, salvage rooms, and pallet storage. The latter, for example, when combined with high humidity provides a perfect environment for booklice development on pallet surfaces, thus conjuring up visions of crawling dust!

Differences in Pest Management Strategy

There are other differences between managing insect infestations in agricultural fields and those in the food industry. One is the interpretation of economic threshold. In agriculture, numbers of insects are often permitted to remain because their presence is of no economic consequence; in the food industry, we frequently strive for complete elimination of an infestation. Further, the use of such biological control agents as parasites and predators can seldom be considered once food ingredients have left the farm, both because substantial numbers of pest insects must be present for the biological control concept to work and because the very presence of parasites and predators constitutes the potential for adulteration. In general, the principles are the same and one simply needs to be concerned with identifying all approaches to control that can possibly work in a given situation and integrating them into a management system.

Location and Construction

First considerations in assessing and planning a pest management program in a food establishment are location and building construction. Such considerations apply not only to food production and manufacturing buildings but to storage facilities and even to restaurants and related food service businesses. Problems frequently arise from a building's proximity to continual sources of reinfestation; e.g., other processing plants, dumping areas, and some types of agricultural production. Often little can be done about such situations other than to emphasize exclusion techniques as an important strategy. Besides mechanical and physical techniques described below, it is frequently desirable to stress open landscaping and weed control, thus reducing harborages close to buildings. These measures should be planned in conjunction with vertebrate animal control programs.

As ideal as older buildings may be in providing suitable environments for insect development, improperly designed new construction may be fully as important. For example, the design of both building and equipment should facilitate cleaning and general maintenance; cracks and crevices should be minimal. All surfaces that could be in contact with food and food containers must be visible for inspection or readily disassembled for inspection and cleaning (Imholt, 1984). Because there are many important elements of construction, they must be thoroughly considered if pest management strategy is to be complete and effective. Two steps are essential. First, experts from different fields must plan together; i.e., the design engineers must, as a minimum, converse with sanitarians, entomologists, and maintenance specialists. Secondly, and equally important, corporate and plant management must make sure design does not get changed in the field nor modified by start-up persons who may be tempted to do so on their own.

Physical and Mechanical Approaches

Physical and mechanical methods are the oldest approaches to insect control; yet, in some cases, they are still the most effective elements of pest management strategy. They consist of direct or indirect measures taken to destroy the insect outright, disrupt normal physiological and behaviorial activities by other than the use of chemicals, or modify the environment to a degree that it is unacceptable or unbearable to the insect (Osmun, 1969). These approaches may be either preventive or corrective. They are based on a thorough knowledge of the ecology of the pest and a realization that there are tolerance limits such as extremes of temperature, humidity, and physical durability beyond which an insect cannot survive. There are also responses to various wavelengths of the electromagnetic spectrum that can attract an insect to a shocking demise.

Temperature: Heat and Cold
Insects vary in their susceptibility to heat, but no insect can survive for long when exposed to temperatures of 60-66°C (140-150°F). Lethal heat can be realistically employed by utilizing a 3-4 hour exposure to 52-55°C (126-131°F) temperatures. This approach is most effective and usually least expensive in processing plants such as bakeries and canneries where existing facilities for heating can be conveniently augmented by space heaters. (See Chapter 15.)

Abundance, distribution, and rate of insect development are affected also by cold. (See Chapter 18.) In northern climates where seasonal sub-zero weather occurs, flour mills and grain bins may remain relatively free of pests. The degree to which cold penetrates a product is principally a function of time.

When outside temperatures reach -22°C (-8°F) or lower, 24-36 hours of exposure
are usually recommended. Cold may be utilized in another manner, that of
storing commodities under refrigeration. Even then, care must be taken to keep
pests such as German cockroaches from congregating in door stripping and outer
layers of insulation.

Humidity and Moisture Content
    Insects must maintain a level of internal water within relatively narrow
limits which, in turn, is strongly influenced by external factors. Most insects
cannot modify their own microclimate and therefore react to adversity either by
moving elsewhere or by dying. High humidity increases the moisture content of
foods, predisposing some to moldy conditions which, in turn, attract specific
insects. Examples of insects which may increase in the presence of high
humidity include booklice, mites, silverfish, mealworms, foreign grain beetles,
and vinegar flies. Similarly, stored grain with a moisture content greater than
12% favors weevil development. Consequently, reductions in humidity and
moisture content are frequently useful elements of pest management.

Light and Near-Visible Radiant Energy
    One of the most useful strategies is the employment of various wavelengths
of light and intensities of electricity to attract and destroy insects. Their
use is based on the photopositive response of many insects, especially to
ultraviolet light. Radiant energy is used in a number of ways: detection of
pest infestation by attraction to light traps; estimation of changing levels of
populations; evaluation of other control measures; the actual control of some
species of insects by electrocutors; and, in the case of insect-free lighting,
the diversion of some insects away from food establishments. Details may be
found in Chapter 8 on insect electrocutors. One word of caution: when insects
are attracted and electrocuted, they usually drop into collection pans beneath
the lights. These pans must be cleaned frequently or else they will become
breeding places for carpet beetles and Trogoderma that, in turn, will move to
processed foods.

Air Movement, Clean Air, and Screening
    The use of high velocity air to exclude flying insects from a building or
room has appeal because it may permit the use of open doorways. However, each
situation must be carefully thought through and engineered. Most certainly such
devices cannot be relied upon as the only approach to control. The addition of
overlapping vertical strips of heavy plastic generally improves the
effectiveness of such installations.
    Often forgotten is the essential nature of proper filters in air
ventilation systems, both in the air intake and the exhaust. Exhaust fans
usually do not run continuously and a calm opening will allow both primary and
secondary pests to enter.
    Windows which can be opened must be screened with suitably small screen
mesh; e.g., 14-16 openings per inch. Doorways need screen doors that swing out!

Protective Packaging
    Since there are many species of insects that are pests of dry foods, effort
has to be made to exclude them from products following processing and
manufacture. No matter how clean the product may be when it enters the channels
of market distribution, the ultimate customer will place the blame on the
manufacturer if an insect is found. One way the product can be protected is by
using insect-resistant containers. This is another element in developing a
total pest management system.
    The success of a food container as an excluder depends on the packaging
material used, the construction of the container, the tightness of the closures,
and of course the attractiveness of the product. The package must be made of
materials that are naturally insect-resistant or treated to resist penetration;

e.g., inclusion of pyrethrins. It must also be structurally designed and fabricated in such a manner that not even a minute crevice is present. Most overwrap materials, in time, can be penetrated by certain chewing insects (e.g., Indianmeal moth larvae), but they do discourage insect entry, especially if product distribution time is short.

Great care must also be taken during the actual manufacture of containers and packing materials and with their storage prior to use. Of particular importance is corrugated cardboard; the channels are ideal harborages for many species of insects, and the bonding materials suitable food. A good practice is for a food processing sanitarian to inspect the premises of container suppliers.

Other Mechanical Devices

Sifting of food materials is another important element of a management system. Depending on the food material, this may be accomplished at appropriate intervals from the farm to the final processor; it reaches a high degree of efficiency in the processing of grain. Properly designed mills incorporate a series of sifters, flotation devices, and other separators to help assure products free of contaminants.

In use for many years in processing plants using grain or grain products are entoletor devices which subject grains to impact by centrifugal force. Those grains with internal infestations are shattered and discharged from the grain stream, and with other materials eggs and adults may also be destroyed.

Trapping devices are useful elements of some management systems. Commonly used are sticky devices of various sorts for the reduction of flies, small beetles, and cockroaches.

BUILDING THE SYSTEM

When considering insect pest management for the food industry, we are immediately confronted with a diversity of problems. One is the difference in products. The foods to be processed or served include inherently dry material such as grain, grain products, and dry blends; high-moisture content fruits and vegetables; liquids such as milk; and products of liquids such as dried milk. Some of these are made more attractive by the addition of nutrients. There are almost as many different ecological situations as there are processed products.

The problem is further compounded by a range of quite different insects, all equipped to survive and flourish in the artificial environments created by man. There are great differences among such pests as rice weevils, the Indianmeal moth, cabinet beetles, and cockroaches. Yet seemingly we build and stock our various food establishments for the convenience of these insects!

In addition to all this, what especially sets the food industry apart from agricultural insect control is the attitude of man himself. Implementation programs must take into consideration the interaction between people and pests affecting their well-being (Sawyer, 1983). Self-imposed, high standards of purity are a way of life. These are admirable, although sometimes extreme, and they often necessitate severe measures of pest control. Economic thresholds give way to consumer alarm over small infestations and thus to the establishment of more severe esthetic thresholds where the goal is frequently the absolute elimination of an insect or insect fragments. It is correct to say that some low levels of insect occurrence can be expected in most foods, and the public must realize that this is unavoidable within the economic constraints of our society. Some reasonable public understanding is necessary in order to avoid wasting food or using pesticides when only trimming and washing may be necessary (Anonymous, 1980).

Pest Recognition and Inspection

Systems of pest management are developed in such a way that they may be preventive, corrective, or a combination of the two. Whatever the case, knowledge of the pest or pest group and its behavior is essential. When an infestation is present, the <u>number one</u> element in developing corrective measures is the identification (usually down to species) of the insect or insects involved. How else can one assess the behavioral and ecological aspects which, in turn, help us choose the appropriate elements for control? Either resident personnel must be able to do this or some professional must be brought in from outside. (See Chapter 5.) Inspection procedures must be frequent and thorough enough to detect the actual insect and evaluate the extent of the problem or potential for a problem. (See Chapter 4.)

The matter of detecting, determining, and subsequently monitoring insects has been, for some species, greatly facilitated by the advent of synthetic pheromones. Some of these chemicals are species-specific; nearly all are group-specific. (See Chapter 7.) Consequently, they are proving to be extremely valuable in the food industry where pests are frequently difficult to locate and their populations hard to assess. The subject of establishment inspection is considered elsewhere in detail (Chapter 4), but it should be remembered that it is an essential ingredient in a pest management program. It is principally through thorough, repeated inspection and monitoring that conditions conducive to insect infestation can be detected. What are most frequently revealed are insanitary conditions. The elimination of such conditions is the second step in pest management because it breaks the life cycle of many pests and prevents further development.

Exclusion of Pests

A pest management system must continuously consider the exclusion of pests. Ways to achieve this strategy were discussed earlier under Mechanical and Physical Controls. The important point is that a food processing or serving establishment can never let up on this matter. There are always trucks unloading and loading; there are open doors and windows, the wrong lights left burning, the arrival of packaging materials, new suppliers of food ingredients, lunch boxes from home, replacement of wooden pallets, and even the arrival of new machinery. As they relate to pest problems, these are not one person's responsibility; they are part of the responsibility of everyone in the company or business. Exclusion is an important part of prevention.

Insecticides

Insecticides remain an essential part of most pest management programs in the food industry although our reliance on them may be lessened by integrating other approaches into the system. Insecticides are advantageous because they are usually immediately effective and relatively inexpensive to use. Many kinds and formulations are available and there is a broad spectrum of application methods; e.g., baits, sprays, dusts, crack and crevice applications, ULV, and fumigants. The latter find their greatest utility in the food industry because fumigant insecticides penetrate and seek out the most remote crevices in large volume structures. The more attention that is paid to selecting a variety of control methods, the more effective insecticides will be when indeed they must be used.

PUTTING IT ALL TOGETHER

Three additional and very important elements must be incorporated into a unified program to achieve a truly effective insect pest management program. They are <u>appropriate people</u>, a <u>systems approach</u>, and <u>adequate evaluation</u>. The persons who must be involved include a sanitarian or professional entomologist together with the plant engineer, quality control supervisor, plant manager, and appropriate corporate persons. Admittedly, not all food industries have this spectrum of personnel; the important point, however, is that all individuals who

could have some concern or responsibility be involved in developing and
implementing pest management strategies. These are not one person operations.
They must be supported by management with an optimal management policy.
Frequently it is prudent to engage a professional consultant or firm that
specializes in food industry problems, both preventive and corrective.
(See Chapter 3.)

The advent of the computer has added a dimension not previously available.
We speak about developing systems of pest management. Often a system of
integrated approaches can be devised without a computer, and thus the lack of a
computer should not be an excuse for not developing a system with all the
elements in their proper place at the proper time. Those who do have access to
such equipment should consider utilizing it to facilitate the entire pest
management program. Computers can be used to develop procedural models based on
our knowledge of pests and their behavior in various environments. It would be
expected that in due time guidance will be available through computer generated
programs to aid the food plant operator. They will facilitate the integration
and timing of control approaches in appropriate pest management systems.

One final point is that of a package consisting of evaluation, record
keeping, and reporting. The topic is expanded elsewhere but it is well to
remember that constant evaluation is essential and that there are too many
facets of a pest management program for one person to keep in his head.
Documentation is the best evidence that elements of the program have been
performed properly and at the correct moment.

This chapter has spoken to the need to consider all possible and reasonable
control elements for the management of insect pests.

REFERENCES

Anonymous. 1979. Integrated pest management in food distribution warehouses.
     National Pest Control Association, ESPC 072101.
Anonymous. 1980. Urban pest management. Nat. Acad. Sc.: 181–185.
Anonymous. 1982. Integrated pest management in food processing plants, Parts I
     & II. National Pest Control Association, ESPC 072119/072126.
Evans, B. R. 1981. Urban integrated pest management. The University of
     Georgia.
Glass, E. H. 1975. Integrated pest management: rationale, potential, needs
     and implementation. Entomological Society of America Special Publication
     75–2.
Imholt, T. 1984. Sanitation Through Design. Avi Publishing Company.
Osmun, J. V. 1969. Physical and mechanical controls in Insect-Pest Management
     and Control. Nat. Acad. Sci. Publ. 1695: 243–281.
Peterson, J. R., and R. B. Mills. 1978. Pest management programs for surplus
     grain. Proc. Symp. Prev. and Contr. of Insects in Stored-food Products:
     369–375.
Sawyer, A. J., and R. A. Casagrande. 1983. Urban Ecology 7:145–157.
Schoenherr, W. 1983. An overview of pest management. Fumigants and
     Pheromones: Issue No. 4, Published by Fumigation Service & Supply and
     Insects Limited, Inc.

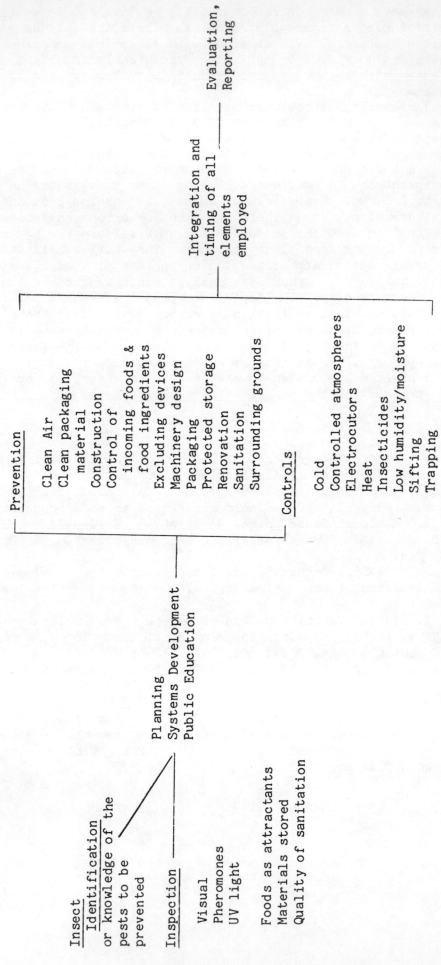

SUMMARY CHART--INSECT PEST MANAGEMENT

Insect
Identification
or knowledge of the
pests to be
prevented

Inspection

Visual
Pheromones
UV light

Foods as attractants
Materials stored
Quality of sanitation

Planning
Systems Development
Public Education

Prevention

Clean Air
Clean packaging
    material
Construction
Control of
    incoming foods &
    food ingredients
Excluding devices
Machinery design
Packaging
Protected storage
Renovation
Sanitation
Surrounding grounds

Controls

Cold
Controlled atmospheres
Electrocutors
Heat
Insecticides
Low humidity/moisture
Sifting
Trapping

Integration and
timing of all
elements
employed

Evaluation,
Reporting

24

THE ROLE OF THE PEST CONTROL OPERATOR (PCO)

Robert C. Yeager

Editor's Comments

Rare is the plant or warehouse that does not need to at least consider insect or pest control. Insects are omnipresent. Every plant has attractions for insects. So the need for some insect control is a given. How should that be provided? Mr. Yeager lists the pros and cons of three options. As a rule of thumb, the larger the plant or facilities, production, etc., the more sense it makes to have your own experts, plant and corporate. Even large companies find it valuable to have a working relationship with PCOs. For smaller companies, an outside PCO is probably the best option. Nonetheless, at least one employee needs to have some knowledge about pest control to interact properly with the PCO including; 1) the initial contractural stages, and 2) some variable spot checking of services rendered. There is little question that a team effort offers the greatest potential rewards.

F. J. Baur

THE ROLE OF THE PEST CONTROL OPERATOR (PCO)

Robert C. Yeager

Rose Exterminator Co.
Copesan Affiliate*
Cincinnati, Ohio 45203

INTRODUCTION

The role for which the professional pest control operator (PCO) has been trained and can serve most effectively and efficiently is one of working closely with whomever is responsible for your sanitation. The PCO can and should help to assure that your operations are in compliance with the laws and regulations of such regulatory agencies as the Federal Food and Drug Administration (FDA) and the Environmental Protection Agency (EPA) as they relate to pests and the contamination they can cause. (Food, Drug, and Cosmetic Act Sections 402(a)(3) and (a)(4)). The above combined with the reduction or absence of consumer complaints related to pests forms the goal of every professional pest control operator in working with concerned industry.

WHO NEEDS PEST CONTROL?

As a PCO it would be easy and reasonably accurate to say "everyone" in answer to the above question. In listing those most in need of pest control, food and its many related industries come first to mind. To stop there, however, would be to overlook the drug industry, the health care industry, etc., in which such pests as insects are also a contamination risk. All of these industries have common aspects of manufacturing such as processing equipment, packaging materials, warehousing, grounds, and landscaping which provide similar attractions to insects thus requiring comparable levels of insect control.

In fact, we need only to reflect on the reasoning behind the publishing of this book and your own experiences to recognize the importance of and need for insect control. Now the question is - who is going to provide the needed pest control services?

OPTIONS

Having decided you need pest control, what are your options? There are three. The advantages and disadvantages of each constitute the main thrust of this chapter.

In-house Personnel

Advantages
Your people will be familiar with the facility and its own unique characteristics. One or more of them may have been involved in the construction

---

*Copesan Services, Inc. was founded in 1958. Its name is derived from a combination of the words COordinated PEst control and SANitation. Today there are several hundred independently owned North American pest control firms working together as Copesan Affiliates to provide pest control services.

of the facility or in the adoption of an existing facility to the needs of your operation. Hopefully such people will also be qualified in the field of pest control.

Since pest control work must at times be performed at unusual hours, your own personnel could be more readily available and have access to parts of the plant where outsiders are not always permitted.

In-plant people might also be able to spot a potential pest problem and correct it in its early stages. Also, it is only natural that you sense a closeness to your own personnel and have a better feeling about their thoroughness and conscientiousness.

Further, there may be a savings in using your own people since they are already on the payroll.

Finally, it provides the freedom of keeping the knowledge of pest problems within the family. Without enlarging on this matter, there are instances where revealing problems to others is difficult. It shouldn't be, but it is.

Disadvantages

There is a real possibility that too much familiarity with one's own facility could develop an acceptance of conditions which may potentially create pest problems. Such conditions might include dead storage, unrepaired cracks, voids and crevices, loose coving and floor tile, snacking at work positions, trash accumulation, possible pest entrances, etc. These situations may not have caused any serious or known problems as yet but have the potential of doing so in a very short time.

To avoid possible problems with the above, time and visible dollars need to be spent to not only have someone intensively trained in pest control but to also take periodic refresher training. This training should be directed at knowing what pesticides to use, how and where to apply them, what to expect, and the hazards involved. Next comes the need to pass the certification examination required by Federal Insecticide, Fungicide, and Rodenticide Act (FIFRA) to permit the purchase of restricted pesticides and a license to use them. Should your operation require the use of fumigants, the need for a license and the training necessary to obtain it is all the more critical. It also means finding the right person within your plant to do this work. Replacement of that person, should that become necessary, could be quite a challenge. Acquiring this training and knowledge is not the disadvantage per se but the pressure and responsibility related to getting it may be.

The need for someone to have the knowledge about handling pesticides will also be important. Proper storage of pesticides is an absolute requirement. The area chosen for this will have to be carefully controlled and be accessible only to designated personnel. The characteristics of the pesticides must be understood as related to vapor toxicity, temperature limit, what to do in case of spillage, leakage, fire protection and precautions, etc.

As we take another look at the person or people who have responsibility for pest control, he, she, or they may have other areas of responsibility. It isn't difficult to envision then, under these conditions, that a conflict of duties causing some element of the operational need to be overlooked or neglected and resulting in a problem of an entirely different nature. This problem alone could negate any hoped for dollar savings in a "do-it-yourself" program.

Without having had pest control experiences in other locations and under varying conditions, there will also be a limited knowledge and/or experience about optional approaches or solutions to stubborn pest problems should they develop. The same absence of experience could surface in trying to use restricted or exotic pesticides should such be required.

Finally, we can't forget insurance. There is a real possibility that you will need additional insurance coverage which could include Product Liability, Blanket Contractual, and Public Liability.

Advantages

The use of a professional PCO can eliminate most if not all the disadvantages listed above. In addition, use of a PCO makes available resource persons both within and outside that organization to assist in the solution of pest or pest related problems.

You can receive frequent, regular, accurate, and complete reports outlining what service and pest activity has taken place since the last report, conditions as related to goals, and possible suggestions as to how improvements can be made.

The increased communications resulting from the regular reports will quickly reflect the distinct advantage of open dialogue and sharing. This feature cannot be over emphasized.

The writer would be in error to imply that the advantages outlined in the two preceding paragraphs would not be possible through a highly trained and certified in-plant person but past experience has strongly suggested that it just doesn't happen.

Your PCO can provide supervisory personnel who will sufficiently acquaint themselves with your operation in order to properly train service technicians at all times. Generally he or she can also provide or arrange for simulated regulatory agency inspections as related to pest control.

Not only will your PCO be acquainted with the proper pesticides for specific needs but the label requirements as well as to mixing, dilution, application, placement, and at a lesser cost in view of quantity purchases.

In addition to being well versed in the control of insect and rodent pests your professional PCO will be familiar with the necessary steps for the control of birds, unwanted vegetation, and some basic suggestions related to landscaping and how they might affect a pest control maintenance program.

As an additional benefit, if your PCO is a member of the National Pest Control Association, he or she can provide ready access to NPCA materials and the additional expertise available from the NPCA staff. Your PCO can also assist in your becoming an associate member of NPCA. Details pertaining to this opportunity are given later in this chapter.

Disadvantages

As a member of the pest control industry, I do not enjoy pointing out any disadvantages to the hiring of a professional PCO especially since most, if not all, these disadvantages can and should be avoided through the careful choice of a PCO. Key points to keep in mind are a well planned and well managed pest control program, good communications, and frequent monitoring. Having said that, let's review the disadvantages which can or have occurred.

There is no question that service businesses can and do occasionally experience an abnormal turn-over in personnel. In the field of pest control this dictates a frequent need for training of new technicians not only regarding a specific facility and its particular pest control requirements but in overall pest control understanding. In the absence of such training the PCO could experience trouble in knowing where to inspect, what to look for, and should a problem be uncovered, a possible uncertainty as to what to do about it. In the presence of such incompetence there could develop on your part a feeling of paying for services not rendered, of seeing the evaporation of professionalism so evident during the initial arrangements and contractual sale.

Having non-company individuals in your plant could lead to possible loss of confidentiality so vital in many operations as relates to new product development and processing conditions.

Before leaving the critiquing arena, it seems fitting to reflect on some comments made by James H. Rutledge during an annual Purdue University Pest Control Conference. Mr. Rutledge is Vice President, Food Protection Services, at Lauhoff Grain Company and a contributor to this book. The thrust of his talk dealt with the basic problems existing between the food industry and the PCO. Among other things, he said:

Food Plant Problems

1. "Too many food companies are too concerned about the cost of services rather than the quality of those services.
2. Some food plants try to transfer total responsibility for pest problems to the PCO. This cannot be done. Industry is still legally responsible for its own sanitation concerns.
3. After having hired a professional PCO, some members of the food industry may tend to ignore the problems of pest control. As a result, a communication gap grows wider and wider between expectations and service rendered until a real problem develops. Pest control services cannot be ignored by management nor are they a substitute for good housekeeping.

Pest Control Industry Problems

1. Some pest control companies, in their eagerness to obtain business, have a tendency to bid too low rather than sufficiently high to cover the quality of service required to achieve quality control.
2. Some members of the pest control industry do not do enough auditing and monitoring of their own services to achieve quality control in their operations.
3. Too many PCO firms have a high rate of turnover and have relied on food plant personnel to familiarize the new technicians with the facility, its uniqueness and its pest control needs. This is not nor should it be the client's problem. Pest control industry supervisors must become sufficiently acquainted with their accounts in order that they can and will train their new technicians prior to account assignment."

We know that there can be advantages and disadvantages to an in-plant pest control program. The same can be said for the use of a professional pest control operator and that such dissatisfaction can ofttimes arise as the result of communications failures, poorly understood expectations, or the manner in which the service is purchased, namely, price vs quality.

IS THERE A THIRD OPTION?

Fortunately there is a third option. It is a combination of both in-house and outside pest control which combines the advantages of both approaches. As a concerned manager, you have the comfort of knowing that there is a team of your people as well as outside professionals working toward a common goal, namely, a plant or facility in which a meaningful sense of pride regarding pest control is felt by all persons involved. This team effort can result in notable reductions in consumer complaints and criticisms from local, state, and federal regulatory agencies.

In June 1982, Pest Control Magazine published the results of a 1981 survey of users of pest control services within the food processing industry. There were 215 replies. Ninety two percent or 198 of those replying have in-house pest control units within their companies. Nearly fifty percent or 108 work in conjunction with outside contracted pest control firms, and about forty two percent or 90 work solely on their own. The remaining eight percent or 17 rely exclusively on an outside PCO for all pest control services. Although there appears to be a definite trend toward in-house pest control, it is also apparent that many companies prefer the comfort and support of a second, objective opinion and pest control certification often lacking in-house.

Of those using a combination of in-house and an outside PCO the quality of service was ranked as good to very good, sixty nine percent or 68, average, twenty five percent or 25, and less than satisfactory, six percent or 6. Unfortunately there was no breakdown on levels of satisfaction reported among the group working solely on their own.

Companies exclusively employing outside PCOs gave somewhat similar ratings, namely, good to very good, fifty nine percent or 9, average, twenty three percent or 4, and less than satisfactory, eighteen percent or 3.

These percentages will vary with the size of the plant involved.

The larger the firm the more likelihood of having a full time qualified person or staff with responsibility for pest control included as part of their sanitation and/or maintenance responsibilities.

From experience we have learned that even the largest firms appreciate having a working relationship or at least a "hot line" for consultation with a PCO. Confirmation of this became evident when certification training classes were offered and the local food industry people were invited. The response was overwhelming. Happily almost everyone attending passed the certification examination. Contrary to what one might think, instead of the "graduates" starting to wear the PCO's hat, the working and consultation relationships deepened considerably.

SO--HOW DOES ONE SELECT A QUALITY PCO
AND SET UP A QUALITY PROGRAM?

1. You will need to have within your operation at least one person with the basic knowledge of what constitutes an acceptable pest control program for your industry. It would even be desirable for that person who is to be the plant contact with the PCO to obtain state certification. This will make possible a much better understanding of pest control "language."
2. Talk with other firms engaged in a similar business to yours. This could give you a starting list of area PCOs from which you may choose the firm which will have the privilege of serving you.
3. Check with the Better Business Bureau and ask questions about specific companies. Usually you will be told if there have been registered complaints and if so, what was done about them.
4. After selecting the firm(s) in which you are still interested, talk with a responsible person within each one. Determine the size and background as relates to your type of business. Are they large enough to handle your account? Have they had sufficient experience in your field of operation? What kind of training program do they have? What is their personnel turnover experience, especially as relates to the technicians and supervisors? Should your interest range sufficiently high, you might even like to visit their place of business.

Once you have become acquainted, it then becomes time to discuss your pest control needs and to begin to develop--

YOUR PEST CONTROL PROGRAM

In doing this, here are a few suggestions, as well as some questions to be answered, which may prove to be helpful in your interacting with your probable PCO.

1. Be totally open regarding the known pest problems. Be explicit about the building(s) to be treated and protected, the nature of your business, the type of products, and the risks involved. Avoid surprises.
2. Have you been involved in any recent regulatory action in which pests were a factor?
3. Do you have pest control service now? Is it in-plant or a professional PCO? What has been your experience? Be frank about it.
4. Provide a very thorough tour of the facility. This will enable not only a review of the known problems but may reveal some additional problem areas and contamination sources such as suppliers, transit, dock handling, storage practices, etc.
5. If possible, provide a schematic of your plant. This will make the tour more understandable and provide a meaningful visual aid in the preparation

of the periodic reports.

6. Discuss clearly the operational schedule and when pest control services can best be applied. Avoid as much as possible "unreasonable" timing such as late nights, Sundays, holidays, etc. although we all understand that such scheduling is necessary and desirable at times. Also be prepared to pay a premium when the PCO is required to work during these times.

7. To maintain open communication, there should be one person from your organization, with appropriate authority and backup, designated as the contact for the PCO.

8. Since your sanitation and the pest control people must of necessity work in close cooperation, clearly establish your sanitation/pest control needs and goals and the role the PCO will play in fulfilling them.

9. You are now ready to request the written proposal and contract. In general it should specify schedules and frequency of routine services as well as emergency and special need responses. Be certain the areas of responsibility are enumerated and approved pesticides and devices to be used are listed. Have spelled out the nature and frequency of reports designed to tabulate areas in need of attention and what was done about them. These reports should also include suggestions for in-plant corrections or repairs should they be appropriate to accomplish your common goals.

10. Finally - agree to a monitoring of the program by not only the PCO but by your people as well. Stay in touch. Good communications is a vital key to the success of this program.

I agree that the above suggestions and approaches are somewhat unconventional but they do demonstrate evidence of interest and concern which will convey to the PCO that you are serious. It will say that you want the very best and that you are willing and eager to strive for it. The effect, generally, is a like amount of effort on the part of the PCO. After all, it's at your desk at which "the buck stops" and you are the one who is responsible for operating a pest controlled plant. Isn't a little extra effort worth it?

Now, as you narrow down the negotiations relating to this program, have the purchasing agent present or represented as may be appropriate. All too frequently the depth of research and effort in striving for the right service and program at a correspondingly right price is lost when the proposal reaches the buyer's desk. This should not be just a buying function but rather addressed to a definite need. Buy the service right - not cheap. The cheapest bottom line isn't always the least expensive.

HOW MUCH IS ALL THIS GOING TO COST?

There is no such thing as a bargain pest control program. First rate pest control service requires knowledgeable problem solving time and time costs money. There is almost no way to cut the time-dollar factor without reducing service. If the reduction in that service is at the expense of quality results, a sincere and worthwhile professional PCO will wisely say thank you but no thank you.

YOUR BEST SOURCE OF CONTINUING
KNOWLEDGE AND TRAINING

The National Pest Control Association, P.O. Box 307, Dunn Loring, Virginia 22027 invites you to become an associate member. Should you like to phone - call (703) 573-8330 and ask for Mrs. Barbara Bonner. You will receive all mailings, be privileged to order any or all of the many membership offerings, at a discount, and be made aware of upcoming conferences, seminars, and various other training opportunities. A very popular and important such opportunity is the Purdue University Conference which is held annually in January.

SUMMARY

1. High quality pest control whether it be preventive or corrective is a necessary service.
2. Diligently follow a clearly designed and workable goal-oriented program. Make it work for you and it will minimize customer complaints as well as unnecessary problems with regulatory people.
3. Put a very high priority on regular inspections, reporting, communications, and monitoring. Without an open sharing of needs, problems, concerns and results, the finest of programs and intentions will be considerably less successful than hoped for.
4. Take full advantage of available high quality training. It is even more helpful to have at least one person in your organization certified in pest control.
5. Consider carefully the three options open to you for pest control--internally handled, an outside PCO, or a combination.

PROTECTION OF AMERICAN FOOD IS IMPORTANT!
WE ALL NEED TO DO OUR PART!

INSPECTION TECHNIQUES

James W. Gentry

Editor's Comments

Periodic, appropriate inspections of a facility, including outside areas, are essential if serious infestation problems are to be avoided. The inspector must be trained, properly equipped, diligent, observant, a communicator, knowledgeable of the plant's raw materials, equipment (processes), construction and previous records on insect control, and have management's interest and support.

The author has used possible evidence of insects as his main target. His information is also of value for other pests (visible and invisible) and for all facets of GMPs.

Keep in mind the importance of behavior of pests when looking for signs of their probable presence.

An extensive checklist on "where-to-look" for insects is provided. It is not intended to be a complete list but should serve as a basis to create the one you need. As an example, if you store bagged edible materials be alert to the possible presence of round holes, usually about 0.1 mm in diameter, in the bags as indicative of insect entry (or exit).

F. J. Baur

INSPECTION TECHNIQUES

James W. Gentry

O'D. Kurtz Associates, Inc.
2411 S. Harbor City Blvd.
Melbourne, FL 32901

INTRODUCTION

Inspections for, and evaluation of, possible insect infestations and other visible pests in food processing plants and storage areas have become more important in recent years since there has been a shift toward integrated pest control. This approach, with its movement away from purely chemical control measures, makes the inspection a primary cost effective method within the integrated program.

The inspection of a facility for insect infestations or potential infestations should be a routinely scheduled part of insect control. Constant vigilance in looking for or searching out insects, insect harborages, improper storage, insanitation, and other situations conducive to insect survival and reproduction is an absolute requirement if the desired control is to be achieved. Inspections should be conducted by qualified personnel. If done in a lackadaisical manner or by unqualified personnel, the inspection loses much of its effectiveness, results in misinformation, or does not produce the desired results. The people conducting the inspection should be specifically trained for this purpose and must be familiar with stored-product pests, their habits, and their reproductive cycles under varying conditions. A knowledge of other insects is necessary in order to recognize insects other than stored-product pests that might find their way into the facility and, if not recognized, possibly create false impressions of infestation of food products.

Many short training courses relating to stored-product insect pests place emphasis on a relatively small number of commonly found insect species. The larger number of "field" or incidental species is often not discussed. Many, such as a variety of non-infesting small moths and beetles, may be confused with stored-product pests. This happens too often and results in the classical case of "a little knowledge is a dangerous thing" whenever these are considered as creating possible infestations. Studies of general entomology quickly reveal that most insects such as thrips, aphids, leafhoppers, plant bugs, gnats, midges, etc. can be placed in a special group or general classification. With this knowledge, one realizes that these insects do not infest stored foods and do not survive very long in a food processing or storage facility where they are out of their natural habitat. These "field" insects are considered as contaminants when they accidentally occur in food. However, recognition will allow the application of appropriate control measures.

INSPECTION

If the inspection is to be efficient and effective, preplanning is required.

34

Preplanning

A good source of information is reports of previous inspections and these should be reviewed if they are available. This gives one a history of what has been found or not found and, if previous inspections were conducted by qualified personnel, allows one to possibly anticipate what to expect. An old facility, for example, poses entirely different problems when compared to a modern, well planned site which was constructed at a recent date under different rules and regulations. If problem areas are listed in prior reports, this allows the inspector to plan some extra time for these areas. Recurring problems that are easily corrected give insight into management's attitude toward Good Manufacturing Practices and sanitation in general.

Other sources of information that may be useful include plans or layouts of the plant, diagrams of equipment and machinery, reports of pesticide applications, etc.

Planning should allow sufficient time for the inspection during appropriate operating hours. Unless there is a valid reason, the inspector should not work beyond normal operating hours. However, one must be ready to work after hours in the event a problem is discovered that requires immediate action.

Proper Equipment

Making sure you have the proper tools is also an essential part of planning. A notebook and pencil are required (pencil is preferred to pens since they are waterproof and notes are not smeared in case of accidental wetting). No matter how much experience the inspector has, one should not rely on memory, but should make extensive notes. Excess information can be discarded but forgotten items cannot be recorded. A flashlight with fresh batteries is also essential and no inspection can be properly conducted without one. This should be a type that throws a strong spot or beam of concentrated light. The flashlight should have a shatterproof lens, and in some situations be explosion proof. An extra bulb and spare batteries should be available. A spatula such as those used in weighing chemicals, or a similar tool, is useful in digging into cracks and crevices. A pair of small forceps for collecting specimens is very useful and small screwcap vials with labels can be used for most specimens. Record desired information on the labels at the time of collection. Again, do not rely on memory at a subsequent time. A small portable vacuum will remove debris from cracks, crevices, and other tight places. A small aerosol container of pyrethrum based flushing agent can be used in locating insects hidden in inaccessible areas.

Special equipment such as blacklights, temperature and humidity measuring devices, etc. may be desirable under certain circumstances. Earplugs, hard hats, safety glasses, safety shoes, hairnets, and other special clothing and/or equipment may also be required.

Personal appearance is also important. The inspector should be neatly dressed, have neatly trimmed hair, and not wear jewelry or trinkets that could fall into products.

The Inspection

Your personal conduct is always critical to a successful inspection. When conducting inspections in other facilities, such as those of contract packers, suppliers, etc. always check with someone who has authority to allow you full access to the property -- the higher the official, the better. Be concise in your conversations and have identification ready in case it is needed. Conclude your introductory session as soon as possible since managers and supervisors are usually very busy people. Always request that someone escort you during the inspection. A representative of the facility can often answer questions, give reasons for certain situations, explain new or unfamiliar equipment, etc., all items that often save considerable time. In the event of an accident or injury, management has a representative present to make sure proper procedures are followed. Most important, this provides an opportunity to make certain points clear, foster cooperation, and to convince management of your ability and integrity. Follow plant rules and other generally accepted rules such as no smoking, drinking, or eating except in designated areas. Do not walk on pallets, bags, cartons, etc. and do not ask employees questions without first obtaining permission to do so.

The Total Inspection

One should keep in mind that the inspection has multiple purposes. The primary purpose, of course, is to maintain the premises and products free of insect and other visible pests. This helps maintain the quality and reputation of the products being produced and strengthens the company image in the retail market. Since the inspectors are assumed to be specifically trained, they should take every opportunity to instruct personnel as appropriate. This, in itself, often has multiple positive results such as allowing others to report problems or potential problems but, more important, strengthens and verifies management's commitment to a pest-free environment. This does not need to be a long, formal process during the inspection. Subtle teaching through explanations and answering questions usually serves this purpose quite well.

Start Inspection Externally

When inspecting other facilities, some observations should be made as you approach. What is the general appearance of the surrounding areas? Is this an old rundown area with dilapidated and/or vacant buildings with garbage in the streets and with abundant pest harborages, or is the location a modern industrial park that is neat and well maintained? Are there animal food or other processing plants in the area that provide insect harborages or could create problems because they operate under different regulations? As you approach the site, a check of general appearances could be helpful later on. Check for open doors, open windows, overflowing garbage receptacles, screening, whether parking spaces are paved and maintained, etc. Check landscape and note the type of shrubbery and its location. Check for pallets against buildings, etc. All of these, and other items you might note, will often give an idea of what might be expected inside. Regardless, the exterior and surrounding areas should be checked in more detail later in the inspection.

While conducting the inspection, one should never hurry. Time must be taken to check completely and thoroughly. Yet, one must not delay unnecessarily. The inspector should also try to cause as little disruption as possible since the inspection itself is a disruption of normal events. Above all, do not make unreasonable demands.

Use All Senses

Once the inspection starts, your most important tools, not mentioned earlier, come into instant use. These are your senses, especially sight. Unfortunately most people have "tunnel vision" and usually cannot see much of what is before them. Fortunately or unfortunately, this "tunnel vision" is a biological fact which requires special training to overcome. Human vision is subjective, selective, and more often than most of us realize, unreliable. Untrained individuals, therefore, usually see what they want to see -- the subject of interest -- without paying attention to anything else because of this subjective selectivity of the eye and brain. This, as good inspectors and photographers know, can be overcome by concentration and experience thereby allowing a more objective view.

Other senses, such as smell and hearing are important. A scent of decaying or sour product alerts one to possible sources of infestations by those insects that are attracted to out-of-condition product. Hearing what others might say, deliberately or inadvertently, is often useful to the inspector. The senses are often used when other pests are included in the inspection. For example, the first clue to bird problems may well be the sounds of bird calls or flight. In some instances touch can be useful in areas that can be reached but not readily seen, but should be used judiciously.

Inspect Everywhere

During the inspection one should check everywhere. A suggested plan, or route, during the inspection is to follow the flow of processing from the entry of raw materials to the point where the finished product leaves the facility. Everything must be checked -- equipment, light fixtures, shelves, cabinets, overhead ledges, cracks, crevices, hollow hand rails, fire hoses, air filters, window ledges, etc. No space should be overlooked. The storage areas, raw materials storage, etc. must be examined. Wall-floor and wall-wall junctions are important. Forklifts, scrubbers, and cleaning equipment or other conveyances are also to be included.

During the inspection be sure to look around you. Do not become distracted by one item to the extent you miss something important. A good habit to form is to look all around you, including overhead, after you have stopped to check something. If you think you have missed something, do not be unwilling to go back and recheck. Do not hesitate to get on your hands and knees, or belly if necessary, to locate problems since pests enjoy the solitude of isolated hard-to-reach places.

The Sanitary Perimeter

The sanitary perimeter offers one of the best sources of information on visible pest evidence, especially in warehouses, and other storage areas. The inspector should carefully search the perimeters with the aid of a flashlight. The floors and walls should be checked for cracks and crevices that may contain product debris and/or insects. The small spatula, or similar tool, mentioned before is useful for digging accumulation from these locations. That portion of the floor under windows or lights is usually more productive since many insects are attracted to light and are found, either dead or alive, in this area. During this portion of the inspection the walls and overhead areas, as well as pallets, and spaces between pallets should also be checked.

Ledges and Flat Surfaces

Any ledges, pallets, slipsheet edges, bag seams, broken bags, etc. should be checked for product residue, insect trails, and insects. Any flat surface may contain dead insects that die from fogging or exhaustion. Pay particular attention to window ledges. Do not forget areas where canned goods are stored. These are often not cleaned and sometimes have a felt-like layer of spider webs, dust, and dead insects that provides ideal sites for insects such as carpet beetles that feed on dead insects. While these pose no threat to canned goods, the adults are often strong fliers and may infest other products some distance away.

Spillage

Spilled product or broken containers should be checked very carefully for insects, especially old product in cracks and between layers on pallets. Many types of spillage, often deep in warehouse stacks, provide ideal breeding sites for a variety of insects. Be alert for discarded food wrappers and drink containers. These can attract a variety of insects and other pests.

Doors and Windows

Frequent checks should be made for open, ill fitting or unscreened doors and windows as these are entry sites for insects and other pests. Always keep in mind that most insects that are classified as stored product pests existed in an external environment prior to adapting to the foods stored by man. Most stored-product pests can, and do, survive outdoors and infestations can occur from the outside.

Rodent Baits and Traps

Since the rodenticides contained in rodent bait are not detrimental to insects, rodent bait is very attractive to insects. This often becomes infested, and on too many occasions the boxes are refilled or replenished with insect infested bait. Careful searches of relatively inaccessible areas often reveal a "lost" bait box with heavy infestations of a number of different insect species. Rodent traps, especially the mechanical types that are enclosed, often contain dead rodents that, if not removed on a regular basis, attract flesh and blow flies and other insects. The traps should be examined in place, then lifted and the area they cover examined since insects frequently hide under these during daylight hours.

Flashlight Technique

As one proceeds, the flashlight beam should be directed over walls at an angle in order to reveal such insects as resting moths. This technique, due to the shadows created, makes insects and insect trails easier to detect. During this examination one can look for broken tiles, cracks in walls, wet areas, etc. that may be attractive to insects.

Light Traps and Electrocutors

Light traps and insect electrocutors should be checked for stored-product pests, especially for dermestids, mites, and if wet or moist, for fly maggots. The type with pull-out drawers is often installed above eye level and the accumulated dead insects behind the drawer may contain fly maggots or other insects. If stored-product pests are found in these traps, more searching for the original source is required.

Old, Outdated, and Returned Products

Old products are always suspect, especially if they remain undisturbed for long periods of time. These are often "on hold" by R&D personnel, and often consist of partial containers or bags of "leftovers" or samples stored in a laboratory cabinet or drawer. Repackaging areas often contain outdated returned goods which are in broken containers. These should always be assumed infested until proven otherwise. These repackaging areas merit special examination in considerable detail. This part of the inspection should not only include the returned goods but every container, bin, or shelf utilized in handling them. These areas are a prime source of infestations due to the broken containers with exposed product. This is especially true when the repackaging area is within the confines of the plant rather than in a separate isolated building.

Equipment

All equipment that is enclosed or that may accumulate product debris in belts, gears, etc. should be checked. This will often require extra effort and usually assistance in removal of panels, etc. to provide access. These include such items as conveyor belts, proofers, various filling machines, grinders, cleaning equipment, etc.

Unused equipment requires special mention. Too often equipment is removed, shut down, or stored without proper cleaning. Under other circumstances it is stored where product debris, especially dust, accumulates on and in this equipment. It is commonly infested with insects if not checked and/or cleaned frequently.

Additional Items and Areas

Other areas and items that require special mention include: electrical circuit boxes, frayed insulation, vending machines, garbage disposal and trash containers, desk drawers, including those of office personnel (they often contain snacks), repair shops, _all_ static areas such as overhead ceiling spaces, closets, under stairs, etc., dirty pallets, ductwork (intakes, exhaust and air-conditioning), filters, stored packaging materials, forklifts (often infested in certain operations), load levelers, tops of offices in warehouses and other overhead spaces, under stick-on labels on burlap bags, and, very important, the roof areas which may have accumulated debris resulting from vents or exhausts that can support insects.

Outdoor Areas

The outdoor areas of the facility should be checked as carefully as the indoor areas since they often provide harborage or potential harborage sites for insects. Such harborage sites as unused equipment, pallets, vegetable and fruit bins, and other items may contain overlooked product residue that will support insects. Also, there are usually small utility buildings, fire control stations, etc., which if they remain undisturbed for some time, are ideal for maintaining low level populations of insects.

One of the primary problems in outdoor areas is dry, dust-like material exhausted from the plant. When this is directed onto railroad tracks, or some other area where it accumulates, good insect breeding sites are created.

Final Inspection Comments

Keep in mind that many insects, especially stored-product pests, may stay completely out of sight, frequently in the product itself. The finding of small numbers, therefore, is often of considerable importance. Even one or two insects on a window ledge may mean a rather extensive infestation in an overhead area, a wall void, or in a product. These findings require an extensive search in order to locate the infestation or to be assured that no infestation exists. Some insects such as moths, cigarette beetles etc. are highly mobile in the adult stage and may be found considerable distances from the point of infestation.

REPORTING

The inspector should give a verbal report at the conclusion of the inspection and follow this with a written report. Both reports should be clear and concise. Any suggestions should be as practical as possible. Certain requirements may be placed on the inspector as far as reporting is concerned and these should be met whenever possible. When the format and contents are left to the inspector, a suggested method of presentation is to make a list of areas with detailed lists of discrepancies, findings, and possible actions. Long narratives require more time and effort to read and make corrections from than a simple list.

SUMMARY

The inspection must be considered and conducted in a serious manner, and have a definitive purpose -- to maintain plant premises and product as free as possible of insects and insect contamination as well as other visual pests. The inspector should be specifically trained for this job and should be familiar with stored-product pests and other insects encountered. The inspection should be preplanned to the extent possible, and basic tools and equipment prepared prior to the inspection. During the inspection, the inspector should check everywhere, including all static areas and equipment. Clear, concise verbal and written reports should be provided.

CHECKLIST

The following "checklist", gathered from various sources, is designed to point out a number of areas in which to look for insects and other pests. Because of the varied nature of individual manufacturing and storage operations, it, of necessity, cannot be totally complete and should, therefore, be modified to individual operations.

Structural Areas

1. Along the wall-floor junctions.
2. In corners at floor level. (These are often not cleaned well and old infested product may be found.)
3. Along wall-wall junctions.
4. In all cracks or holes in floors or walls.
5. On window sills and ledges.
6. Inside control panels, switchboxes, etc.
7. On the floor and in corners of any pits.
8. Along overhead rafters and pipes.
9. In areas above false ceilings, or in other voids.
10. Inside light fixtures.
11. Behind sliding fire doors.
12. Around rim and in baskets of floor drains.
13. Around sump pumps.
14. Under stairways.
15. On the floor and in corners of fire-protection (dry-valve) rooms.
16. Behind peeling baseboards.
17. Behind handrail supports or uneven surfaces.
18. In small depressions in block or cement walls.
19. All duct-work.
20. Sink-holes under buildings built on slabs.
21. Between layers of roofing, especially where there are open spaces.
22. Between the folds of fire hoses and in the seams of cloth covers for or inside the cabinets for these hoses.
23. On the floor and in the corners of any pits for scales, bucket elevators, etc.
24. In bottom of elevator pits.

Equipment

25. In dead spots inside equipment where material can accumulate.
26. Any voids, cracks, holes, etc. in or around equipment, e.g., false bottoms, areas where legs meet the floor, holes from missing bolts, area under equipment close to floor, etc.
27. In hollow handrails and equipment supports.
28. Unused equipment in storage.

Cleaning Related Items

29. In and under wet mops, particularly if stored on the floor.
30. In accumulations inside vacuum cleaners.
31. Between broom bristles, especially at the base of the bristles, where it is difficult to clean.
32. In accumulations in trash cans and behind plastic liners for trash containers. (Be sure containers in break rooms, lunch rooms, locker rooms, and restrooms are checked.)
33. In accumulations of floor sweepings.

34. In, behind, and under trays of insect electrocutors.
35. Inside and under mechanical rodent traps.
36. On rodent glue boards, in pheromone traps, etc.
37. In rodent bait and under bait stations.
38. In and under spider webs.

Production/Materials Areas

39. In and around damaged packages of edible material.
40. The interior spaces of pallets and on pallet contact surfaces.
41. Under and along edges of slip sheets.
42. Along the seams/folds of paper or burlap bags.
43. Under stick-on labels on burlap bags.
44. In hollow wooden or metal doors.
45. In and around areas used for storage of packaging materials.
46. In empty containers waiting to be filled.
47. In spills of any edible materials (don't forget about dextrins/starchy materials used as glues).
48. In tailings from sifters, particularly for edible materials.
49. In the materials removed from individual packages, e.g., cans, bottles, etc., when cleaning them.
50. In and around areas used for reclaim or salvage, particularly of edible materials.
51. Under lids, tarps, etc. used to cover bulk containers of edible material.

Miscellaneous Areas

52. In and around areas used for long-term or dead storage, including down equipment and above-office storage.
53. In any accumulation of wetted material(s).
54. Under and behind any items stored on the floor adjacent to walls, e.g., boxes, pipes, boards, dunnage, etc.
55. Inside pipes stored on racks or on floors along walls. Always check inside any cut off pipes that are not capped.
56. In loose insulation.
57. Inside desk drawers, especially in desks where food is stored or eaten.
58. Inside, under, and on top of storage cabinets.
59. Around the wheels of any mobile equipment, e.g., forklifts, mechanical sweepers, scrubbers, etc.
60. Inside and on top of employee lockers.
61. Inside the bottom of stand-type hollow ashtrays.
62. Between the folds of firehoses and in the seams of the cloth covers for the hose holders.
63. In the soil of potted plants (usually found only in offices and laboratories but sometimes in lobby/waiting rooms).
64. Under, behind, and if possible, inside vending machines.
65. In, behind, and under refrigerators and microwaves in lunchrooms.
66. In cabinets under sinks.
67. Inside pockets of company provided uniforms. (Also check in folds of jackets, etc. left hanging for long periods).
68. Under shelves, cabinets, etc. mounted close to floor.
69. Behind pictures and other decorative wall hangings.
70. In bins used to collect dirty clothes.
71. In food accumulations, particularly behind false walls, in trucks or railcars.
72. In the door channels and under slip sheets in railcars.

73. In the rooms of heating, air conditioning and ventilating units.
74. In areas used by outside contractors to store equipment.
75. In and around seals (curtains) for truck doors.
76. In and around any areas with off odors.

Outdoor Areas

77. In material in plugged or clogged gutters.
78. In any accumulations on the roof, particularly those around exhaust vents.
79. Inside all small sheds, fire-control houses, etc.
80. Around areas of standing water.
81. In bird nests, particularly those built on the main structure.
82. Around areas under lights.
83. In areas of vegetation adjacent to main structures.
84. Under piles of trash and debris.
85. In and around piles of pallets or accumulations of old or unused equipment.
86. In soil contaminated by food product(s).
87. Inside smoke vents.
88. Inside the bases for silos of edible products.
89. In and around garbage compactors/dumpers.
90. In and around areas used to store sawdust/wood chips for boiler fuel.
91. In areas under dock levelers.
92. All railroad sidings.

INSECT PEST IDENTIFICATION

George T. Okumura

Editor's Comments

Proper pest control is predicated on knowing what you are trying to control hence species identification is paramount. Knowing the species in a potential problem can help you decide what the risks are, what actions are indicated, and how prompt these actions need to be. This chapter illustrates a few of the common and important look-alikes, and it provides an extensive list of references.

Be aware that size of individual insects can vary as a function of maturity and food ingested.

Should insects represent a significant continuing potential risk to operations and products, you would be wise to have at least one informed individual on your staff and to have on file a list of individuals you can approach, as needed, for prompt, accurate species identification.

F. J. Baur

INSECT PEST IDENTIFICATION

George T. Okumura

Okumura Biological Institute (OBI)
6669 14th Street, Sacramento, California 95831

INTRODUCTION

One of the major problems in the food industry relative to sanitation is the live insect infestations and the presence of insect fragments in raw and finished products. In order to treat and prevent further problems of this nature, the pests involved must first be identified before control or eradication measures can be applied. Current regulations regarding labeling of pest control substances designed to protect consumers require application sites or locations be listed. Species identification as well as site conditions will be essential prior to selection of the proper insecticide.

Different pests, no matter how much they may look alike, occasionally do not respond to control measures in the same manner. One may be more resistant to a specific chemical than another; thus, the choice of control substance depends on an accurate identification. Also, certain species are more vulnerable to treatments in a particular life stage, while others are better treated during a different stage of development.

Any person charged with the responsibility for sanitation in the various aspects of the food industry should be aware of the many pest species likely to infest the products manufactured or handled. A case in point involved a California shipper who freighted several carloads of rice to a Midwest breakfast cereal company. The shipment was later found to be heavily infested with a sap beetle, which was determined to be Glischrochilus quadrisignatus. The cereal company wanted to reject the entire shipment on this basis, but it was pointed out that this beetle does not occur in California but does occur in the Midwest. Upon further investigation it was learned that the carloads of what had originally been dry rice had been exposed to rain during the period when their hatches were opened to inspect the grain at the destination point. This resulted in decomposition of the grain and mold formation, both of which were attractive to the beetles. The consignee (cereal company) was shown to be in error; therefore, they were liable.

One must also be able to differentiate between the look alike species and be able to identify those causing specific problems, or a person must know where to obtain the information to do this (i.e., appropriate literature sources, professional identifiers, private consultants, and/or specialty training programs).

Some of the more commonly used references are given at the end of this chapter. The professional identifiers (insect taxonomists or systematists), though few in number, are associated with the agricultural departments of some state governments, the U.S. Department of Agriculture and its cooperative extension services (at some land grant universities), and a few other colleges, universities, and museums. Some private consultants may be able to make the proper identifications. The food industry may prefer to have their own personnel trained by taking appropriate college classes or to have them attend instructional sessions tailored specifically to the industry's needs. Such sessions are provided in lecture series by some professional entomologists.

Before going to battle, it is important to know something about the habits of your enemy. Insects over the course of their usually relatively short lives go through developmental changes referred to as metamorphosis. The more primitive forms--such as silverfish, firebrats (Thysanura), cockroaches (Orthoptera), termites (Isoptera), earwigs (Dermaptera), and psocids (Psocoptera)--to name some that might be encountered in a warehouse situation, undergo gradual transformation from an egg stage through successive nymphal stages. The youngest nymphs are essentially like the adults, except for smaller size, lack of wings in certain groups where wings are present on adults, and sexual immaturity. This developmental process is known as incomplete metamorphosis.

The vast majority of insects, however, undergo dramatic changes during their development or life cycle. From eggs, they hatch into small larvae (usually worm-like in appearance), feed and grow larger through successive larval stages, enter a relatively immobile stage called the pupa, where amazing transformation occurs resulting in the emergence of winged, sexually mature adults, bearing no resemblance to the earlier stages. This developmental sequence is called complete metamorphosis. The following large groups of insects (Orders), which contain species known to impact on the food industry, develop through complete metamorphosis: beetles (Coleoptera); moths (Lepidoptera); flies (Diptera); and bees, wasps, and ants (Hymenoptera).

An example of how this kind of knowledge can be important in resolving problems arising with food sanitation is given below:

A woman decided to file suit against the processor of a particular package of pizza mix after having discovered moth larvae in the product. Her claim against the retail store and the manufacturer of the pizza mix alleged that upon opening the package, six months subsequent to its purchase, she became ill at the sight of the "worms." It was explained to the lawyer of the would-be plaintiff that this larva, identified as the Indianmeal moth (Plodia interpunctella), undergoes complete metamorphosis, requiring about six weeks to complete development from egg to adult when exposed to an average temperature of 80° F. Had the package been infested at the time of purchase, there would have to be remnants of pupal and adult stages present in the mix six months later. Since none were found, the product must have become infested more recently while in storage at the woman's home. The suit was dropped.

IMPORTANCE OF BEHAVIOR

The behavior of two similarly appearing insects may differ significantly and, therefore, be a factor in preparing the appropriate eradication or control procedures. For example, consider a warehouse in which the storage at one end is infested by a beetle pest but that at the opposite end is not. One should be more concerned about taking immediate and more comprehensive control measures if the responsible insect is the red flour beetle (Tribolium castaneum), since it can fly and could quickly infest the other areas. A similar looking and related pest, the confused flour beelte (Tribolium confusum), does not fly and cannot spread as quickly, thus permitting more time for control.

A person trained to identify the pests most likely to be found infesting certain food products would be able to disregard incidental non-pest species which are occasionally discovered in the vicinity of the products. A competent sanitarian could utilize detection traps on the site, such as black light electrocutor traps, baited sticky traps, sex pheromone traps, etc., by determining on the basis of captured specimens which products are in danger or are already infested by certain species. For instance, if a detection device was trapping bean weevils (Acanthoscelides obtectus), one should know to look for the potential infestation site only where beans are stored, since this is the only host for bean weevils, thus eliminating the costly and time consuming effort of searching all available products in the storage areas.

DESCRIPTIONS OF SOME SELECTED PESTS

Grain beetles - The sawtoothed grain beetle (Oryzaephilus surinamensis) and the merchant grain beetle (O. mercator) are widely distributed in the U.S. and Canada and look very much alike. The sawtoothed species is a major pest of grain, cereals, flour, and other stored products while the merchant grain species is associated more often with dried fruits, nuts, seeds, and dry pet foods. Adults of both species are about 2.5 mm long, have thoraxes with six tooth-like projections on each side, and are flattish, slender, and brownish in color. To separate the two, one must compare the size of the tubercle behind the eye. If the width of the tubercle is less than half the vertical diameter of the eye, it is the merchant grain beetle while in the sawtoothed grain beetle the tubercle is greater than half the vertical diameter of the eye. The merchant grain beetle flies while the sawtoothed beetle does not.

Flour beetles - Two other closely related and similar appearing pests are the confused flour beetle (Tribolium confusum), which does not fly, and the red flour beetle (T. castaneum), which does fly. The adults are about 3.2 mm long, reddish brown, and have longitudinal ridges on the wing covers. Examining the head of a beetle ventrally, the space between the eyes of the confused flour beetle is nearly three times the longitudinal diameter of the eye, and the terminal three segments of the antennae become gradually larger; while in the red flour beetle the space between the eyes is much smaller and the terminal three segments of the antennae are abruptly enlarged and nearly of equal size. The confused flour beetle is reputed to be the most common and damaging pest of flour in the U.S. Both species infest flour and cereals.

The cigarette beetle (Lasioderma serricorne) and the drugstore beetle (Stegobium paniceum) are small beetles (2.5 mm long), oval, reddish brown to light brown, and have the head hidden from view when observed from above. The cigarette beetle has the antennae serrate with the segments of uniform size throughout, and lacks striations on the wing covers; the drugstore beetle has the last three segments of the antennae much larger than the other segments, and the wing covers are striated and conspicuously pubescent. Both species are cosmopolitan in distribution and can be found in a great variety of stored foods, tobacco, and drugs.

The warehouse beetle (Trogoderma variabile) is another small oblong beetle, about 3 mm long, and brown-black with brownish orange, wavy lines running across the wing covers from side to side. The fuzzy looking larvae are covered with arrow-shaped and rat-tail-like hairs which when ingested can cause illness to humans. This beetle is a major pest of stored foods and other animal and vegetable products in many parts of the world. Other species of Trogoderma are serious pests, particularly the khapra beetle (T. granarium), a species currently being eradicated from the U.S. by the government. If a khapra beetle is found on a premise, that establishment must shut down for one to two months or until further critical inspection and fumigation by methyl bromide. In the past, misidentification has occurred and unnecessary treatment was applied. Therefore, it behooves one to seek out an expert to have the Trogoderma group identified correctly.

The cadelle (Tenebroides mauritanicus) is a cosmopolitan species and is considered a serious pest of flour mills, granaries, warehouses, ships, and stores, where they mainly infest grains, cereals, and nuts. These elongated and relatively large beetles (5-10 mm) are generally flattened, shiny black to brown, have deep striations on the wing covers, and dense and coarse punctations on the head and thorax.

46

There are three most important beetle pests of stored grain. One, the lesser grain borer (Rhyzopertha dominica)--originally from the tropics but now throughout the U.S.--is about 2.4 mm long, elongate in shape and dark brown to black in color, with a serrated antennal club of three segments. The other two pests are weevils, distinguishable by their long slender snouts: the rice weevil (Sitophilus oryzae) is about 3.2 mm long and reddish brown to nearly black, top of the thorax having circular punctures, and the wing covers having four yellow-orange spots; the granary weevil (S. granarius) is slightly larger (about 4.8 mm long), brown to brownish black, top of the thorax having elongated punctures, and no yellow-orange spots on the wing covers. The latter species cannot fly and prefers cooler temperature climates, while the rice weevil can fly and prefers warmer regions.

No less important that the beetles as pests of food products in storage are certain species of moths. Unlike the beetles, whose adult and larval stages both cause feeding damage, only the larval stages of moths are capable of feeding on the stored food products.

Moth pests of grain - The European grain moth (Nemapogon granella) is a widespread pest of all kinds of grains in storage, as is the Angoumois grain moth (Sitotroga cerealella). The adults of both are small (about 10-12 mm in wing spread), the former species being light gray, densely speckled with black, while the Angoumois grain moth is yellowish brown, often with a single black dot near the center of the wing, and the hind wing is notched on the apex. The larva of the Angoumois grain moth feeds within kernels of grain while the other feeds on the surface.

Phycitine moth pests - Included in this group are five of the most common, more generalized pests of stored products. The first four listed appear much alike, being basically pale brownish-gray in color, with a few darker wavy transverse bands across the forewings and having a wing spread ranging from about 18 to 25 mm. A closer examination of internal structures (genitalia specifically) may be required to make a positive identification of the exact species. The larvae are usually creamy white, often with a pinkish cast and with a brownish head. They generally feed on the surface, spoiling the products, not only by their feeding, but by webbing material together with their droppings (frass) into an unsightly and disagreeable mess. The species included here are: The Mediterranean flour moth (Anagasta kuehniella), tobacco moth (Ephestia elutella), almond moth (Cadra cautella), and raisin moth (C. figulilella). A fifth phycitine species in this same category of general stored product pests is the Indianmeal moth (Plodia interpunctella). This moth is in the same size range as those above but differs significantly in general color patterns. It has forewings which are whitish-gray on the basal (inner) one-third and deep red-brown on the apical (outer) two-thirds.

Storage areas often are infested with one or more species of cockroach. They are generally scavengers, foraging at night on most anything organic. Due to their habits, they could potentially spread disease organisms to stored foods or otherwise contaminate them with their droppings, egg cases, and offensive odors. The following species of cockroaches are found indoors in North America: The American cockroach (Periplaneta americana); Australian cockroach (P. australasiae); Oriental cockroach (Blatta orientalis); German cockroach (Blattella germanica); and brown-banded cockroach (Supella longipalpa). The largest of these is the American cockroach, being 38 to 51 mm long; its prothorax is dark and surrounded peripherally by a whitish color. The mature Australian cockroach is about 32 mm long and has yellow markings on the thorax and yellow streaks at the base of the wings. The Oriental cockroach is about 25 mm long and has an overall dark color. The most commonly encountered cockroach in the U.S. is the German cockroach, which measures about 13 mm in length and is recognizable by a pair of parallel dark vertical stripes on the prothorax.

Though only slightly larger than the German, the brown-banded differs by having transverse dark bands on the wings.

The following pests are usually associated with food products which have deteriorated into poor condition. The flat grain beetle (Crytolestes pusillus), as the name implies, is flat, reddish brown in color and distinctive in being one of the smallest stored product pests (1.5-2 mm long). Even smaller still (about 1 mm long on average) are the psocids (Psocoptera). They are white to pale gray in color. Psocids apparently prefer to feed on the mold formed on the products when kept in high humidity. A large group of non-insects, the mites, have only a few members which can be found on occasion associated with food such as: the grain mite (Acarus siro); cheese mite (Tyrolichus casei); and dried fruit mite (Carpoglyphus lactis). Mites on the whole are some of the most difficult pests to identify, mainly due to their very small size, requiring special handling and microscopic techniques.

SUMMARY

The above brief discussion of some of the most common and most widely distributed species of known pests of stored food is obviously meant only to touch on some of the more important potential pests a food sanitation official may encounter. Other species of beetles and moths, plus some insects of other Orders, can cause serious problems in certain localities and under circumstances conducive to their survival. Therefore, a general familiarization with a few of these will not be adequate for someone who must protect products in storage. Much more extensive training is necessary. Additionally, this person must know what literature is helpful and what experts are available to help in making troublesome identifications (i.e., Trogoderma spp., mites), or for other consultations.

In any insect problem relative to pest control, customer complaints, lawsuits, insect damage, or contamination, etc., the first important step is the identification of the insect. After correctly identifying the insect, one can obtain from a literature search necessary detailed information such as its economic status, behavior, medical implication, control, and other pertinent facts to assist in resolving the problem.

Insects must go through developmental stages before reaching adulthood, the reproductive stage. Complete and incomplete metamorphoses are the two generalized types of development. The former type of metamorphosis involves drastic changes through four stages: egg, larval, pupal, and adult, which is the case for insects such as the beetles and the moths. Incomplete metamorphosis involves three gradual stages: egg, nymph, and adult; examples of insects with this type of life history are cockroaches, psocids, and silverfish. Mites are not classified as insects, and they go through egg, larval, and nymphal stages before becoming adults.

The major groups of pests in the food industry are the beetles (Coleoptera) and moths (Lepidoptera). Some of the more important and common beetles are the sawtoothed grain beetle, merchant grain beetle, confused flour beetle, red flour beetle, cigarette beetle, drugstore beetle, warehouse beetle, flat grain beetle, cadelle, lesser grain borer, rice weevil, and granary weevil. The moths include European grain moth, Angoumois grain moth, Mediterranean flour moth, tobacco moth, almond moth, raisin moth, and Indianmeal moth. Cockroaches often cause problems. The species most involved with food industry are the American cockroach, Australian cockroach, Oriental cockroach, German cockroach, and brown-banded cockroach. Mites, psocids, and silverfish are usually considered secondary or minor pests.

SELECTED REFERENCES

Aitken, Audrey D. 1963. A key to the larvae of some species of Phycitinae (Lepidoptera, Pyralidae) associated with stored products, and of some

48

related species. Bul. Entomol. Res., 54 (part 2):175-188, illus.

Arnett, Ross H., Jr. 1963. The Beetles of the United States: A Manual for Identification. Washington, D.C.: Catholic Univ. of Amer. Press, 1112 pp., illus.

Borror, Donald J., and Dwight M. Delong. 1970 (3rd ed.). An Introduction to the Study of Insects. New York: Holt, Rinehart, and Winston, 812 pp., illus.

Campbell, W. V. 1969. Stored Grain Insects and Their Control in the Middle Atlantic States. Bul. No. 75, South. Coop. Series, North Carolina Agr. Exp. Sta., Raleigh, 30 pp., 32 figs.

Chu, H. F. 1949. How to Know the Immature Insects. Dubuque, Iowa: Wm. C. Brown Co., 234 pp., 631 figs.

Comstock, J. H. 1949. An Introduction to Entomology. Ithaca, N.Y.: Comstock Pub. Co., 1064 pp., 1228 figs.

Dillon, Elizabeth S., and S. Lawrence. 1961. A Manual of Common Beetles of Eastern North America. Evanston, Ill.: Row, Peterson, 884 pp., 544 figs.

Dunstan, G. Gordon, and Thomas R. Davidson. 1967. Diseases, Insects, and Mites of Stone Fruits. Pub. 915, Canada Dept. Agr., Ottawa, 48 pp., 30 figs.

Ebeling, Walter. 1959. Subtropical Fruit Pests. Div. Agr. Sci., Univ. of California, Berkeley, 435 pp., 160 figs.

Ebeling, Walter. 1975. Urban Entomology. Univ. of California, Div. of Agr. Sci., 695 pp., illus.

Essig, E. O. 1926. Insects of Western North America. New York: MacMillan, 1036 pp., 766 figs.

Essig, E. O. 1951. College Entomology. New York: MacMillan Co., 900 pp., 308 figs.

Falcon, Louis A., et al. 1967. Light Traps and Moth Identifications: An Aid for Detecting Insect Outbreaks. Leaf. 197, Agr. Exp. Sta., Univ. of California, Berkeley, 16 pp., 4 pls.

Frost, S. W. 1942 (reprint: Dover Pubs., New York). Insect Life and Insect Natural History (orig. entitled General Entomology). New York: McGraw-Hill Book Co., 526 pp., illus.

Helfer, Jacques R. 1962. How to Know the Grasshoppers, Cockroaches, and Their Allies. Dubuque, Iowa: Wm. C. Brown Co., 351 pp., 579 figs.

Holland, W. J. 1914 (reprint: Dover Pubs., New York). The Moth Book. New York: Doubleday, Page & Co., 479 pp., 263 figs.

Jaques, H. E. 1951. How to Know the Beetles. Dubuque, Iowa: Wm. C. Brown Co., 372 pp., 865 figs.

Jaques, H. E. 1951. How to Know the Insects. Dubuque, Iowa: Wm. C. Brown Co., 205 pp., 411 figs.

Jeannel, R. 1960. Introduction to Entomology. London: Hutchinson, 344 pp., 150 figs.

Kono, Tokuwo, and Charles S. Papp. 1977. Handbook of Agricultural Pests: Aphids, Thrips, Mites, Snails, and Slugs. California Dept. of Food & Agr., Sacramento, 205 pp., illus.

Kurtz, O'Dean L., and Kenton L. Harris. 1963. Micro-analytical Entomology for Food Sanitation Control. Washington, D.C.: Assoc. Official Agr. Chemists, 576 pp., 825 figs.

Lutz, Frank E. 1948. Field Book of Insects. New York: G. P. Putnam's Sons, 510 pp., 100 pls.

Mallis, Arnold. 1983. Handbook of Pest Control. New York: MacNair-Dorland Co., 1101 pp., illus.

Matheson, Robert. 1951. Entomology for Introductory Courses. Ithaca, N.Y.: Comstock Pub. Co., 629 pp., 500 figs.

Merrill, L. G., Jr., et al. 1956-61. Economic Insects of New Jersey. Bul. 293, 295, 296, 305, 306, 321, 354, Ext. Serv., Rutgers Univ., New Brunswick, 112 pp., illus.

Metcalf, C. L., and W. P. Flint. 1962. Destructive and Useful Insects. New York: McGraw-Hill Book Co., 1087 pp., illus.

Munro, J. W. 1966. Pests of Stored Products. London: Hutchinson & Co., Ltd.,

234 pp., illus.

Okumura, George T. 1961. Identification of Lepidopterous Larvae Attacking Cotton with Illustrated Key. Bur. Ent. Spec. Pub. No. 282, Dept. Agr., State of California, Sacramento, 80 pp., 52 figs.

Okumura, George T. 1972. Warehouse Beetle a Major Pest of Stored Food. Nat. Pest Cont. Opr. News, 32(1):4, 4 figs.

Peairs, L. M. 1948. Insect Pests of Farm, Garden, and Orchard. New York: John Wiley & Sons, 549 pp., 648 figs.

Peterson, Alvah. 1948 (4th ed., 1962). Larvae of Insects: Part I. Lipidoptera and Hymenoptera. Ann Arbor, Michigan: Edwards Brothers, Inc., 315 pp., illus.

Peterson, Alvah. 1951 (4th ed., 1960). Larvae of Insects: Part II. Coleoptera, Diptera, Neuroptera, Siphonaptera, Mecoptera, Trichoptera. Ann Arbor, Michigan: Edwards Brothers, 416 pp., illus.

Randolph, N. M., and C. F. Garner. 1961. Insects Attacking Forage Crops. Bul. 975, Agr. Ext. Serv., Texas A & M, College Station, 26 pp., 44 figs.

Ross, Herbert H. 1966. A Textbook of Entomology. New York: John Wiley & Sons, 539 pp., 401 figs.

Smith, Marion R. 1965. House-infesting Ants of Eastern United States: Their Recognition, Biology, and Economic Importance. Tech. Bul. No. 1326, U.S. Dept. Agr., Washington, D.C., 105 pp., 50 figs.

Swain, Lester A. 1948. The Insect Guide. Garden City, N.Y.: Doubleday & Co., 261 pp., 175 figs.

Swan, Lester A., and Charles S. Papp. 1972. The Common Insects of North America. New York: Harper & Row, Pub., Inc., 750 pp., 1422 figs.

Wigglesworth, V. B. 1964. The Life of Insects. Cleveland, Ohio: World Pub. Co., 359 pp., 164 figs.

Wright, J. M., and J. W. Apple. 1959. Common Vegetable Insects. Cir. 671, Coop. Ext. Serv., Univ. of Illinois, Urbana, 38 pp., illus.

IMPORTANT BEHAVIORAL ASPECTS OF SELECTED INSECTS
(A Table)

Fred J. Baur

Editor's Comments

The preceding chapter covered insect pest identification. This chapter summarizes, in tabular form, some information on insect behavior scattered throughout the literature and some unpublished. Both infesting and environmental insect species are listed. Species judged likely to be encountered in plants were selected for illustration purposes and the value of the information presented. Fleas and mites were added to the environmental portion of the table since occasionally employees encounter these in the home.

It is hoped that this introductory effort will stimulate entomologists to improve on its completeness and accuracy.

Remember, behavior is the key to the selection of optimum control measures and to the urgency of action.

F. J. Baur

IMPORTANT BEHAVIORAL CHARACTERISTICS OF SELECTED INSECTS
(A Table)

Fred J. Baur
The Procter & Gamble Company
6090 Center Hill Road
Cincinnati, Ohio 45224*

INTRODUCTION

In any problem, species identification is paramount. Insect behavior, habits, and capabilities vary so much from species to species that the risk and control steps can be understood properly only if the species is accurately identified. Would you prefer to have in your plant an insect that only crawls or one which both crawls and flies? Is it important that the insect might like to hide in your packaging materials or perhaps penetrate them and enter a finished product? A better understanding of insect behavior will help put and keep concerns in perspective and optimize avoidance and control efforts.

The purpose of each column in the table should be self-evident. The information is believed to be reasonably accurate. It will, at least, give you some idea of variability in behavior and a comparison between species. Be aware that the experts do not always agree on behavior or capabilities; hence, in some instances, ranges are listed, and in others more than one opinion is listed. The presence of only one figure or range tends to indicate unanimity. The writer took the liberty of rounding off most figures to reflect the accuracy of the data. The data are for favorable conditions generally judged in the 75-90°F (24-32°C) and the 50-60% RH ranges, with ample food. Variations in the real world can be extreme.

Differences of opinion also arose with the four yes or no answer columns for flying, reaction to light, found in light traps, and the ability to penetrate packaging film. The positions taken reflect the majority opinion; as examples, (1) the sawtoothed grain beetle is said to not fly, (2) the sawtoothed grain beetle can penetrate a package although this ability is very weak, and (3) the Indianmeal moth is sufficiently attracted to light to be found in an electrocutor tray. The available information on attractiveness of electrocutors to flying infesting-type insects is sparse, hence the column dealing with "found in light traps" is not listed. You can infer that if the species flies and is attracted to light, it probably can be found in the trays of insect electrocutors. Most studies have dealt only with the environmental-type insects, predominantly flies.

A few additional remarks are in order on the important aspect of penetration of packages. The aim to simplify or generalize by a yes or no answer may lead the reader astray. This ability is subject to wide variability. The key variables are: the chemical nature, the thickness, and the toughness of the film; the condition of the end seals, if present; and, of course, the insects' innate capability to chew or make holes, whether adult or larval stages. Use of the word "yes" says there is some ability. Very few species can penetrate essentially all flexible films. The cadelle, cigarette beetle, and lesser grain borer are species that can. Foil, polypropylene, and polyester tend to be the most resistant films. If your need requires that you be sure, consult the appropriate literature starting with chapter 23 of this book.

*Present address: 1545 Larry Avenue, Cincinnati, Ohio 45224

Remember, the information in the table is presented to give a comparative picture on a few species as a guide to proper concern. This is a first attempt hence far from complete. There could be inaccuracies, and the non-uniform handling, at times, reflects the variability in available information. It is hoped that this effort will stimulate an improvement, one that will also expand on the behavioral aspects covered. An expansion should include items such as reactions of insects in a product or spill to heat and to disturbance. Insects vary in their tendencies to rise to the surface under such exposures. A large undertaking is required and needed.

IMPORTANT BEHAVIORAL CHARACTERISTICS OF SOME INSECTS FOUND IN PLANTS (PRIMARILY FOOD)

Product-Infesting Species

Species	Size in.	Size mm	Type(s) of food	Approx. no. of eggs per female (lifetime)	Approx. life cycle[a] (days)	Approx. adult lifetime (days)	Able to fly[b]	Attracted to light[b]	Able to penetrate flexible packaging materials[c]	Other comments
Angoumois grain moth	1/2-2/3	13-17	grains, cereals	40, 400, 50-300	35, 40-65	7-10, 2-14	Yes	Yes	Yes	More in southern U.S.; night fliers
Black carpet beetle	1/8-3/16	3-5	animal products, grains, cereals	100	180-365	15-30, 30-60	Yes	Yes	Yes	Found in cracks in warehouse floors; adults very destructive, larvae very resistant
Cadelle beetle	1/3	8	grain (germ), rice, flour, nuts, fruits	1,000	65-135 (270-410 if hibernates)	360-720	Yes	?	Yes	Frequently in grain elevators
Cigarette beetle (see Drug store beetle)	1/10-1/8	2-3	wide variety of dried vegetable products, tobacco, cocoa, nuts	20-100	30-55	14-28, 14-42	Yes	Yes	Yes	Strong flier and penetrator; likes warm areas
Coffee bean weevil	1/8	3	coffee, corn	50	30-35, 30-70	85-135	Yes	?	Yes	Leaps on window panes
Confused flour beetle (see Red flour beetle)	1/7	3.5	grains and grain products; omnivorous feeder	450	40	365, 540	No	No	Yes	Many consider most serious pest; confused with red flour beetle
Drug store beetle (see Cigarette beetle)	1/10-1/8	2-3	Omnivorous; vegetable products, condiments	20-100	60-210	14-28, 14-42	Yes	Yes	Yes	Likes warmer temperatures; serrated back

54

Name			Food						Remarks
Flat grain beetle (rusty grain similar)	1/16	1.6	grain, processed food, corn, dates	240, 100-400	45-65	365	Yes	Yes	Yes (weak)	Likes moist, moldy conditions
Granary weevil	1/8-3/16	3-5	Compressed farinaceous materials	50-250	30-50	210-240	No	No	Yes (weak)	Bores with mandibles; plugs holes; northern U.S.; can go 2-3 weeks without eating
Indianmeal moth	5/8	16	grain, flour, corn, nuts, dried foodstuffs	100-300	40-55, 25-130, 25-305	5-13, 7-25	Yes	Yes	Yes	Night fliers; larvae like to wander; check frass; may overwinter as larvae
Lesser grain borer	1/8	3	grains, milled grains	300-500	30	ca. 180	Yes	Yes	Yes	Primarily south and central states; larvae very active
Meal moth	1	2.5	cereals, cereal products, dried vegetable matter	200-400, 120-400	40-55, 35-70	7	Yes	Yes	Yes	Webbing binds up food being eaten
Mealworm (yellow)	>1/2	>13	grain, wet and moldy grain and refuse	280, 500	120-730, 520	60-90	No	No	Yes	Likes dark; tremendous flight range; attracted to souring odors
Mediterranean flour moth	<1	<25	flour, nuts, chocolate, dried fruits, stored foods	120-680, 100-400	30-40, 55-65, 45-130	7, 9-14	Yes	Yes	Yes	Night fliers; flour mill pest; webbing clogs machinery; may overwinter as larvae
Merchant grain beetle (see Sawtoothed grain beetle)	1/7	3.5	grain and grain products	150, 260	30-40, 25-100	125	Yes	Yes	Yes (weak)	Sawtoothed look-alike; tubercle less than 1/2 diameter of eye

a Egg to adult, favorable conditions.
b If can fly and are attracted to light, probably will be found in insect electrocutors.
c Paper, polyethylene, or comparable. Both larval and adult stages of beetles have mandibles (chewing parts), only larval stages for moths.

55

(continued on next page)

IMPORTANT BEHAVIORAL CHARACTERISTICS OF SOME INSECTS (Continued)

Product-Infesting Species (Continued)

Species	Size in.	Size mm	Type(s) of food	Approx. no. of eggs per female (lifetime)	Approx. life cycle[a] (days)	Approx. adult lifetime (days)	Able to fly[b]	Attracted to light[b]	Able to penetrate flexible packaging materials[c]	Other comments
Red flour beetle (see Confused flour beetle)	1/7	3.5	stored grain and grain products	350, 450	40–90	365, 380	Yes	Yes	Yes (weak)	Serious pest; differentiate from confused flour beetle by antennae and eyes
Rice moth	1/2–7/8	13–24	rice, nuts, chocolate	100–200, 40–400	40, 45–140	7–14, 2–26	Yes	Yes	Yes	Very dense webbing
Rice weevil	1/10–1/8	2–3	grain borer (like granaries)	300–400	30	90–185, 120–150	Yes	Yes	Yes (weak)	Southern states
Sawtoothed grain beetle (see Merchant grain beetle)	1/10–1/8	2–3	grain and grain products, dried fruits, chocolate, sugar	280, 300, 50–250	20–80, 30–50	135, 180–300, 180–1,100	No	No	Yes (weak)	One of most important; difficult to eradicate
Spider beetle (hairy)	1/10–1/7	2–3.5	variety, including grain and grain products	40–120	90–120	?	No	?	Yes	Likes 35–40°F; looks like a spider
Warehouse beetle	1/8	3	most anything; likes protein	10–45, 100	40–50	10, 28–42	Yes	Yes	Yes	Larvae like dark and hibernate (diapause) as long as 3 years

56

Species	Description	Habitat	Type(s) of food	Approx. no. of eggs per female (lifetime)	Approx. life cycle[a] (days)	Able to fly	Attracted to light	Found in light traps	Able to penetrate flexible packaging materials	Non-chemical control	Chemical control (label permitting)	Other comments
Ant, carpenter	0.25-0.5 in. (6-13 mm) long; reddish brown or black with a constricted waist and 2 pairs of wings with forewings larger than hind wings	Indoors in wood that is moist, decayed, or previously damaged by termites, such as ceilings, walls, floors, window sills, and doors; outdoors in dead, decaying trees	Dead insects, carrion, household foods, and other organic debris	Variable with food supply, queen dynasty, size of colony	Varies according to cast, season, and temperatures	Reproductives	Not highly	Yes	Yes	Protect wood with varnish, paint, etc.; replace wood if damage is extensive	Apply a residual spray or dust inside holes and to surface of wood; after a few days, plug holes with plastic wood, putty, etc.	Widely distributed throughout U.S.; dwell in and excavate wood; tend to bite
Ant, crazy	About 0.06 in. (2 mm) long; dark brown to black; body slender with long, 12-segmented antennae and one node on the petiole[d]	Outdoors in trash, cracks in trees, rotten wood, and in soil under objects; indoors in walls behind baseboards, in rotten wood, and underneath houses	Omnivorous; seeds, honeydew sap, fruits, household foods, and dead and live insects	Variable with food supply, queen dynasty, size of colony	Varies according to cast, season, and temperatures	Reproductives	Not highly	Yes	Yes	Good housekeeping and cleanliness	Indoors via baits, dusts, and spot treatment with residual sprays; outdoors via baits, granules, dusts, and sprays	Generally found throughout U.S.; will get into plants; sugar main attractant

[d]Attachment between abdomen and thorax.

(continued on next page)

57

IMPORTANT BEHAVIORAL CHARACTERISTICS OF SOME INSECTS (Continued)

Environmental (Household) Species (Continued)

Species	Description	Habitat	Type(s) of food	Approx. no. of eggs per female (lifetime)	Approx. life cycle[a] (days)	Able to fly	Attracted to light	Found in light traps	Able to penetrate flexible packaging materials	Non-chemical control	Chemical control (label permitting)	Other comments
Ant, little black	Less than 0.1 in. (<3 mm) long; jet black; 2 nodes on petiole	Indoors in woodwork or masonry of buildings; generally outdoors beneath rocks in lawns, and soil	Omnivorous; insects, honeydew, nectar, pollen, and household foods	Variable with food supply, queen dynasty, size of colony	Varies according to cast, season, and temperatures	Reproductives	Not highly	Yes	Yes	Good housekeeping and cleanliness	Indoors via baits, dusts, and spot treatment with residual sprays; outdoors via baits, granules, dusts, and sprays	Generally found throughout U.S.; will get into plants; sugar main attractant
Ant, pharaoh	Less than 0.1 in. (<3 mm) long; light yellow to reddish; 2 nodes on the petiole; clubbed antennae with last 3 segments enlarged	Outdoors in lawns and gardens; indoors in wall voids, subfloor areas, attics, cracks and crevices behind baseboards and plaster, and in furniture	Omnivorous; plants, household foods, and dead and live insects; likes fatty foods	Variable with food supply, queen dynasty, size of colony	Varies according to cast, season, and temperatures	No (adults have wings)	Not highly	No	Yes	Good housekeeping and cleanliness	Baits preferred; sprays and dusts tend to fractionate colony	Generally found in the U.S. but not in California; will get into plants; sugar main attractant

58

Pest	Description	Habits	Food							Nonchemical control	Chemical control	Remarks
Booklice (psocids)	Small, soft-bodied; about 0.04 in. (1 mm) long; both winged and wingless forms	Outdoors on bark and foliage; indoors on books and papers, and in humid areas	Molds, fungi, cereals, pollen, dead insects	60-100, 200	20, 30-60, 110	Some species (common ones no)	No	No		Reduce moisture and eliminate mold formation; clean infested products; clean up spilled foods	Spot treat infested surfaces	Dense webbing from winged psocids; common species difficult to detect due to size and lack of color; check packaging materials
Centipedes	Many segments and legs—one pair legs per segment; brownish, flattened arthropods; 1 in. (25 mm) or more long	Active at night as predators of insects and spiders; usually associated with damp, dark places; indoors may be found in closets and bathrooms	Insects and spiders	Varies with numerous species; example is 45	1,095	No	No	No		Remove individuals from building; eliminate breeding areas outdoors	Residual sprays to and near foundation and walls and around steps; dust crawl spaces	Remove accumulations near and in hiding and breeding areas; damp locations; under accumulations
Cockroach, American	1-1.5 in. (25-38 mm) long; reddish brown with yellow band around edge of pronotum; slender cercif with last segment twice as long as wide	Outdoors in palms and decaying trees; indoors throughout buildings but commonly around pipes, sewer drains, and wet and damp areas	Omni-vorous; food-stuffs, starch, waste refuse, soap, residues, etc.	95-225	180-400, 365-1,095	No	Glides (at least)	Yes	Yes	Daily sanitation of food areas; prevent entrance by sealing all cracks and openings	Residual sprays or dusts indoors with perimeter treatments outside	Called palmetto bugs, shadroaches; likes warm, dark, moist places

eHardened body wall plate on back—prothorax.
fPair of appendages at end of abdomen.

59

(continued on next page)

IMPORTANT BEHAVIORAL CHARACTERISTICS OF SOME INSECTS (Continued)

Environmental (Household) Species (Continued)

Species	Description	Habitat	Type(s) of food	Approx. no. of eggs per female (lifetime)	Approx. life cycle[a] (days)	Able to fly	Attracted to light	Found in light traps	Able to penetrate flexible packaging materials	Non-chemical control	Chemical control (label permitting)	Other comments
Cockroach, Australian	0.75–1.5 in. (19–38 mm) long; reddish brown with yellow markings on pronotum and a yellow streak on lateral wing margins	Outdoors in gardens and greenhouses; indoors throughout houses and buildings	Anything organic; therefore, any stored food products, grains, etc.	500	250–440, average 350	Yes	No	Yes (probably)	Yes	Prevent entrance into houses and buildings by sealing all cracks and openings; inspect all plants brought inside; daily housekeeping	Residual sprays or dusts indoors with perimeter treatments outside	Closely resembles American, but smaller; prefers warmer climates
Cockroach, brown	1–1.5 in. (25–38 mm) long; dark reddish brown with yellowish margin of the pronotum; cerci "blunt," triangular, and less than twice as long as wide	Outdoors in palms and decaying trees; indoors throughout houses and buildings	Anything organic; therefore, any stored food products, grains, etc.	240–290	340–350	Yes	No	Yes (probably)	Yes	Daily housekeeping; sanitation of food areas; prevent entrance into houses and buildings by sealing all cracks and openings	Residual sprays or dusts indoors with perimeter treatments outside	Also can be confused with American, mainly in Southeast; much less common than American

Cockroach	Description	Habitat	Diet							Control	Treatment	Remarks
Cockroach, brown banded	0.5–0.6 in. (13–16 mm) long; dark brown; 2 pale brown bands traversing the wings	Widely distributed throughout buildings; most abundant in food areas	Omni-vorous; food stuffs, starch, etc.	175	Yes	No	90–280, 95–280	Yes	Yes	Daily sani-tation of food areas; prevent entrance by sealing all cracks and crevices	Residual sprays or dusts	Behind and beneath shelves, cabinets, etc.; under sinks, tables; not near water; runs extremely fast
Cockroach, Florida	1.6–1.9 in. (41–48 mm) long; deep reddish brown to black; forewings extend not further than the mesonoium[9]	Mostly out-doors under logs, palmetto leaves, and under bark of dead pine trees, but occasionally will enter houses and other buildings	Anything organic; there-fore, any stored food products, grains, etc.	480–960	No	No	About 160	No	Yes	Daily housekeep-ing; sani-tation of food areas; prevent entrance into houses and buildings by sealing all cracks and openings and by in-specting all firewood coming into the house	Residual sprays or dusts indoors with perimeter treatments outside	Sometimes called "wood" roach; infrequently in homes
Cockroach, German	About 0.6 in. (16 mm) long; pale brown or tan; 2 parallel dark streaks on the pronotum	Can be found in most areas of the plant, but most abundant in moist and warm (humid) locations	Omni-vorous; food-stuffs, starch, etc.	260	No (can glide)	No	50–60, 55–125, 90–300	No	Yes	Prevent entrance by sealing all cracks and openings; inspect all receipts; daily sani-tation of food areas	Residual sprays or dusts; crack and crevice treatment most effective	High reproduc-tivity; runs rapidly; odoriferous; shuns light; most common indoor cock-roach species in the U.S.

9Hardened wall plate on back—mesothorax.

(continued on next page)

61

IMPORTANT BEHAVIORAL CHARACTERISTICS OF SOME INSECTS (Continued)

Environmental (Household) Species (Continued)

Species	Description	Habitat	Type(s) of food	Approx. no. of eggs per female (lifetime)	Approx. life cycle[a] (days)	Able to fly	Attracted to light	Found in light traps	Able to penetrate flexible packaging materials	Non-chemical control	Chemical control (label permitting)	Other comments
Cockroach, Oriental	1–1.25 in. (25–32 mm) long; dark brown, almost black; female wings rudimentary; tarsih lack an arolium[i]	Outdoors in leaf debris and under boards and stones; indoors throughout buildings, especially in dark, damp, and sewer pipe areas, etc.	Omnivorous; foodstuffs, starch, etc.	200	360, 300–800, 350–740	Yes (rarely)	No	Yes (probably)	Yes	Daily moisture control; prevent entrance by sealing all cracks and openings	Residual sprays or dusts indoors with perimeter treatments outside	Called water bugs; usually in damp basement areas
Cockroach, smoky brown	1–1.5 in. (25–38 mm) long; uniformly brownish black or mahogany	Outdoors in woodpiles and decaying trees; indoors throughout houses, buildings, and greenhouses	Anything organic; therefore, any stored food products, grains, etc.	480–960	About 360	Yes	No	Yes (probably)	Yes	Daily housekeeping; sanitation of food areas; prevent entrance by sealing all cracks and openings	Residual sprays or dusts indoors with perimeter treatments outside	Frequently confused with American; mostly in the South

62

(continued on next page)

Pest	Description	Where found	Food						Control		Remarks
Cricket, field	Almost 1 in. (25 mm) long; usually black with wings that project back like pointed tails	Fields, near buildings, openings around foundations	Wide variety of grains, other insects	40–170, 150–400	78–90, 60–>100, 65–100, about 365	Yes	Yes	Yes	Exclusion	Residual sprays or baits	Widely distributed; frequently indoors but prefers outside; attracted to outdoor lights
Cricket, house	About 0.75 in. (19 mm) long; light yellowish brown; three dark bands on the head and long thin antennae	Warm places like near fireplaces; can be found in bakeries; will conceal themselves in cracks and crevices	Soft dough and cereal products, fruits/vegetables	40–170, 100	60–>100, 265–320	Yes (most often walks, runs, or jumps)	Yes	Yes (less than field species)	Exclusion	Residual sprays or dusts	Usually outdoors; garbage dumps
Earwigs	Beetle-like; short; 0.5–1 in. (13–25 mm) long; pincer-like appendages at tip of abdomen	Usually found in soil or under debris; mulched flower beds	Scavenges on insects, etc., such as aphids, mites, scales	30, 45	40–100, 90–140	No	No	No	Remove individuals; eliminate breeding areas outdoors	Residual sprays to and near foundation walls and around steps; dust crawl spaces	Active at night
Fleas	Small; 0.06 in. (2 mm) long; reddish brown, wingless; bodies compressed laterally; legs long and adapted for jumping	On pets; can be in carpets, upholstered furniture; in cracks and crevices	Prefer blood meal from pets; if not available, humans; larvae feed on organic matter on premises	Up to 450	14–730	No	No	No	Thoroughly vacuum house or steam clean	Apply pesticides or dusts, sprays, or dips to infested animals; treat premises with residual sprays, both indoors and outdoors; apply repellents to legs and arms when entering infested area	Immature stages can be dormant for months, if undisturbed

hPart of legs.
iPad-like structure between claws.

IMPORTANT BEHAVIORAL CHARACTERISTICS OF SOME INSECTS (Continued)

Environmental (Household) Species (Continued)

Species	Description	Habitat	Type(s) of food	Approx. no. of eggs per female (lifetime)	Approx. life cycle[a] (days)	Able to fly	Attracted to light	Found in light traps	Able to penetrate flexible packaging materials	Non-chemical control	Chemical control (label permitting)	Other comments
Fly, fruit	About 0.1 in. (3 mm) long; brownish black to brownish yellow; have a feathery bristle on the antennae	Fermenting or rotting fruit and vegetable material and in garbage cans	Tomato products, fruit juices	500	8-10	Yes	Yes	Yes	No	Sanitation and destruction of breeding sites; tight-fitting garbage containers	Residual and space sprays; baits	Pupal stage hard to kill; hard outer shell and low respiration; breeds rapidly
Fly, house	About 0.25 in. (6 mm) long; dull gray; thorax marked longitudinally with 4 dark stripes; abdomen pale and fourth wing vein is angled	Warm organic material such as animal and poultry manure, garbage, decaying vegetables and fruits, and in piles of moist leaves and lawn clippings	Wastes and refuse, almost any organic matter	350-900	6-10	Yes	Yes	Yes	No	Sanitation and destruction of breeding sites; tight-fitting garbage containers; screens on windows and doors	Larvacides; residual and space sprays; baits	Needs to regurgitate to eat
Millipedes[J]	Many-segmented and many-legged; two pairs of legs/segment; 1-1.5 in. (25-38 mm) long	Usually found under stones, mulch, etc., where humidity is high	Decaying vegetable matter	Varies with numerous species	120, 365	No	No	No	No	Remove individuals; eliminate breeding areas outdoors	Residual sprays to and near foundations, walls, and around steps; dust crawl spaces	Entry in fall most likely; crawl up walls and drop from ceilings

Pest	Description	Habits	Food							Control (cultural)	Control (chemical)	Remarks
Mites,j household	About 0.03 in. (0.8 mm) long or less; several species; some come in from outdoor vegetation (clover mites), others may be found in dust and debris (dust mites), others from birds, rodents, etc.	Found in homes; some may attack humans in the absence of normal hosts like birds, rodents, or insects; others are harmless (clover mites); others cause allergies	Human skin debris	Varies with species	5–7, 12–20	No	No	No	No	Eliminate roosting areas for birds; control rodents and insects	Spray surfaces over which mites crawl; reduce severe infestations with space sprays	Can cause allergy; not likely to infest products; grain mites similar in behavior
Pillbugs/ sowbugsj	Wingless, oval, gray arthropods; approximately 0.5 in. (13 mm) long; body segments like armored plates	Slow-moving, crawling; live in areas of excessive dampness and humidity, usually very cool areas like basements; most active at night; found under trash, boards, decaying vegetation, mulch, leaf litter	Decaying vegetable matter	25–30, 60–120	720	No	No	No		Remove individuals; eliminate breeding areas outdoors; control moisture	Residual sprays to and near foundation walls, walls, around steps, and damp areas; dust crawl spaces	Pillbug rolls into ball; sowbug cannot become a ball; has two tail-like appendages

jNot insects.

(continued on next page)

IMPORTANT BEHAVIORAL CHARACTERISTICS OF SOME INSECTS (Continued)

Environmental (Household) Species (Continued)

Species	Description	Habitat	Type(s) of food	Approx. no. of eggs per female (lifetime)	Approx. life cycle[a] (days)	Able to fly	Attracted to light	Found in light traps	Able to penetrate flexible packaging materials	Non-chemical control	Chemical control (label permitting)	Other comments
Silverfish/ firebrats	Slender, wingless insects covered with scales; 0.3-0.5 in. (8-13 mm) long; gray; 3 long tail-like appendages at the tip of abdomen	In bathrooms, under rocks, beneath bark of trees, in nests of insects, birds, and mammals	Feed on materials high in protein, sugar, and starch; usually feed on book-binding starch, on clothing, glue and paste, cereals, wheat, flour, paper, etc.	50-100, 100	60-120, 80-85, 90-720	No	No	No	Yes	Good housekeeping	Spot-treat infested surfaces; dust wall voids and crawl spaces	Found near boiler rooms, water pipes, etc.; silverfish like high moisture, firebrats low moisture; silverfish shiny, like darkness; firebrats like temperatures above 100°F

Termites[k]

Species	Size	Type(s) of food	Approx. no. of offspring per queen	Approx. life cycle[l] (years)	Approx. life cycle[m] (years)	Able to fly[n] (reproductives)	Attracted to light[n]	Ability to penetrate	Other comments
Dampwood	1 in. (25 mm) (winged); 3/4 in. (19 mm) (soldiers)	wood, cellulose, corrugated paper	several thousand	1-5 or more	up to 25-30	Yes	Yes	Yes	High-moisture wood required; low risk to industry; no contact with soil needed
Drywood	1/2 in. (13 mm)	wood, cellulose, corrugated paper	several thousand	1-5 or more	up to 25-30	Yes	Yes	Yes	Found in perimeter southern states; tropical; undecayed wood of low moisture; no soil contact
Subterranean	1/2 in. (13 mm) (winged); 1/4 in. (6 mm) (soldiers)	wood, cellulose, corrugated paper	millions	1-5 or more	up to 25-30	Yes	Yes	Yes	Throughout U.S.; colonies in ground; enter buildings via foundation cracks, joints, utility openings

[k]Key points of differentiation from ants: no waist, the two pairs of wings equal length, straight antennae.
[l]Egg to winged alate or reproductive, favorable conditions.
[m]Primary reproductive.
[n]Could, therefore, be found in light traps.

67

USE OF PHEROMONES AND FOOD ATTRACTANTS
FOR MONITORING AND TRAPPING STORED-PRODUCT INSECTS

Wendell E. Burkholder

Editor's Comments

An important developing control device for monitoring purposes is the pheromone trap. These traps are inexpensive, very sensitive, and can be species specific.

Food attractant traps have been around much longer, many homemade, but little used.

Anyone who has used glue boards for rodent control has recognized the value of the sticky or adhesive type traps for insects.

Traps have significant utility in a total program to monitor species and numbers and some utility in total insect management for both stored-product and environmental insects.

For background information, the integral parts of any trap are 1) the structure or housing, 2) an attractant, and 3) a means of capturing or retaining the attracted insects.

Be aware that FDA on January 4, 1984, commented to Zoecon Corp. relative to the use in the food industry of pheromone traps (see Food Chemical News of February 6, 1984). The gist of FDA's response was that they see no need to change their policy concerning the presence of insects in food plants and that use of the traps should be done responsibly and as part of the "overall program to keep insect control within allowable limits" (emphasis added).

F. J. Baur

THE USE OF PHEROMONES AND FOOD ATTRACTANTS FOR MONITORING AND TRAPPING STORED-PRODUCT INSECTS

Wendell E. Burkholder

Stored Product and Household Insects Laboratory
Agricultural Research Service, U.S. Department of Agriculture
Department of Entomology, University of Wisconsin
Madison, Wisconsin 53706

INTRODUCTION

Recent demonstrations have shown that traps using pheromones and food attractants are useful in detecting low levels of insect infestations in grain storages, warehouses, and food processing facilities. By continued monitoring, important information can be obtained pertaining to species present, pest location, and population levels. Specific control steps may then be applied when necessary. In this way efficiency is increased, lower costs result, and less product is damaged or contaminated when compared to conventional procedures. The uses of pheromones and new trapping procedures are not difficult. However, the timing and placement of traps as well as interpretation of the results require a good understanding of insect biology and behavior.

In this chapter I will discuss: 1) the biology and behavior of some of the major storage pest insect species relative to sex pheromones, food attractants, and control methodology; 2) traps that have been developed recently that, with pheromones and food attractants, show promise for greatly improving pest management procedures; and 3) trap placement and interpretation of trap catch.

BIOLOGY AND BEHAVIOR AS THEY RELATE TO CONTROL METHODOLOGY

There are two basic reproductive patterns: Adults are either short-lived (< 1 month) and require no feeding for reproduction; or are long-lived (> 1 month) and must feed for reproduction.

The Short-Lived Adults

The first reproductive type is exemplified by a number of dermestid species. Dermestid adults are usually nourished by their reserve fat and are able to produce their sex pheromones and reproduce efficiently by coordinating male and female sexual activity periods. Pheromone release by many female Trogoderma spp. is restricted to a relatively narrow time span during the day and coincides with the male activity period.

Selection would favor exposing the reproductive individuals to predation for as short a time as possible. Dermestid adults are often found singly, for example, in mud dauber nests. At certain times of the day the female may be seen on the outside of the mud dauber nest in a calling position with the abdomen raised from the substrate at a 45° angle. Pheromone is released at this time. The male crawls to the outside of a similar nest and elevates his antennae in analert, searching position at the same time of day. On perceiving the pheromone the male flies to the female and mates with her. The preoviposition period lasts only 1 day. The female generally deposits eggs in small batches in different locations over several days. Eggs are frequently deposited in the cracks in food packages or in animal nests. It appears that survival is enhanced by distributing the eggs in widely dispersed habitats. It is obvious that the adults must depend on an effective communication system for continued beetle propagation. These adults therefore depend on an effective sex pheromone.

The pheromones of several Trogoderma species incorporate isomers of 14-methyl hexadecenal (Table 1). The aldehyde, which is highly active, is produced by those species that exhibit female calling in which the abdomen is raised. The alcohol component is also produced by these species but is less volatile and is active only by contact or over a very short distance. The alcohol is the primary pheromone component of some Trogoderma species in which a different calling behavior has been observed. When these species "call" males, the females not only raise their abdomens, but also rub their hind legs in a fanning motion over the ventral abdominal tip.

The Attagenus spp. (carpet beetles) also release pheromones during calling (Barak and Burkholder, 1978). The pheromones of two species are known (Table 1). Anthrenus flavipes (furniture carpet beetles) females behave in a similar fashion and release pheromone while in a calling posture very early in the morning (Burkholder et al., 1974). The pheromone has been identified (Fukui et al., 1974) (Table 1). The hide beetle (Dermestes maculatus) females produce a sex pheromone which attracts males (Abdel-Kader and Barak, 1979). Levinson et al. (1978) reported on a male-produced pheromone in the same insect that is secreted by a setiferous area on the ventral surface of the abdomen. This substance promotes recognition of sexually mature males by females.

Several of the bruchid seed beetles (Bruchidae) have pheromones. Hope et al. (1967) first demonstrated the existence of a male-produced sex pheromone in the bean weevil, Acanthocelides obtectus (Table 1). Preliminary evidence for a female sex pheromone which attracts males was reported for Callosobruchus chinensis (azuki bean weevil) by Honda and Yamamoto (1976), and the cowpea weevil Callosobruchus maculatus by Rup and Sharma (1978). Tanaka et al. (1981) found that C. chinensis has a copulation release pheromone composed of several synergistic components that was distinct from the female sex attractant. It is produced and released by both sexes, but affects only the male. Qi and Burkholder (1982a) examined the biology and behavior of C. maculatus, including pheromone production and release, effect of mating on release, calling behavior, gland location, and pheromone isolation. They reported that females began releasing the pheromone soon after emergence and continued production for one week. An unusual calling behavior was described which was synchronized with pheromone release. During calling, the metathoracic legs fan and rub the elevated abdominal tip. Mating reduced pheromone release by females. However, male response was not inhibited by mating.

The common stored-product anobiid beetles are the cigarette beetle, Lasioderma serricorne and the drugstore beetle, Stegobium paniceum (Anobiidae). These beetles have female-produced sex pheromones and do not require feeding for egg maturation (Coffelt and Burkholder, 1972; Kuwahara et al., 1975). Both pheromones have been identified (Chuman et al., 1979; Kuwahara et al., 1978) (Table 1).

In stored-product moths long-distance sex pheromones are produced by the female. Each species has a specific time in which the females release pheromone. For example the Indianmeal moth Plodia interpunctella is usually active in the late evening and the Mediterranean flour moth Anagasta kuehniella is active in the early morning (personal communication, J. Sargent). However, synthetic pheromones may stimulate some males to fly at other times.

Table 1. Major Pheromone Components of the Short-Lived Adult
Stored-Product Beetles

DERMESTIDAE

Trogoderma inclusum   ♀   (Z)-14-Methyl-8-hexadecen-1-ol and
and T. variabile   (Z)-14-methyl-8-hexadecenal
(warehouse beetle)   (Burkholder and Dicke, 1966; Rodin
                             et al., 1969; Cross et al., 1976).

Trogoderma glabrum   ♀   (E)-14-Methyl-8-hexadecen-1-ol and
                             (E)-14-methyl-8-hexadecenal
                             (Burkholder and Dicke, 1966; Yarger
                             et al., 1975; Cross et al., 1976).

Trogoderma granarium   ♀   92:8 (Z:E)-14-Methyl-8-hexadecenal
(khapra beetle)   (Levinson and BarIlan, 1967; Cross
                             et al., 1976).

Attagenus unicolor   ♀   (E,Z)-3,5-Tetradecadienoic acid
(= megatoma)   (Burkholder and Dicke, 1966;
(black carpet beetle)   Silverstein et al., 1967).

Attagenus   ♀   (Z,Z)-3,5-Tetradecadienoic acid
elongatulus   (Barak and Burkholder, 1977a,b;
                             Fukui et al., 1977).

Anthrenus flavipes   ♀   (Z)-3-Decenoic acid (Burkholder et
(furniture carpet   al., 1974; Fukui et al., 1974).
beetle)

BRUCHIDAE

Acanthoscelides   ♂   (E)-(-)-Methyl-2,4,5-tetradeca-
obtectus   trienoate (Hope et al., 1967;
(bean weevil)   Horler, 1970).

ANOBIIDAE

Stegobium paniceum   ♀   2,3-Dihydro-2,3,5-trimethyl-6
(drugstore beetle)   (1-methyl-2-oxobutyl)-4H-pyran-4-one
                             (Kuwahara et al., 1975, 1978).

Lasioderma   ♀   4,6-Dimethyl-7-hydroxy-nonan-3-
serricorne   one (Coffelt & Burkholder, 1972;
(cigarette beetle)   Chuman et al., 1979).

Table 2. Major Pheromone Components of Some Stored-Product
         Moths

GELECHIIDAE

Sitotroga   ♀   (Z,E)-7,11-Hexadecadien-1-ol acetate
  cerealella   (Keys and Mills, 1968; Vick et al.,
  (Angoumois grain   1974).
  moth)

PYRALIDAE

Ephestia   ♀   (Z,E)-9,12-Tetradecadien-1-ol acetate
  elutella   (Brady, 1973; Brady and Nordlund,
  (tobacco moth)   1971; Brady and Daley, 1972; Brady
                              et al., 1971a,b; Kuwahara and
Plodia   ♀   Casida, 1973; Kuwahara et al.,
  interpunctella   1971a,b).
  (Indianmeal moth)

Cadra cautella   ♀
  (almond moth)

Anagasta kuehniella   ♀
  (Mediterranean
  flour moth)

Cadra figulilella   ♀
  (raisin moth)

The pheromones of stored-product moths have been identified from several
important species (Table 2). Recent studies indicate secondary components may
improve trap catch for some species (Vick et al., 1981). Since the Indianmeal
moth pheromone attracts four additional species of stored-product moths, it is
quite useful for detection purposes.

The Long-Lived Adults

Beetles that require adult feeding to reproduce usually live for a number of
months. In contrast to the insects discussed above, these beetles often utilize
a male-produced aggregation pheromone. Our studies (Khorramshahi and Burkholder,
1981) with Rhyzopertha dominica (Bostrichidae), the lesser grain borer, have
demonstrated a strong male and female response to the male-produced aggregation
pheromone. The pheromone is responsible for the typical sweet odor that is
associated with lesser grain borer infested grain. The 2-component pheromone was
isolated, identified, and synthesized by Williams et al. (1981) (Table 3).
Subsequent studies in my laboratory have demonstrated that the pheromone is most
effective when used in the natural ratio of one part of component 1 and two parts
of component 2. High concentrations (30 mg) of the pheromone in the natural
ratio attracted significantly fewer insects in a plastic grain-probe trap than
the pheromone-free control. This suggests a dispersal or epideictic effect of

Table 3. Major Pheromone Components of the Long-Lived Adult
          Stored-Product Beetles.

BOSTRICHIDAE

| Rhyzopertha <br> dominica <br> lesser grain <br> borer | ♂ | 1-Methylbutyl (E)-2-methyl-2-pentenoate (dominicalure 1) and 1-methylbutyl (E)-2,4-dimethyl-2-pentenoate (dominicalure 2) (Khorramshahi & Burkholder, 1981; Williams et al., 1981). |

TENEBRIONIDAE

| Tribolium <br> castaneum
 (red flour <br> beetle) | ♂ | 4,8-Dimethyldecanal (Suzuki and Sugawara, 1979). |
| Tribolium <br> confusum
 (confused flour <br> beetle) | ♂ | |

CUCUJIDAE

| Cryptolestes <br> ferrugineus
 (rusty grain <br> beetle) | ♂ | (E,E)-4-8-dimethyl-4,8-decadien-10-olide (ferrulactone I) (3Z,11 S)-3 dodecen-11-olide (ferrulactone II) (Borden et al., 1979; Wong et al., 1983). |

---

the aggregation pheromone at high concentrations. Newly emerged insects migrate away from crowded conditions. However, older adults appear to become habituated to the pheromone.

The Sitophilus weevils (Curculionidae) have been studied extensively in our laboratory. The male rice weevil, Sitophilus oryzae, produces an aggregation pheromone that attracts both males and females (Phillips and Burkholder, 1981). Males of the granary weevil, Sitophilus granarius, and the maize weevil, Sitophilus zeamais, also produce aggregation pheromones (Faustini et al., 1982, Walgenbach et al., 1983). The chemical identification of the Sitophilus weevil pheromones was reported by Schmuff et al. (1984). These studies have demonstrated that maximum pheromone production occurs when food is present. Thus, not only the presence of mates is signaled, but also the availability of food. These beetles survive for less than a week without food and water, so, the importance of signaling the presence of both is obvious.

The pheromones of the Tribolium flour beetles (Tenebrionidae) have been studied intensively. Ryan and O'Ceallachain (1976) discovered a male-produced aggregation pheromone which attracts both sexes of Tribolium confusum. They also isolated a female-produced pheromone which is attractive to male T. confusum. The role of the sex pheromone is likely to be twofold, sex recognition and sexual stimulation. The existence of a sex pheromone in Tribolium is not surprising. I believe sex pheromones will be discovered in many species in which only aggregation pheromones are presently known. They are probably effective over a very close range or by contact after aggregations occur. The aggregation pheromone of Tribolium castaneum and T. confusum was identified by Suzuki and Sugawara (1979) (Table 3).

The Tribolium flour beetles possess odoriferous glands on the abdomen and thorax which emit benzoquinones (2-methyl- and 2 ethyl-1, 4-benzoquinone) when disturbed. These pungent secretions function as defensive chemicals and provide protection from predators (Roth 1943; Alexander and Barton 1943; Loconti and Roth 1953; Happ 1968; Tschinkel 1975; Markarian et al. 1978) and inhibit fungal (Englehardt et al. 1965) and bacterial growth (DeCoursey et al. 1953). Faustini and Burkholder (unpublished) have demonstrated that the quinones of T. castaneum, under certain stress conditions as overcrowding, dissolve the aggregation pheromone and act as epideictic (dispersal) pheromones.

The rusty grain beetle, Cryptolestes ferrugineus, (Cucujidae) is a pest of stored grain. Borden et al. (1979) reported that both sexes respond to the odor of mixed sex beetle populations as well as to the odor of males. The aggregation pheromone was reported by Wong et al. (1983) (Table 3).

The sawtoothed grain beetle, Oryzaephilus surinamensis, and the merchant grain beetle, O. mercator (Cucujidae), are cosmopolitan insects. Pierce et al. (1981) presented data on attractiveness of volatiles from beetles that suggest both sexes produce an aggregation pheromone. The nature of the attractants was not reported.

TYPES OF PHEROMONE AND FOOD ATTRACTANT INSECT TRAPS

Corrugated Paper Insect Traps

The use of multilayered corrugated paper with pheromones for trapping stored-product insects was reported by Burkholder (1976) and Barak and Burkholder (1976). The corrugations are attractive to the insects as hiding sites similar to cracks in walls or floors. In our early studies we routinely used an insecticide in the traps to kill the insects. Either the corrugated paper was treated with a residual insecticide or a volatile insecticide (Vapona) was used. Due to concerns relating to pesticide use in food plants, we developed a modified corrugated trap that does not contain an insecticide (Burkholder, 1984; Barak and Burkholder, 1984) (Fig. 1). The trap contains a plastic dish with an oil lure that also kills the insects. Both larvae and adults crawl or fall into the dish containing 20 to 30 drops of the oil. The insects appear to die by suffocation. The primary constituents of the oil lure are extracts of wheat germ oil (Nara et al., 1981; Nara and Burkholder, 1983), oat oil (Freedman et al., 1982; Mikolajcek et al., 1983), and mineral oil. Several volatile components including octanoic acid have been identified from wheat germ oil. These volatiles initiate aggregating activity in Trogoderma larvae (Nara et al., 1981). Mikolajczak et al. (1984) have reported on a number of different compounds in oats that attract sawtoothed grain beetles. The most active compounds were (E)-2-nonenal and (E,E)-2-4-nonadienal. Other active compounds were hexanal, heptanal, octanal, (E)-2-heptenal, and 2-furaldehyde. Subsequent studies have resulted in further identifications of highly active attractants in oat oil. This trap is unusual in that the lure is also the killing agent.

The trap also accommodates pheromone-treated rubber septa. Up to four different pheromone-treated septa may be used at one time. Oil lures are especially effective for Trogoderma spp. larvae, adult sawtoothed grain beetles,

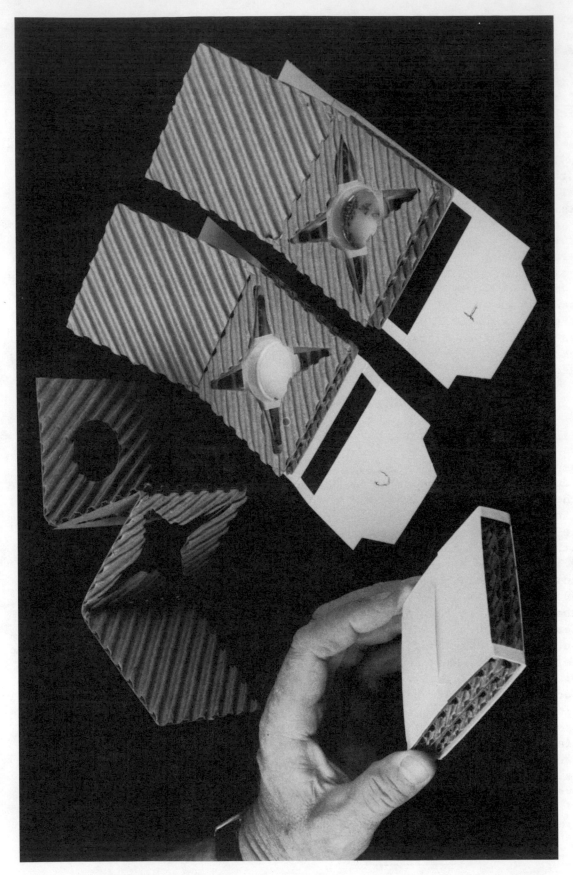

Fig. 1. View of corrugated paper trap (9 x 9 x 2 cm) with a plastic dish containing an absorbent paper pad and approximately 30 drops (1 ml) of an oil lure that also serves as a killing agent. Pheromone-treated rubber septa are placed in the V-notched sections of the trap. (Burkholder, 1984; Barak and Burkholder, 1984)

Oryzaephilus surinamensis, and _Tribolium_. The combination of food lures and pheromone lures has greatly improved the usefulness of the trap. Traps with pheromones and attractants are available from the Zoecon® Corporation, P. O. Box 10975, Palo Alto, California 94303 under the STORGARD^TM label.

Grain-Probe Insect Traps

The use of perforated grain-probe insect traps with pheromones is a relatively new procedure. A perforated brass trap was described and tested by Loschiavo and Atkinson (1967, 1973), Loschiavo (1974, 1975), and Barak and Harein (1982). These traps did not utilize pheromones or other lures. During the past several years a plastic trap was designed that has a number of advantages over the metal version. Major drawbacks of the metal trap were high cost and lack of commercial availability, probably due to expense and difficulty in mass production. Other concerns were: 1) need for greater effectiveness, 2) need for a larger trap, and 3) need for easy application of pheromone and food lures. A plastic material was selected to meet the requirements of low cost, high impact resistance and strength, and ease of machining. Plastic materials are also easily adapted for mass production. The plastic selected, Lexan®, has the added advantage in that it is transparent. Another advantage of the 2.5 cm tube selected was that the 3.2 mm walls allowed the machining of sloping holes. The trap has 186 holes that are 2.79 mm in diameter that slope at a 50° angle (Burkholder, 1984) (Fig. 2). Larger holes may be used for large grains such as corn. The upper end of the trap is closed with a plastic plug that is perforated to allow the insertion of one or several 3.2 mm polyethylene tubes of various lengths. Rubber septa containing pheromones or food lures may be attached to each polyethylene tube. Also the tubes may be used directly as attractant-releasing devices by sealing attractant compounds inside them. A rope attached to the top of the trap is useful in securing and recovering the trap.

Insects fall or crawl into the trap and drop through the plastic funnel into the lower part of the device. A clear plastic tube insert is provided to make removal of the insects easier. The tube is marked in a graduated scale (ml) in order to estimate by volume large numbers of the trapped insects. A thin film of white petrolatum (U.S.P.), mineral oil (food grade), or the oil lures used in the corrugated paper traps (1 ml) may be used inside the inserted tube to promptly kill trapped insects. The oil lures are especially useful in luring a wide variety of insects, including larvae. Care should be taken to prevent the oil from getting on the insect entry holes. A quick-release lower plug allows easy access to the trapped insects. A metal device has been designed to fit over the trap to facilitate pushing it into the grain (Fig. 3). This device can be attached to the standard Seedburo T-handle and threaded extensions available for the deep-probe grain sampler. It is also useful as a tool to quickly remove the lower plug of the trap.

The trap works remarkably well without pheromones or attractants and appears to be superior to the metal traps (Table 4). The aggregation pheromones of both the lesser grain borer and _Tribolium_ have been used successfully with the trap. When compared with the control, 3-4 times as many insects were caught in the traps during field studies with _Tribolium_ pheromone (R. Cogburn, unpublished) and the lesser grain borer pheromone (R. Cogburn, unpublished and R. Mills, personal communication). Our studies have demonstrated the importance of pheromone concentration when using the grain-probe traps. High concentrations (30 mg) of the pheromone are repellent to the lesser grain borer when compared to the control or to lower concentrations (1, 3, or 10 mg). A similar effect occurs with the _Tribolium_ pheromone. At the present time, studies are being conducted in a number of locations throughout the world. Experience in California rice storages with the _Tribolium_ aggregation pheromone has indicated that as many as

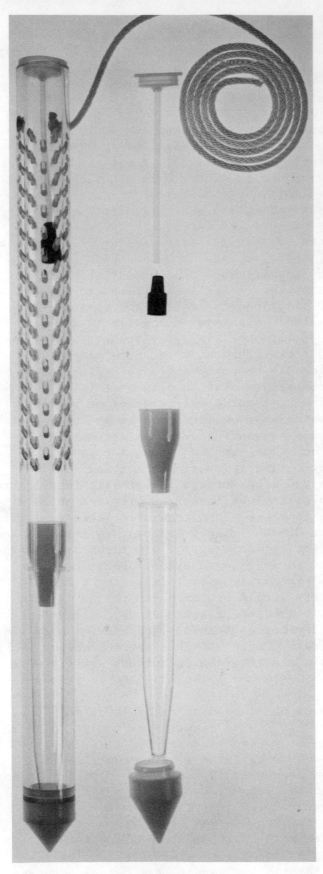

Fig. 2. (LEFT) View of plastic grain-probe insect trap, on left the assembled trap (2.5 x 38 cm); on the right the component parts (top to bottom) - plastic end cap with tube for dispersing attractants or for attachment of rubber septa; the rubber septa; plastic funnel; plastic graduated test tube; plastic snap-off cap. (Burkholder, 1984)

Fig. 3. (BELOW) View of metal device (9.5 x 3.1 cm) that fits over the plastic grain-probe trap and is attached to a Seedboro T-handle with threaded extension to facilitate pushing the trap into the grain. (Burkholder, 1984)

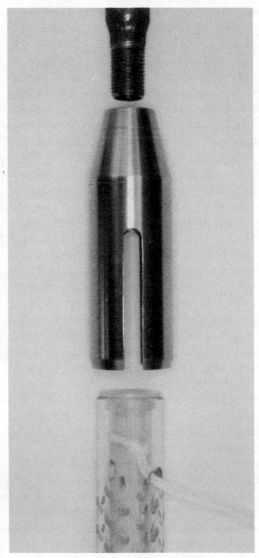

Table 4. Comparison of Granary Weevils (_Sitophilus granarius_)
Captured in Probe Traps in 14.5 kg Containers of Wheat*.

Plastic Trap	Metal Trap**
19%	3.2%

* 100 insects per container (5 replicates), insects allowed to
   settle for 24 h followed by 48 h test.
** Modification of Loschiavo trap (Barak and Harein, 1982).

15 _Tribolium_ may be caught in a trap after three days when conventional sampling
failed to detect the insects. Studies in progress will undoubtedly add to the
information on time-temperature effects, type of grain, trap placement, pheromone
concentration effects, and other ecological information. In addition, our
studies (Qi and Burkholder, 1982b; Kramer et al. 1984) have demonstrated the
usefulness of probe traps in detecting insects in oil or repellent-treated grain.
     Time is a significant factor in the use of these grain-probe traps. The
longer the trap is in place the greater the sensitivity. In this way, the traps
will detect infestations when grain trier samples do not. Also insect migrations
in grain are well known and the traps will detect such movements. Although the
trap was designed to be used with pheromone and food attractants it has some
usefulness without them. These traps may be used continuously from the day the
grain is stored until it is marketed. The plastic grain-probe traps are
virtually indestructible but should be removed when liquid fumigants are used.
     The plastic grain-probe trap is manufactured by Walco Mfg. Co., 282
Jefferson St. Oregon, Wisconsin and is marketed by the Zoecon Corporation, Palo
Alto, California under the STORGARD^TM label. Pheromones and attractants for use
in the traps are also available from them.

Adhesive Traps

     Perhaps the most common method of trapping insects, especially moths, has
been to apply adhesive (sticky) material to an environmentally protected surface
such as cardboard or plastic. Cogburn et al. (1984) utilized 2 inverted plastic
pie plates separated by a 4.8 cm diameter PVC pipe. Adhesive was applied to the
upper surface of the lower plate. The advantage of this trap is durability, low
cost, and ease in which it can be either hung inside or staked outdoors. The
trap is successful in capturing both moths and beetles, especially the lesser
grain borer and _Sitotroga cerealella_, the Angoumois grain moth. A variety of
adhesive traps is commercially available.

Funnel Traps

     A variety of funnel traps has been used for trapping both stored product
moths and beetles. Shapas and Burkholder (1978) utilized a small funnel attached
to a bottle baited with pheromone to trap adult _Trogoderma_ males in field
studies.

International Pheromones LTD. is currently marketing a funnel trap for stored-product moths. Insects are killed by an insecticide inside the trap. While the traps usually have a larger insect capacity than adhesive traps the insecticide restricts the use of these traps in food plants.

The grain probe trap may be modified to be used in head-space areas of bins, in warehouses, and in outdoor settings. Modifications include adding a funnel with a 2.5 cm diameter hole in the narrow end. This allows the funnel to be slipped over the probe trap so that flying insects will fall or crawl into the trap's holes. A rain or dust cover is placed over the trap. When a large reservoir is required the bottom end plug and tube insert is replaced with a .4 liter plastic bottle. Cogburn (personal communication) has successfully used the probe trap attached to posts and trees to catch lesser grain borers.

TRAP PLACEMENT AND INTERPRETATION OF TRAP CATCH

Placement of the corrugated paper traps in warehouses is dependent in part on the size of the warehouse and on available supporting posts or other places where there is little or no traffic. It is suggested that they be placed approximately 16 m apart in a general grid system so that all areas are evaluated. Our studies in California (Smith et al., 1983) with Trogoderma pheromone in cardboard traps demonstrated that in one warehouse greater trap catches occurred near exterior walls with doors. In another warehouse, the largest trap catch was inside and away from the walls, indicating an infestation or potential site of an infestation. There was no significant difference between traps placed 1.5 m above the floor or on the floor.

In general, the traps should not be placed near open doors or windows. This is to prevent attracting insects into a facility from outside. Moth traps are usually more effective if placed at near ceiling levels (personal communication, J. Sargent).

In trapping studies with dominicalure pheromones outside rice bins in Texas, Cogburn et al. (1984) found that more lesser grain borers were caught in traps placed near the concrete supporting slab than at 2 or 6 m above the slab. In our studies (Williams et al. 1981) pheromone treated corrugated cardboard traps worked well when placed on storage room floors or as pheromone treated adhesive traps suspended 2.4 m above the floor.

The dominicalure pheromones have been utilized by R. J. Hodges and colleagues (personal communication) to determine the presence of both lesser grain borers and Prostephanus truncatus, the larger grain borer, in farm maize stores in Tanzania. They placed pheromone treated corrugated paper traps either in the farm storage house or outside on a storage platform. The dominicalure pheromone may therefore be used for simultaneous monitoring of the two species in areas such as Kenya and other countries where the larger grain borer is especially threatening to the maize crops, both in the field and in storage.

Ideally several grain-probe traps should be used in a bin. Traps should be pushed several inches below the surface at intervals of 2-3 m starting from the center and radiating at 90° angles giving priority to the North, and followed by the South, East, and West quadrants. Additional traps may be pushed down into the grain up to approximately 4 m. Traps should be checked daily if insects are suspected or at least weekly under average conditions. If 1 to 8 insects are caught in 7 days or less, further bin evaluation and possible treatment should be considered. These traps provide a continuous monitoring system.

Some insects such as Tribolium are often near the surface of the bin while other species are usually more generally distributed. Trapping near suspected or known warm areas is recommended because infestations are much more likely.

Indianmeal moths have been successfully trapped utilizing either rubber septa, hollow fibers, or plastic materials baited with pheromone. Vick et al. (1981) reported that 10 mg of ($\underline{Z},\underline{E}$)-9,12-tetradecadien-1-ol acetate, Z9,E12-14:acetate or a mixture of 4 mg of the acetate and 6 mg of ($\underline{Z},\underline{E}$)-9-12-tetradecadien-1-ol was effective in trapping the Indianmeal moth. Stockel and Sureau (1981) reported satisfactory results in trapping the Angoumois grain moth with pheromone amounts between 300 and 900 ug/rubber septa dispensers.

Traps, if protected, may be placed outside of warehouses to catch migrating insects. The traps may therefore intercept insects before they have a chance to move inside a warehouse. Trapping should begin when temperatures reach approximately 19°C. May through October are usually the months of most intensive use except in heated areas or in semi-tropical or tropical areas where year-round trapping is necessary.

Pheromone lures should be refrigerated or protected from heat prior to their use. Lures should be replaced according to the manufacturers' recommendations. The traps should be inspected at least once a week and the trap catch counted, identified, and recorded. Trapped insects should also be removed from the corrugated paper traps to prevent dermestids such as Trogoderma from feeding on them. It is important to keep an accurate map of trap locations as most insect traps, including light traps, have the potential for providing food for dermestid beetles.

The interpretation of the trap catch is not as difficult as it may appear. One or several insects in a trap usually indicates the presence of a small infestation or accidental entry of stray insects. Repeated catches over a period of several weeks indicates the likelihood of an infestation. Large numbers of insects (10-30) usually indicate a serious problem. If the traps are placed on a grid system the first week's catch may pick up stray insects in a random pattern. In subsequent weeks the trap catch will likely indicate more precisely the problem areas by pin-pointing active insect sites. Repeated trapping may offer partial control by eliminating some of the offending insects as well as impairing reproduction.

Several success stories of finding elusive insect infestations in food plants and warehouses have been received. The result has been an estimated savings of many thousands of dollars in product loss and reduced insecticide usage because of a quick and efficient localized treatment. Follow-up trapping has been recommended to monitor the successful elimination of insect infestations in these instances.

DISCUSSION

Pheromone traps can be expected to attract insects from a relatively large area and therefore should uncover insect activity before other conventional sampling methods. The corrugated paper or probe trap, with or without pheromone, may be used along with other sampling procedures to provide information on the insect population. It is important to know what the population is to provide a basis for subsequent monitoring studies. The total picture of insect biology, behavior, and ecology also needs to be considered in the interpretation of subsequent monitoring. Plant managers and sanitarians, as well as commercial pest control personnel, have a new lower threshold level to consider. It may be possible to control insects much easier now because of the early and accurate information provided by the traps.

Early detection and trapping of insects can theoretically provide some control. If the density of insects can be kept below tolerance levels by trapping, at least partial control may be achieved. Moreover time is allowed for other environmentally sound insect control practices such as improved sanitation, aeration of grain, and correcting construction faults.

Pheromone traps can be expected to catch migrating insects. For example, studies by Williams and Floyd (1970) with maize weevils suggested that migrating weevils may not fly directly to the corn in the field, but could fly to some other host, prior to the time the corn becomes susceptible. The male-produced aggregation pheromone of the maize weevil could be used to trap these insects before they have a chance to infest corn in the fields. In addition, pheromone traps could be used to intercept insects before they migrate from the field to storage bins, warehouses or other storage facilities.

The use of a grid system in trap placement assures uniform coverage of a warehouse or bin. Distribution, both horizontally and vertically, may be required with traps for flying insects or with grain-probe traps.

The corrugated paper insect traps have also been useful in trapping certain insects and mites that are environmental problems. The traps are especially effective in trapping adult moth flies (Psychodidae), also referred to as drain flies, filter flies, or sewerage flies. Other insects that are caught in the traps are miscellaneous small flies, fungus gnats, silverfish, and cockroaches.

Recent advances in trap design along with success in developing broad-spectrum food lures for stored-product beetles have stimulated commercial development of new insect monitoring systems. The many years of basic research in this area have resulted in some commercial success.

SUMMARY

Pheromones and food attractants have been combined with improved traps for monitoring and trapping a wide variety of stored-product insects. They are powerful tools, and when handled properly, can aid in the effective management of pest insects. The traps, when placed on a grid system, are effective in continuously and efficiently sampling over a large area. It offers a savings in time spent on visual and in manual searches. Pheromones will soon be available for other important stored-product insects such as the grain weevils. Pheromones along with newly designed traps offer a bright future for detecting and monitoring stored-product insects. Pesticide application costs should be reduced because of better timing of applications. The objective is to avoid unnecessary pesticide applications. Advances in other areas such as improved aeration systems and biological controls will aid in this effort.

ACKNOWLEDGEMENT

This research was supported by the College of Agricultural and Life Sciences, University of Wisconsin, Madison, and by a cooperative agreement between the University of Wisconsin and the ARS, USDA. I thank Janet Klein, Joel Phillips, Catherine Walgenbach, and Alan Barak for their help and suggestions. The assistance and contributions of Dave Walford, Walco Mfg., Oregon, Wisconsin during the development of the plastic grain-probe insect trap is greatly appreciated. Mention of a proprietary product or company name in this paper does not constitute an endorsement by the USDA.

LITERATURE CITED

Abdel-Kader, M. M., and A. V. Barak. 1979. Evidence for a sex pheromone in the hide beetle, Dermestes maculatus (DeGeer) (Coleoptera: Dermestidae). J. Chem. Ecol. 5:805-813.

Alexander, P., and D. H. R. Barton. 1943. The excretion of ethylquinone by the flour beetle. J. Biochem. 37:463-465.

Barak, A. V., and W. E. Burkholder. 1976. Trapping studies with dermestid sex pheromones. Envirn. Entomol. 5:111-114.

Barak, A. V., and W. E. Burkholder. 1977a. Behavior and pheromone studies with Attagenus elongatulus Casey (Coleoptera: Dermestidae). J. Chem. Ecol. 3:219-237.

Barak, A. V., and W. E. Burkholder. 1977b. Studies on the biology of Attagenus elongatulus Casey (Coleoptera: Dermestidae) and the effects of larval crowding on pupation and life cycle. J. Stored-Prod. Res. 13:169-175.

Barak, A. V., and W. E. Burkholder. 1978. Interspecific response to sex pheromones, and calling behavior of several Attagenus species (Coleoptera: Dermestidae). J. Chem. Ecol. 4:451-461.

Barak, A. V., and W. E. Burkholder. 1984. A versatile and effective trap for detecting and monitoring stored product Coleoptera. Prot. Ecol. (In press).

Barak, A. V., and P. K. Harein. 1982. Trap detection of stored-grain insects in farm-stored, shelled corn. J. Econ. Entomol. 75:108-111.

Borden, J. H., M. G. Dolinski, L. Chong, V. Verigin, H. D. Pierce, Jr., and A. C. Oehlschlager. 1979. Aggregation pheromone in the rusty grain beetle, Cryptolestes ferrugineus (Coleoptera: Cucujidae). Can. Entomol. 111:681-688.

Brady, U. E. 1973. Isolation, identification and stimulatory activity of a second component of the sex pheromone system (complex) of the female almond moth, Cadra cautella (Walker). Life Sci. 13:227.

Brady, U. E., and R. C. Daley. 1972. Identification of a sex pheromone from the female raisin moth, Cadra figulilella. Ann. Entomol. Soc. Am. 65:1356.

Brady, U. E., and D. A. Nordlund. 1971. Cis-9,trans-12 tetradecadien-1-yl acetate in the female tobacco moth Ephestia elutella (Hubner) and evidence for an additional component of the sex pheromone. Life Sci. 10:797.

Brady, U. E., D. A. Nordlund, and R. C. Daley. 1971a. The sex stimulant of the Mediterranean flour moth Anagasta kuehniella. J. Georgia Entomol. Soc. 6:215.

Brady, U. E., J. H., III, Tumlinson, R. B. Brownlee, and R. M. Silverstein. 1971b. Sex stimulant and attractant in the Indian meal moth and in the almond moth. Science 171:802.

Burkholder, W. E. 1976. Application of pheromones for manipulating insect pests of stored products. Pages 111-122, in, T. Kono and S. Ishii (eds.) Insect pheromones and their applications, Japan Protection Association, Tokyo.

Burkholder, W. E. 1984. Stored-product insect behavior and pheromone studies: Keys to successful monitoring and trapping. Proc. Third International Working Conference on Stored-Product Entomology (Manhattan, KS). pp. 20-33.

Burkholder, W. E., and R. J. Dicke. 1966. Evidence of sex pheromones in females of several species of Dermestidae. J. Econ. Entomol. 59:540-543.

Burkholder, W. E., M. Ma, Y. Kuwahara, and F. Matsumura. 1974. Sex pheromone of the furniture carpet beetle, Anthrenus flavipes (Coleoptera: Dermestidae). Can. Entomol. 106:835-839.

Chuman, T., M. Kohno, K. Kato, and M. Noguchi. 1979. 4,6-dimethyl-7-hydroxy-nonan-3-one, a sex pheromone of the cigarette beetle (Lasioderma serricorne F.). Tetrahedron Letters 25:2361-2364.

Coffelt, J. A., and W. E. Burkholder. 1972. Reproductive biology of the cigarette beetle, Lasioderma serricorne. I. Quantitative laboratory bioassay of the female sex pheromone from females of different ages. Ann. Entomol. Soc. Am. 65:447-450.

Cogburn, R. R., W. E. Burkholder, and H. J. Williams. 1984. Field tests with the aggregation pheromone of the lesser grain borer (Coleoptera: Bostrichidae) Envirn. Entomol. 13:162-166.

Cross, J. H., R. C. Byler, R. F. Cassidy, Jr., R. M. Silverstein, R. E. Greenblatt, W. E. Burkholder, A. R. Levinson, and H. Z. Levinson. 1976. Porapak-Q collection of pheromone components and isolation of (Z)- and (E)-14-methyl-8-hexadecenal, potent sex attracting components, from females of four species of Trogoderma (Coleoptera: Dermestidae). J. Chem. Ecol. 2:457-468.

DeCoursey, J. D., A. P. Webster, W. W. Taylor, Jr., R. S. Leopold, and R. H. Kathan. 1953. An antibacterial agent from Tribolium castaneum (Herbst). Ann. Entomol. Soc. Am. 46:386-392.

Englehardt, M., H. Rapoport, and A. Sokoloff. 1965. Comparison of the content of the odoriferous gland reservoirs in normal and mutant Tribolium confusum. Science 150:632-633.

Faustini, D. L., W. L. Giese, J. K. Phillips, and W. E. Burkholder. 1982. Aggregation pheromone of the male granary weevil, Sitophilus granarius (L.). J. Chem. Ecol. 8:679-687.

Freedman, B., K. L. Mikolajczak, D. R. Smith, Jr., W. F. Kwolek, and W. E. Burkholder. 1982. Olfactory and aggregation responses of Oryzaephilus surinamensis (L.) to extracts from oats. J. Stored Prod. Res. 18:75-82.

Fukui, H., F. Matsumura, M. Ma, and W. E. Burkholder. 1974. Identification of the sex pheromone of the furniture carpet beetle, Anthrenus flavipes LeConte. Tetrahedron Lett. 40:3563-3566.

Fukui, H., F. Matsumura, A. V. Barak, and W. E. Burkholder. 1977. Isolation and identification of a major sex-attracting component of Attagenus elongatulus (Casey) (Coleoptera: Dermestidae). J. Chem. Ecol. 3: 541-550.

Happ, G. M. 1968. Quinone and hydrocarbon production in the defense glands of Eleodes longicollis and Tribolium castaneum (Coleoptera: Tenebrionidae). J. Insect Physiol. 14:1821-1837.

Honda, H., and I. Yamamoto. 1976. Evidence for and chemical nature of a sex pheromone present in azuki bean weevil, Callosobruchus chinensis L. Proc. Symposium on Insect Pheromones and their Applications (Nagaoka and Tokyo) p. 164.

Hope, J. A., D. F. Horler, and D. G. Rowlands. 1967. A possible pheromone of the bruchid Acanthocelides obtectus (Say). J. Stored Prod. Res. 3:387.

Horler, D. F. 1970. An allenic ester produced by the male dried bean beetle, Acanthoscelides obtectus (Say). J. Chem. Soc. C:859.

Keys, R. E. and R. B. Mills. 1968. Demonstration and extraction of a sex attractant from female Angoumois grain moths. J. Econ. Entomol. 61:46.

Khorramshahi, A., and W. E. Burkholder. 1981. Behavior of the lesser grain borer Rhyzopertha dominica (Coleoptera: Bostrichidae): Male-produced aggregation pheromone attracts both sexes. J. Chem. Ecol. 7:33-38.

Kramer, K., R. W. Beeman, W. E. Spiers, W. E. Burkholder, and T. P. McGovern. 1984. Susceptibility of stored product insects to alkyl ketones and derivatives. J. Kansas Ent. Soc. (In press).

Kuwahara, Y. and J. E. Casida. 1973. Quantitative analysis of the sex pheromone of several physitid moths by electron-capture gas chromatography. Agric. Biol. Chem. 37:681.

Kuwahara, Y., H. Fukami, R. Howard, S. Ishii, F. Matsumura, and W. E. Burkholder. 1978. Chemical studies on the Anobiidae: Sex pheromone of the drugstore beetle, Stegobium paniceum (L.) (Coleoptera). Tetrahedron 34:1769-1774.

Kuwahara, Y., H. Fukami, S. Ishii, F. Matsumura, and W. E. Burkholder. 1975. Studies on the isolation and bioassay of the sex pheromone of the drugstore beetle Stegobium paniceum (Coleoptera: Anobiidae) J. Chem. Ecol. 1:413-422.

Kuwahara, Y., H. Hara, S. Ishii, and H. Fukami. 1971a. The sex pheromone of the Mediterranean flour moth. Agric. Biol. Chem. 35:447.

Kuwahara, Y., C. Kitamura, S. Takahashi, H. Hara, S. Ishii, and H. Fukami. 1971b. Sex pheromone of the almond moth and the Indian meal moth: Cis-9,trans-12, tetradecadienyl acetate. Science 171:801.

Levinson, H. Z., and A. R. BarIlan. 1967. Function and properties of an assembling scent in the khapra beetle Trogoderma granarium. Riv. Parasitol. 28:27-42.

Levinson, H. Z., A. R. Levinson, T.-I. Jen, J. L. D. Williams, and G. Kahn. 1978. Production site, partial composition and olfactory perception of a pheromone in the male hide beetle. Naturwissenschaften 65:543-545.

Loconti, J. D., and L. M. Roth. 1953. Composition of the odorous secretion of Tribolium castaneum. Ann. Entomol. Soc. Am. 46:281-289.

Loschiavo, S. R. 1974. Laboratory studies of a device to detect insects in grain, and of the distribution of adults of the rusty grain beetle, Cryptolestes ferrugineus (Coleoptera: Cucujidae), in wheat filled containers. Can. Entomol. 106:1309-1318.

Loschiavo, S. R. 1975. Field tests of devices to detect insects in different kinds of grain storages. Can. Entomol. 107:385-389.

Loschiavo, S. R., and J. M. Atkinson. 1967. A trap for the detection and recovery of insects in stored grain. Can. Entomol. 99:1160.

Loschiavo, S. R., and J. M. Atkinson. 1973. An improved trap to detect beetles (Coleoptera) in stored grain. 105:437-440.

Markarian, H., G. J. Florentine, and J. J. Pratt, Jr. 1978. Quinone production of some species of Tribolium. J. Insect Physiol. 24:785-790.

Mikolajczak, K. L., B. Freedman, B. W. Zilkowski, C. R. Smith, Jr. and W. E. Burkholder. 1983. Effect of oat constituents on aggregation behavior of Oryzaephilus surinamensis (L). Agric. and Food Chem. 31:30-33.

Mikolajczak, K. L., B. W. Zilkowski, C. R. Smith, Jr., and W. E. Burkholder. 1984. Volatile food attractants for Oryzaephilus surinamensis (L.) from oats. J. Chem. Ecol. 10:301-309.

Nara, J. M., and W. E. Burkholder. 1983. Influence of molting cycle on the aggregation response of Trogoderma glabrum (Coleoptera:Dermestidae) larvae to wheat germ oil. Environ. Entomol. 12:703-706.

Nara, J. M., R. C. Lindsay, and W. E. Burkholder. 1981. Analysis of volatile compounds in wheat germ oil responsible for an aggregation response in Trogoderma glabrum larvae. Agric. and Food Chem. 29:68-72.

Phillips, J. K., and W. E. Burkholder. 1981. Evidence for a male-produced aggregation pheromone in the rice weevil, Sitophilus oryzae (L.). J. Econ. Entomol. 74:539-542.

Pierce, A. M., A. C. Oehlschlager, and J. H. Borden. 1981. Olfactory response to beetle-produced volatiles and host-food attractants by Oryzaephilus surinamensis and O. mercator. Can. J. Zool. 59:1980-1990.

Qi, Yun-Tai, and W. E. Burkholder. 1982a. Sex pheromone biology and behavior of the cowpea weevil Callosobruchus maculatus (Coleoptera: Bruchidae). J. Chem. Ecol. 8:527-523.

Qi, Yun-Tai, and W. E. Burkholder. 1982b. Protection of stored wheat from the granary weevil by vegetable oils. J. Econ. Entomol. 74:502-505.

Rodin, J. O., R. M. Silverstein, W. E. Burkholder, and J. E. Gorman. 1969. Sex attractant of female dermestid beetle Trogoderma inclusum LeConte. Science 165:904.

Roth, L. M. 1943. Studies on the gaseous secretion of Tribolium confusum Duval II. The odiferous glands of Tribolium confusum. Ann. Entomol. Soc. Am. 36:397-424.

Rup, P. J., and S. P. Sharma. 1978. Behavioural response of males and females of Callosobruchus maculatus (F.) to the sex pheromones. Indian J. Ecol. 1:72-76.

Ryan, M. F., and D. P. O'Ceallachain. 1976. Aggregation and sex pheromones in the beetle Tribolium confusum. J. Insect. Physiol. 22:1501-1503.

Schmuff, N., J. K. Phillips, W. E. Burkholder, H. M. Fales, C. Chen, P. Roller, and M. Ma. 1984. The chemical identification of the rice weevil and maize weevil aggregation pheromones. Tetrahedron Lett. 25:1533-1534.

Shapas, T. J., and W. E. Burkholder. 1978. Patterns of sex pheromone release from adult females, and effects of air velocity and pheromone release rates on theoretical communication distances in Trogoderma glabrum. J. Chem. Ecol. 4:395-408.

Silverstein, R. M., J. O. Rodin, W. E. Burkholder, and J. E. Gorman. 1967. Sex attractant of the black carpet beetle. Science 157:85.

Smith, L. W. Jr., W. E. Burkholder, and J. K. Phillips. 1983. Detection and control of stored food insects with traps and attractants: the effect of pheromone-baited traps and their placement on the number of Trogoderma species captured. Natick Technical Report TR 83/008:13 pp.

Stockel, J., and F. Sureau. 1981. Monitoring for the Angoumois grain moth in corn. In: "Management of Insect Pests With Semiochemicals" E. R. Mitchell ed. pp 63-73.

Suzuki, T., and R. Sugawara. 1979. Isolation of an aggregation pheromone from the flour beetles, Tribolium castaneum and T. confusum (Coleoptera: Tenebrionidae). Appl. Entomol. Zool. 14:228-230.

Tanaka, K., K. Ohsawa, H. Honda, and I. Yamamoto. 1981. Copulation release pheromone, erectin, from the azuki bean weevil (Callosobruchus chinensis L.) J. Pestic. Sci. 6:75-82.

Tschinkel, W. R. 1975. A comparative study of the chemical defensive system of tenebrionid beetles. I. Chemistry of the secretions. J. Insect Physiol. 21:753-783.

Vick, K. W., J. A. Coffelt, R. W. Mankin, and E. L. Soderstrom. 1981. Recent developments in the use of pheromones to monitor Plodia interpunctella and Ephestia cautella In: "Management of Insect Pests With Semiochemicals" E. R. Mitchell, ed. Plenum Press, N.Y. pp 19-28.

Vick, K. W., H. C. F. Su, L. L. Sower, P. G. Mahany, and P. C. Drummond. 1974. (Z,E)-7,11-Hexadecadien-1-ol acetate: The sex pheromone of the Angoumois grain moth, Sitotroga cerealella. Experientia 30:17.

Walgenbach, C. A., J. K. Phillips, D. L. Faustini, and W. E. Burkholder. 1983. Male-produced aggregation pheromone of the maize weevil, Sitophilus zeamais, and interspecific attraction between three Sitophilus species. J. Chem. Ecol. 9:381-341.

Williams, R. N., and E. H. Floyd. 1970. Flight habits of the maize weevil, Sitophilus zeamais. J. Econ. Entomol. 63:1583-1588.

Williams, H. J., R. M. Silverstein, W. E. Burkholder, and A. Khorramshahi. 1981. Dominicalure 1 and 2: Components of aggregation pheromone from male lesser grain borer Rhyzopertha dominica (F.) (Coleoptera: Bostrichidae). J. Chem. Ecol. 7:759-781.

Wong, J. W., V. Verigin, A. C. Oehlschlager, J. H. Borden, H. D. Pierce, Jr., A. M. Pierce, and L. Chong. 1983. Isolation and identification of two macrolide pheromones from the frass of Cryptolestes ferrugineus (Coleoptera: Cucujidae). J. Chem. Ecol. 9:451-474.

Yarger, R. G., R. M. Silverstein, and W. E. Burkholder. 1975. Sex pheromone of the female dermestid beetle Trogoderma glabrum (Herbst). J. Chem. Ecol. 1:323-334.

INSECT ELECTROCUTOR LIGHT TRAPS

David Gilbert

Editor's Comments

    This chapter presents a detailed, pragmatic review,
with historical perspective, of insect electrocutors. The
literature has long needed such an effort.
    The author puts the use of light traps into proper
perspective in a total insect control program. They are
unquestionably of value as monitoring tools and, depending
on the plant, insect species, and control needs, they can
also be of significant value in controlling the numbers of
flying species. Obviously, to be caught the insects have to
be able to fly and not be repelled by light (see chapter 6).
    The author has drawn greatly from the efforts of his
father, who has been the pioneer of the development and use
of insect electrocutors.
    Covered in this chapter are types of traps and lamps,
proper selection and placement, and maintenance. Don't
forget, as part of maintenance, to empty the trays
periodically and check for numbers and species (monitor).

                                              F. J. Baur

FIGURE 1. This modern industrial light trap features two Sylvania (40-watt, conventional phosphor) blacklight lamps, an aluminized steel reflector, a removable insect-catch tray, easy maintenance design (no tools required), and corner-mount option for strategic installation.

INSECT ELECTROCUTOR LIGHT TRAPS

David Gilbert

Don Gilbert Industries, Inc.
P. O. Box 2188
Jonesboro, Arkansas 72402-2188

INTRODUCTION

Over the last two decades, insect electrocutor light traps have come to be recognized as a valuable insect control tool. Today, there are an estimated 200 thousand industrial design light traps in use. Most of the advanced design traps are found in food and pharmaceutical plants. Many supermarkets, restaurants, and hospitals are also using them, but the question of how they are best put to use and their true value has been somewhat of a mystery.

Television ads for consumer oriented, 15 watt, single-lamp traps have promised to clear an entire acre or more of insects from backyards, pools, and patios for so long that they have fostered a lingering misperception that any little trap hung up anywhere in any facility might (as if by magic) be expected to clear the premises of flying insects almost instantly. There is real and significant value to the use of light traps in our food and drug industry facilities, restaurants, supermarkets, hospitals, etc.; but just what is it, how is it done, and which trap designs do it best?

A Short History of Light Traps

Our early ancestors learned long ago that insects were drawn to flame: first the campfire, then the candle, and then the lantern. The electric light

FIGURES 2 and 3. Artist's renderings of the first "electric-light" trap and the first "electrocutor" light trap.

was first used to trap flying insects soon after Edison invented it in the 1880's. The first light trap was an electric bulb hanging over a wash tub filled with fluid. It drowned and sometimes poisoned its victims.

The idea of electrocuting flying insects attracted to a light was conceived in the early 1900's. The "granddaddy" of today's traps featured a standard 20 watt incandescent bulb and a wood-framed electrocuting grid and hung (outdoors) from a post or chain. Improvements came slowly. For years light traps remained very similar to this early 1900's design and were only used outdoors.

The conventional blacklight phosphor still used today was developed in 1938. The wooden frame was replaced with metal in the 40's. Larger outdoor traps with two 48 inch long, 40 watt blacklight lamps appeared in the 50's. Outdoor traps have more or less turned out to be an evolutionary dead end (see section entitled "Indoor/Not Outdoor"), but the 40 watt lamps have proved effective over the years and are used today in the most advanced industrial designs.

The first escape-deterrent trap design with a self-contained removable insect-catch tray came on the market in 1968 and won a Putnam (Food Processing Magazine) award that year for contribution to the food industry. This indoor ceiling-hung trap was effective and killed a lot of night-flying insects, but was not very effective against day-flying insects (flies, etc.).

In 1969, the same designer discovered that light traps were significantly more effective against day-flying insects if they were placed down low, close to the floor and came out with the first wall-mount trap to do just that. With the back of the trap flat against the wall, the importance of reflector panels, materials, and theory became apparent. The corner-mount design came along in 1976 providing more versatility in the strategic installation of traps. Today, there are a number of different light trap designs for various strategic purposes.

THE VALUE OF LIGHT TRAPS

Sanitation is the basis of every good pest control program, chemicals play an important part, proper lighting is essential, as are a number of other preventative procedures; but not even a combination of these procedures can keep all the flying insects out of any facility. While light traps won't instantly mesmerize and magically eliminate every flying insect in a plant, they do eliminate a good percentage of those fliers which invariably escape other procedures, and as such, are an effective, integral part of the overall pest control system. They also function as a monitor, keeping constant check on the type and number of fliers present.

Detecting a Problem

In many companies, management is not aware of the numbers of flying insects that are actually present in their building. Placing a light trap in any type of facility (including those as critical as food plants, pharmaceutical laboratories, and hospitals) too often reveals, not only flies, but a good many other flying insects in surprising numbers. Some insects just naturally hide from sight. Others are so small we don't notice them until they pile up in the catch tray of a trap. While finding insects in your facility is hardly what you would call good news, "knowing you have a problem" is often said to be "half the solution." These traps can give you an enlightening look at insects that often go unnoticed.

Light Traps as a Method of Control

People who catch more insects in their facility than they had previously imagined were there are usually convinced that light traps are an effective means of control, indeed more than 100% effective it would seem. Evidence to the contrary comes from the common observation that a fly will sometimes buzz around a light trap and then fly away in the other direction or at times pays no attention to a trap at all. An inconsistent response to light appears to be part of the nature of the fly and is sometimes considered to be sufficient evidence that light traps don't work well against them.

While it is true (discounting negatively phototactic insects and non-fliers) that flies are one of the hardest, if not the hardest insect to catch, by knowing and using the insects' nature and habits, effective systems of control can be designed (see section entitled "Designing For Flies"). Furthermore, designing traps and systems well enough to control flies provides a more effective overall system for controlling all types of flying insects.

Light Traps as a Monitor

By keeping an eye on the insect catch, light traps also function as an early warning monitor system. For example, if a breakdown in sanitation or chemical procedures occurs, then more insects than usual will show up in the catch trays of the traps. This gives warning that something is not being taken care of properly and something can be done before the problem gets further out of hand.

Species identification can also lead to the solution of a problem. For instance, an unusual insect, one you are not familiar with, may show up in numbers sufficient to cause concern. Identifying the species of the insect and its habits (particularly those of propagation) may lead to the source and solution of the problem. Refer to chapters 5 and 6 on insect identification and behavior, respectively.

HOW TO USE LIGHT TRAPS

Indoor/Not Outdoor

Sunlight contains large quantities of all the wavelengths of electromagnetic energy which both humans and insects see as light. This includes all of the wavelengths which attract insects to light traps. For that reason, light traps are not effective outside during daylight hours. At night, outdoor traps will attract and kill a lot of insects; but while there may be some psychological comfort in hearing all those insects being zapped, the truth seems to be that there are just too many insects in the great outdoors for a trap to be of any real value there. Exterior night flier problems can be more economically and effectively controlled with the strategic use of outdoor lighting. (See section entitled "A Different Sort of Light Trap.") Inside a building, where the number of insects can be limited, is where the true value of light traps is found.

Area Coverage

The most often asked question about light traps is: "How much area will these traps cover?" It is erroneous to state that a light trap will control "X" square feet; especially without knowing the type of insects involved, the degree of control desired, or the environmental and structural conditions of the particular facility. The distance at which an insect responds is affected by lamp type and trap design, of course, but also by the visual acuity and nature of the particular insect(s).

Generally speaking, most flying insects cannot see or respond to a light trap more than 100 feet away. Flies will respond at 20 to 25 feet (Gilbert, 1967-84) with a significant increase in response at about 12 feet (Pickens, 1984), but by nature frequently move about in search of food, water, breeding materials, a cooler or warmer spot, etc. and inadvertently search out light traps in the process. From practical experience, we know that placing traps about 50 feet apart throughout a facility can be used as a general rule of thumb for flies and most other flying insects (Gilbert, 1967-84). However, this is only a general rule of thumb. Because of differing control requirements, the optimum number of and distance between traps will vary from facility to facility and from room to room.

It would be nice if there were an easy 1-2-3 way to explain exactly how many traps to use and where to place them, but there are too many variables involved. There are hundreds of thousands of different species of flying insects, more than 350 species of flies alone. There are also environmental and structural differences to consider. A facility with a fly-prone dumpster or railroad track area and grassy or tilled fields near big open bay doors and no physical barriers between there and the process area is a different situation than a relatively clean facility with a production area secluded away behind double-door entryways, chambered corridors, and narrow passageways. Generalizations can be made, however (about behavior differences between day and night-flying insects, for example), which provide a basic understanding and guide for the reasonable application of light traps.

Light Traps as Part of a System

Light traps should not be considered a replacement for good sanitation, proper lighting, or other preventative procedures. The installation of a well designed light trap system may allow you to reduce or perhaps even eliminate some current procedures (an undesirable chemical, for example) and still maintain the same level of control, but light traps should be used as only one of your integrated pest management tools in order to achieve the best control.

The first step in installing a light trap system is to limit, as much as possible, the number of flying insects the traps will have to deal with. Given time, light traps can handle the heftiest of temporary surges, but not a constant overflow. If poor sanitation or inadequate chemical procedures allow on-site breeding grounds to develop or if improper lighting consistently draws in an unnecessarily large number of night-flying insects, then even the most efficient, functionally designed, and strategically located traps can be overpowered. Use all of the insect control tools at hand and as masterfully as possible.

Strategic Installation

The strategic placement of light traps within a facility can dramatically increase their effectiveness. The basic idea is to place the traps, as much as possible, in the insects' path and at those points where they are most likely to respond. You need at least a general idea of the types of insects you are dealing with, how many there are, where they are coming from, and where they are going when they enter your facility. However, because of the difficulty in catching flies (day-fliers) and the more than adequate performance of "fly designed" systems against other species of fliers, the fly is the primary target in this endeavor. The fly's infamous reputation as a transporter of filth and bacteria is also not to be ignored.

Designing for Flies (Day-flying Insects)

Low Flying Creatures
Install traps down low, if you expect to catch a fly. It is in their nature, like most day-flying insects, to be near the floor when they are active and responsive to the traps. They spend a great deal of their time skimming surfaces as if constantly searching for something. Tests show that a trap placed near the floor is four to ten times more effective than one placed eight feet high (Gilbert, 1968; Pickens et al., 1969, 1975). Though many, perhaps most other flying insects (especially night-fliers), are usually found at higher

FIGURE 4. Install light traps down low (below 5') for day-flying insects (flies) and 8-10 ft. high for night-fliers. Install traps low to catch both types.

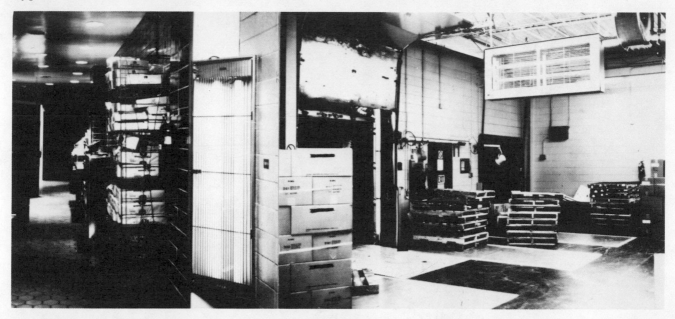

heights in a building nearer the warmth and lights, they are, as a general rule, much more responsive to ultraviolet light than day-fliers (flies) and come down to traps significantly better than day-fliers go up to one. The improvement in performance is worth considerable effort to install traps down low (putting a steel beam in the concrete to protect it from lift trucks, for example).

Ceiling-hung traps are necessary, however, in certain circumstances. An obvious example is at a loading dock door which is deluged with night-flying insects. This situation calls for large rectangular traps mounted 8 to 10 feet from the floor and perpendicular to the door so as not to draw insects in from outside, but to trap them as they enter. At loading docks or perhaps in other high insect traffic areas, both high and low traps may be useful, but in most cases low traps will handle both day and night-fliers (see Fig. 4 above).

Periodic Response to Light
As mentioned earlier, flies appear to be inconsistent in their response to light. This is due to their photo-periodic nature. They go through intermittent periods of attraction to and disinterest in light (Beck, 1968). If we watch a fly in a house, for example, we notice that every so often it goes to a lamp or a window, perhaps not immediately, and it doesn't always stay riveted there; but every now and then it will try to fly right into the light. At other times the fly may be more interested in looking for food, water, a playmate, a cooler or warmer spot...who knows?

What we do know is that flies respond to light traps in a very similar manner. At times it may seem as if the traps are not working, with a fly buzzing nonchalantly in front of a trap; but eventually the fly will enter a responsive period. In tests where 100 flies have been released into a 25' x 25' room, an average of 12 to 13 flies were caught in the first minute and a half, 60 more were caught over the next two to three hours, and essentially all were in the trap in 6 to 7 hours with a noted increase in responsiveness at sundown (Gilbert, 1969; Pickens, 1969). What this variable, but dependable, response pattern tells us is: don't bunch traps up in one area, but scatter them throughout a facility; so that when the fly is ready to respond, there will be a trap nearby to take care of it. The peak response at sundown and secondary peaks at various other times (sunrise, etc.) tell us the traps should be operated around the clock (24 hours a day).

FIGURE 5. This imaginary 100,000 sq. ft. food plant has more than its share of typical problems. Using the floorplan to highlight probable breeding areas and sketch in probable insect paths helps to determine the best locations for traps.

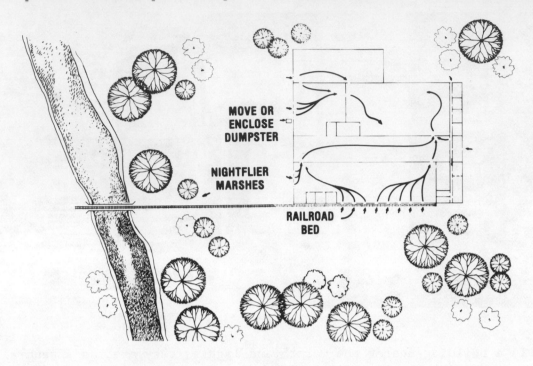

MOVE OR
ENCLOSE
DUMPSTER

NIGHTFLIER
MARSHES

RAILROAD
BED

Using the Floorplan to Map Strategy

The best way to begin designing a light trap system is with a floor plan of the facility spread out in front of you. Consider the types of insects that might be encountered and where they are likely to be coming from. Start by locating and marking all:

(A) dumpsters (flies),
(B) railroad tracks subject to food particle spillage (flies),
(C) grassy or tilled fields and standing water (night-fliers), and
(D) any other probable breeding areas of flying insects (include stored product areas, if applicable).

Then highlight all entrances. Determine the areas inside the facility which need the most protection and the path an insect would have to follow to reach those "critical" areas from each entrance and sketch them in. Traps will be placed progressively at key points along these paths. Locate any narrow passageways (doorways) along these paths where insects will be bottlenecked and highlight them. Keep in mind that insects will move toward food odors. Expect heavy fly traffic from dumpster areas. Note expected traffic from all insect sources.

Placing the Traps

Entrances: The First Line of Defense

Consider placing the first traps in an area 12 to 25 feet inside the entrances. A fly's eyes must adjust to sudden changes in light intensity much like a human's eyes do. At that time, a fly is about as responsive to a light trap as it will ever be. Also, any insects that can be eliminated here means that there will be that many less to deal with later on, closer to production areas. Place these traps where insects will get the best/longest look at them.

FIGURE 6. A floorplan layout of light traps demonstrating several possible locations for double-door entrance foyers which give insects more time to respond to light traps. The use of foyers makes this system superior to that in Figure 7.

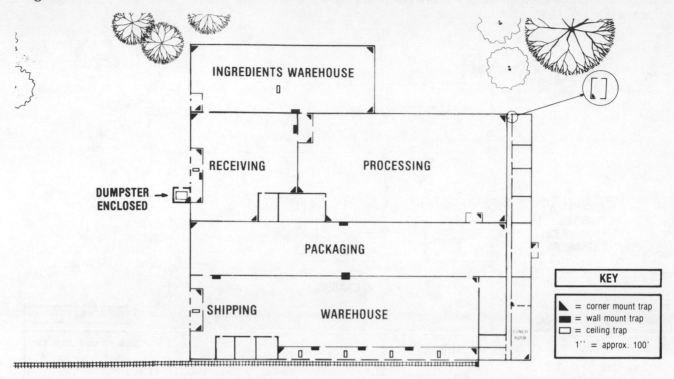

Of course, the best way to keep insects out is to seal up the cracks and keep all the doors closed all the time. This should be done as much as possible, but people must get in and out and deliveries must be made. Consider constructing "double-door" entries and installing a light trap in the foyer between them. Here, the insect is given significantly more time to respond to the trap while trying to negotiate the second door (see Fig. 6 above).

A few companies have even installed "triple-door" entries with a trap in each of the two foyers. Yet, no matter how elaborate the entrance trap locations, a few insects are inevitably going to get through. These must be handled by other traps strategically placed between here and the area(s) to be protected.

Production Areas: End of the Path

At least one physical barrier (wall) between outside entrances and any production area is a must if good control of flying insects is expected. Once the insects get into an open product area, they are harder to trap. Product odors compete with the effectiveness of light traps. Just how much is determined by what odor and which insects are there and if the insect is hungry or ready to lay eggs at the time. Temperature variances around products and machinery can also interfere with insect flight patterns. Half of the insects that get into this area may end up in a trap and the other half in your product.

Try to catch insects before they get into these areas. Locations previous to the production area should take precedence in placing traps especially if a high degree of control is not necessary or budgeting allows for only a limited number of traps. If on the other hand, the intention is to install the best system possible, then go ahead and install a number of traps in the production area and catch as many as you can. In any case, consideration should be given to placing at least one trap in (and/or adjacent to) each production area as a monitor. The entrance to a production area is a good spot for a light trap or double-door foyer where the trap can function as both a monitor and a control unit (see Fig. 6).

95

FIGURE 7. A floorplan layout of light traps without double-door entrance foyers. Keep in mind that there are a number of things which do not show up on floorplans (light, odor, wind, temperature, type and number of insects, degree of control desired, etc.) that will alter the optimum number and/or location of traps.

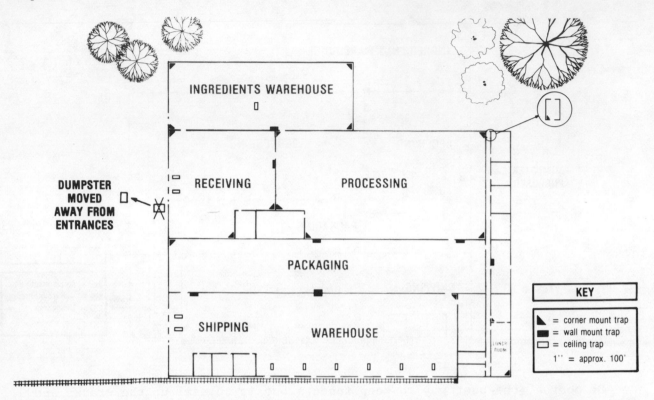

Bottlenecks

Bottlenecks are narrow passageways (doorways) which insects must pass through before moving further along into the facility. Here, traps can be set up in ambush, certain to get good exposure to each and every insect as it attempts to negotiate the bottleneck. As with entrance locations, every insect caught here is one less to deal with later on. Strip or air curtains placed in doorways are useful in deterring insects to some extent, providing more time for them to respond to the traps. Air curtains must be powerful (1680 fpm at 24" above the floor), well adjusted, maintained, and not subject to counteracting winds. By looking at the pathways you have already sketched on your floorplan, you can get an idea how crucial a particular bottleneck is in protecting an area(s) by the number and density of insect paths which funnel through it. Some bottlenecks can be as strategically crucial as entrances or even more so.

Stairwells

Stairwells are excellent places for light traps. They are often the best spot to trap insects on their way to another floor, but more importantly, a stairwell which has an outside entrance functions as a ready made double-door entrance foyer.

Corners

Traps that fit in corners have a number f advantages. Because of temperature variations or perhaps odors trapp d there, flies sometimes develop a fondness for a particular corner in a facilit . A corner also provides a good place to install traps out of the way of lift trucks and people, but what is most important is catching insects. Don't stick a trap in a corner just to get it out of the way. Still, a corner near an entrance or bottleneck can make an excellent trap location even better. Corner mounted traps also have an advantage in many circumstances of being visible from all points in a room while a trap mounted flat on a wall would have blind spots. Always keep corners in mind. Use them whenever reasonably possible.

Areas Off the Insect Pathways

Don't completely forget these areas. Keep the fly's periodic response in mind. Spread the traps throughout the facility while still staying in the fly's way as much as possible. Decisions between these two sometimes conflicting objectives usually depend on whether overall control or the protection of a specific area is of greater importance.

Monitoring

The catch of light traps should be noted as the information can be quite useful. A general increase in the number of insects caught hints that on-site breeding grounds may have developed. An increase in a particular species of insect or the number caught in one particular trap is an even stronger signal of on-premise breeding. If breeding grounds have developed, knowing which trap has the largest increase can help you locate the problem.

A light trap used specifically for monitoring purposes should be placed in the position of greatest exposure to the area to be monitored and left on for a time period of at least 24 hours which would, of course, include both sundown and sunup (high response periods for many species).

Stored-Product Insects

If the target insect is a stored-product pest, then the source of that insect should be treated as their primary entrance. Since the source of the insect and the product to be protected are most often one and the same, traps should be placed in close proximity to the product to lure away and kill as many as possible before they can propagate and multiply the problem. A lamp which produces energy in the green rather than the ultraviolet part of the spectrum may be helpful (see section entitled "A Lamp for Stored-Product Insects").

Where Not to Place Traps

Don't place ceiling-type light traps in open product (production) areas. Use wall-mount light traps only. Though every trap you use should have a catch tray, an insect flying at high speed or one subject to a good gust of wind could still fall out and into whatever is beneath it.

Don't install light traps in a potentially explosive environment; operating rooms where oxygen is used or rooms in cereal plants which have excessive dust particles, for example. Light traps do produce a small electrical arc to kill insects and could possibly ignite such substances. There are light traps without electric grids that (though not nearly as effective) may be of some use in these circumstances.

Don't install a trap so it shines directly out a glass door since it may congregate insects there like horses at a starting gate.

Other Factors to Consider

Competing Light

Light traps must compete with the existing lighting inside a building for the attention of insects. Trap effectiveness will not be diminished in rooms lit with daylight fluorescent or incandescent lighting. The ultraviolet lamps of light traps have a significant advantage over both of these, but trap efficiency is reduced by the use of mercury vapor lighting. Mercury vapor emits a high amount of the ultraviolet energy which insects find so attractive. The use of mercury vapor will not necessarily negate light trap effectiveness, but can noticeably reduce it (especially in the case of night-fliers).

Sodium vapor lighting, on the other hand, contains only a very small amount of ultraviolet energy and using it for indoor lighting will noticeably increase the effectiveness of light traps. Its use is highly recommended. Cost efficient operation is another reason to consider sodium vapor. Color rendition can be a bit of a problem under low pressure sodium vapor light, but this problem is more or less absent with high pressure sodium vapor. Special lighting can then be used in specific areas where high quality color rendition is necessary.

Sunlight from windows will compete for insects' attention to an even greater extent than mercury vapor. Efforts should be made to screen out sunlight where control is needed and consideration given to placing the traps in the darker parts of rooms. If excessive sunlight is unavoidable, trap effectiveness will be limited to the shadier hours.

Odors

Odors are primary attractants to food, breeding materials (often one and the same), and sexual partners. It is no wonder then that insects have a highly developed sense of smell and that dumpsters and production areas are the most attractive spots. It is practically impossible to keep flies away from garbage areas. Insects can track an attractive odor to its source from a good distance away, but as many as possible should be kept away and uncontrolled breeding must be avoided. Keep dumpster areas located as far away from entrances and as clean as possible. Purge any cracks and crevices in the concrete beneath the dumpster with a residual chemical to prevent "fly factories" from developing in the garbage residue found there.

Propensity to Flight

While it is true that flies are highly attracted to the odors of barnyards, animal carcasses, and dumpsters, you can't depend on them to stay there. "The restlessness and mobility of flies is one of their most important vector [germ spreading] attributes for it helps to expose them to a vast array of pathogenic organisms and to the food we eat." "Fly traffic between kitchens and privies has . . . been well documented by the studies of the house fly, Drosophila, and other species by Shura-Bura (1952) in the Crimea, Pimentel and Fay (1955) in Texas, Zaidenov (1960) in Chita, and Murvosh and Thaggard (1966) in the Bahamas" (Greenberg, 1973). In placing light traps, you can depend on insects to move back and forth along odoriferous pathways from dumpsters to production areas and other scent producing areas.

Temperature

Insects are cold blooded. They do not have the human capacity to generate body heat. The favorite fly temperature is around 82 degrees F and five degrees one way or the other can make a noticeable change in a fly's behavior. Like a snake to pavement, they will seek a warmer spot when it gets too cold (or a cooler one when it gets too hot). Many day-fliers cease to fly at about 48 degrees and are immobile (don't crawl) below 45 degrees F. Knowing this provides another clue as to which areas they might frequent or avoid and, consequently, where a trap may or may not be needed.

Wind Currents

Flying insects are lightweight and can be moved about by small amounts of wind. The gust of wind you feel when you open a door of some facilities (caused by positive air pressure) may be strong enough to keep some insects out while a negative pressure may be enough to pull some in. Keep in mind that flying insects have a tendency to follow a wind rather than fight it.

On the other hand, some are strong enough fliers to overcome fairly stiff breezes, especially if there are odors on the wind. Flies have been known to travel over four miles a day against a wind of up to 10 mph (Yerington and Warner, 1961). Even air curtains are not much good against flies unless they are high powered (1680 fpm at 24" above floor), extremely well adjusted (no gaps), carefully maintained, and not subject to counteracting winds.

Counter-clockwise Flight Pattern

Most flies seem to follow a counter-clockwise route when exploring new areas (Pickens, 1984). If there are no other overriding factors or considerations, placing traps to the fly's right as he enters a new area is a good idea.

A SUMMARY OF STRATEGIC INSTALLATION

(1) Don't expect magic from light traps. Use them as only one tool against flying insects. Prevent constant overload of insects with good sanitation, proper use of lighting and chemicals, and then complete the system with strategically installed traps to control normal flying insect traffic and temporary surges.

(2) Light traps will be four to ten times more effective against day fliers (flies) if we place them down low as close to the floor as possible.

(3) The fly is periodic in its response to light traps, so make sure not to bunch the traps up in one area. When the fly is responsive we want a trap nearby to take care of him. Spacing them 50 ft apart is only a general rule of thumb.

(4) Know the path insects must take to reach your most critical areas and place traps in their way.

(5) Consider entrances and bottlenecks for prime trap locations. Keep stairwells and corners in mind as well.

(6) Place at least one trap in each production area as a monitor and perhaps more for control purposes.

(7) Do not use ceiling-hung traps in production (open product) areas, or any electrocutor light traps in explosive areas or where attractant light shines directly out a glass door which is used at night.

(8) Take note of existing lighting. Light traps must compete with it for the attention of insects. Sodium vapor lighting is the best to use in this respect, incandescent and fluorescent are O.K., while mercury vapor is the worst.

(9) Learn as much as you can about the nature and habits of flying insects. Keep the effects of odors, temperature, wind, light, etc. in mind as you lay out a light trap system.

(10) You might consider buying your light traps from a company that provides system design services. They have experience. Make good use of these services, but combine it with a bit of your own reasoning. No one can fine tune your system the way you can, working with it on a day to day basis. If you notice insects congregating in a particular place, perhaps where the expert put no trap, then put one over there and see for yourself if it works better.

A DIFFERENT SORT OF LIGHT TRAP

Insect electrocutor light traps are not effective outdoors and yet some of the worst flying insect problems develop there around the outside of a facility. There is, however, a different sort of light trap that can solve this problem. It is the strategic use of outdoor lighting that will keep a majority of night-flying insects trapped away from a facility's entrances.

Sodium vapor lighting contains a very small amount of the ultraviolet energy which insects find so attractive. Clear-mercury vapor, on the other hand, contains a great amount of it. By using sodium vapor lighting for all lighting near the building and putting up large clear-mercury vapor lamps 150-200 feet away from the building, the majority of insects can be concentrated out where they are no problem.

Be careful not to locate any lighting directly over entrances. Use sodium vapor out away from the entrance and pointing back to light the area. There are two types of sodium vapor lighting: high and low pressure. The inability to distinguish colors beneath low pressure sodium vapor can be a problem, but this problem is more or less absent with the use of high pressure sodium vapor.

HOW TO SELECT A LIGHT TRAP

There are significant differences between $200 - $500 industrial light trap designs and the cheaper, consumer oriented backyard bug traps. The size and quality of both the attractant lamps and reflector surfaces are the foremost and most obvious differences. A superior transformer may be used to charge the electrocuting grid or a higher quality ballast to provide greater lamp efficiency. The quality and durability of all component parts, materials, and workmanship; the ease of insect removal and lamp replacement; the availability and quality of service and various other design features will also contribute to the higher cost and superior performance of a light trap. However, though the best traps are never the cheapest, neither are they always the highest priced.

Attractant Lamps

The type of blacklight lamp used in a trap is important, of course; but any claim to have a super one-of-a-kind insect attractant lamp is presently unfounded. Though research continues in this area and a more effective lamp does remain a possibility, there are currently (October 1984) no private label or special wavelength lamps which are not commercially available to everyone. The best lamp (the Sylvania F40/350BL) is currently available from several sources: your local electrical supplier, the lamp manufacturer, and some light trap manufacturers to name a few. You should shop around for the best price. Hopefully, the following discussion can explain some of the jargon used in reference to attractant lamps so you will be better able to judge the merits of a new improved lamp, if and when one becomes available.

The Attraction

No one really knows what an insect sees, but our best avenue to understanding the attraction of insects to light and coming up with new innovations is in seeing how measurable lamp characteristics correspond to insect response patterns. Therefore, the theoretical attractiveness of a lamp is determined by the wavelengths of energy emitted, the intensity of energy emitted, the size of the luminous area, and the reflection of energy from

FIGURE 8. The energy spectrum. Humans see wavelengths of energy from 380 to 770nm as light. Flies respond best to wavelengths between 330 and 350nm.

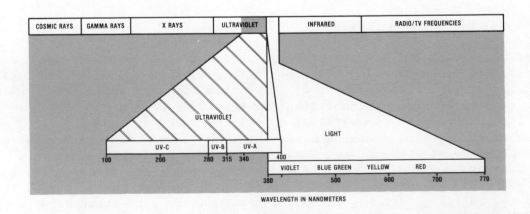

WAVELENGTH IN NANOMETERS

surfaces around it (Killough, 1961). In choosing a lamp, however, theory should be supported by practical tests.

Available Lamps

You will find a number of different brandnames on ultraviolet lamps, but there are currently only three basic types or phosphors to choose from. The phosphor is the coating on the inside of the glass tube which surrounds and filters the cathode produced mercury arc producing the lamp's various wavelengths of light (energy). The three currently available phosphors are the Sylvania (conventional) phosphor (barium silicate, lead activated), the Philips phosphor (strontium floral borate, europium activated), and the G.E. phosphor (strontium borate, europium activated).

In addition to the phosphor, the wattage of the lamp and the quality of the ballast will affect the energy intensity the lamp produces. Hartsock (1961) reported that although insect attraction increases with increasing intensity (brightness), the rate of increase steadily declines. There is a point of diminishing return. The size of a lamp's luminous area and its ability to light up a reflective surface must also be taken into consideration. Tests indicate that at least 40 watts are needed to do a respectable job of catching flying insects and that a total of 80 watts (two lamps) has proved best (Gilbert, 1968-84). The most successful industrial light trap designs use two, 40 watt, straight-tube, blacklight lamps each of which is 48" long and an inch and a half in diameter.

Searching for the Best Phosphor

Most of the theoretical "which lamp is best" talk centers around which phosphor does the best against flies because all three phosphors work comparatively well against other insects, especially night-fliers which respond more consistently to light and to a wider range of wavelengths. The problem with the theoretical talk has been that the published research on the subject is rather cryptic and sparse, hard to decipher, and the whole subject highly subject to misinterpretation. As a result, there is a lot of misinformation out there passing as fact.

Practical Tests

In these tests the different lamps are placed either in a trap in a test-chamber for a specific amount of time or side by side in identical traps and interchanged at regular intervals to see which lamp catches the most flies. The lamps are tested at various ages (stages of depreciation) as well. In both types of tests and at various stages of depreciation the results are the same. Of the three lamps currently available, the Sylvania F40/350BL (conventional phosphor) lamp consistently comes out on top (Gilbert, 1967-84) (see Table 1).

Theory Agrees with Practical Tests

As it turns out, the practical test results do support current scientific theory. Research (Thimijan and Pickens, 1973) established the theoretical response curve for flies, the peak of which is between 330 and 350 nanometers. When this theoretical curve is compared with the spectral energy distribution

TABLE 1. The three currently available 40 watt straight-tube blacklight lamps.

RANK	LAMP	PHOSPHOR	ENERGY PEAK
1st	Sylvania F40/350BL	(barium silicate, lead activated)	352nm
2nd	Philips F40T12BL	(strontium floral borate, europium activated)	365nm
3rd	General Electric F40BL	(strontium borate, europium activated	370nm

FIGURE 9. House fly attraction of equal radiant energy bands expressed as relative energy value. Based on 400 house flies exposed to each band (Thimijan and Pickens, 1973).

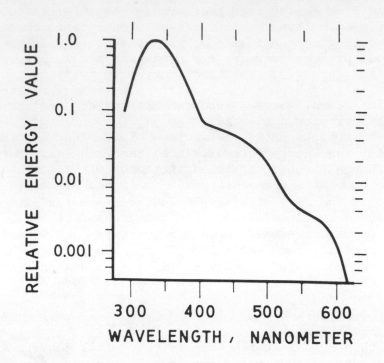

curves of the various lamps, it is found that even though the G.E. F40BL has more energy (intensity) at its peak of 370nm than the Sylvania has at its 352nm peak, the intensity of the G.E. lamp declines drastically from its peak and has practically no energy below 350nm where the fly is most responsive (see Fig. 9 above). The Sylvania F40/350BL has the largest amount of energy between 330 and 350nm of any available lamp and grades out on top in comparison of the curves as well as in the practical tests.

Claiming the Peak

The (Thimijan and Pickens, 1973) fly response peak is the most recent and reasonably acceptable theory. Research (Cameron, 1938) which incorrectly found 365.6nm (still quoted by some) to be the most attractive wavelength to houseflies is attributable to the lack of test instruments in 1938 which could deliver sufficient intensity at lower wavelengths. Thimijan and Pickens' equipment was considerably more advanced than Cameron's and their findings are also backed by the results of practical tests.

The BLB Myth

BLB (blacklight blue) lamps should not be used in light traps. Pickens found that BLB lamps are less effective (Pickens, 1984). They also cost approximately three times as much. The only difference in BLB and standard BL (blacklight) lamps is that the BLB contains a cobalt blue filter which removes energy from the visible end of the spectrum (above 420nm), removing approximately 20% of the insect attractant energy at the same time. This makes it more expensive and perhaps more effective for other purposes, but less effective in attracting flying insects.

The misunderstanding that exists concerning BLB lamps is due to an erroneous interpretation of Thimijan and Pickens (1969). BLB lamps are not more, but rather less effective in catching flies (or any other known flying insect) than the standard BL lamp.

Table 2. Non light-trappable flying insects

sawtoothed grain beetle	cadelle beetle
yellow mealworm	merchant grain beetle
fruit-piercing moth	larger grain borer
granary weevil	coffee been weevil
confused flour beetle	spider beetle

A Lamp for Stored-Product Insects

Laboratory tests (Marzke et al., 1973) suggest that replacing one of the two ultraviolet lamps in a light trap with a green (Sylvania F40T12G) lamp may be of value in dealing with many stored-product pests. The green lamp produces most of its energy between 500 and 560nm with a peak around 530nm. The combined spectral curves of the green and ultraviolet lamps resemble the typically wide ranging response curve of stored-product pests and some flies (Pickens, 1983; Mazokhin-Porshniakov, 1960). However, Kirkpatrick et al. (1970) found BL lamps alone to be as effective as BL plus green. Sufficient practical tests and on-site experience are lacking and have yet to prove or disprove the theoretical evidence.

Trappable Insects

A list of all the light-trappable flying insects would fill an entire book by itself. It is much easier to include a list of negatively-phototactic insects. Insects not included in this list are more or less responsive to light traps. Note that there is some question as to the inclusion of the confused flour beetle and all the insects listed in the right hand column, but if these insects are at all trappable, it is to no significant extent (see Table 2 above).

Trap Design

Reflective Surfaces

The surfaces in light traps which reflect energy from the blacklight lamps are very nearly as important as the lamps themselves in attracting insects. Generally speaking, the larger the reflecting area, the larger the attraction, but only if a suitable material has been used. If a suitable material is used, the whole reflecting surface, rather than just the lamps, becomes the attractant. The success of larger reflecting areas is believed attributable to the stimulation of a larger percentage of the multifacets in the eyes of insects (Pickens, 1984).

Aluminized Steel

The combination of specular and diffuse ultraviolet reflection patterns of ultraviolet by aluminized steel makes it the superior material for this purpose. Anodized aluminum, with slightly higher diffuse reflection patterns, but considerably lower specular reflection intensity, rates second in both theory and practical tests (Gilbert, 1975-76).

Directional Control vs Two-sided Attraction

Some light trap designs are open on two sides and have no reflector panel while others have a reflector panel which shuts off light from one side of the unit, but reflects a great deal more out the other. Whether this combination of both lamp and reflected light out of one side of a trap is superior to non-reflected lamp light out of both sides of a trap depends on the area and way in which the trap is used. A two-sided trap will be visible from more areas, but a trap with a good reflector will be more attractive within its illuminated hemisphere of visibility.

FIGURE 10. The large aluminized steel reflector panel (behind the lamps and grid) of this wall mount trap is a most effective design feature.

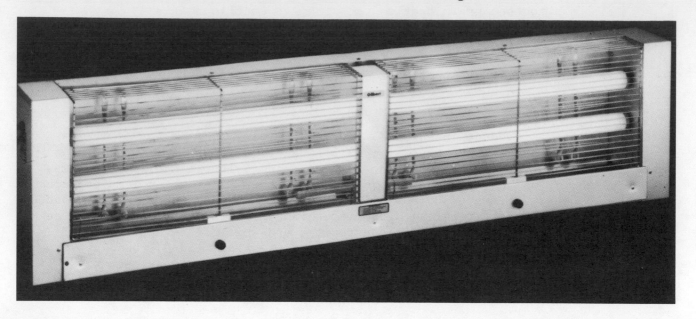

Wall-Mount Traps

Research (Gilbert, 1968; Pickens, Morgan, and Thimijan, 1969) supported the design and use of wall-mounted light traps for strategic installation at lower heights for improved performance against day-flying insects. Previous to 1968, all light traps were made to hang from the ceiling (or to sit on a post out in the yard). Low installation still requires the use of a wall or corner-mounted trap. A rare exception is when a ceiling-type trap can be hung low beneath a table or conveyor belt and still be visible enough to be effective.

Corner-Mount Traps

A trap with the corner-mount option has its advantages. Insects sometimes develop a fondness for a particular corner; corners often provide for a superior strategic placement in getting out of the way of lift trucks and people while staying in the insects' way; and in many circumstances corner mounting allows a trap to be visible from all points in a room when a flat wall mounted trap would have blind spots (see Fig. 11).

Recessed Installation

Some trap designs have an inner body which is removable from the outer casing of the trap. This allows the trap to be mounted into and flush with the wall taking up less space and giving a more attractive appearance and can still be serviced if need be. If necessary, an entirely new trap (inner body) could be slipped into the old casing.

Ceiling-Hung Traps

Ceiling-hung traps are primarily used against night-flying insects which have a tendency to be at higher heights than day-fliers, nearer the warmth and lights up there. They are sometimes used for fly control in high traffic areas or in hospitals where scuffling feet or aseptic floors can sometimes send flies higher than they are normally found.

The Electrocutor Element

The electrocutor grid is comprised of oppositely charged parallel metal wires. The electrical charge is generated by a special high voltage transformer designed to deliver the precise voltage with true current limiting features.

FIGURE 11. Optional corner-mount. The 45-degree angled sidewalls and flat back of this light trap design allow it to be installed either flat against the wall (like a typical wall-mount trap) or snugly in a corner.

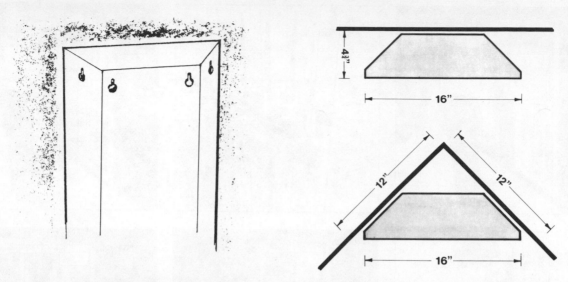

The quality (lifespan) of this smaller type of transformer, delivering 3500 volts at 9 milliamperes current, is dependent on how well the secondary is insulated from moisture.

Usually, the electrical circuit will include a capacitor (for storage) which provides extra voltage at the instant the insect's body closes the circuit to help prevent insects from sticking to the grid. The grid itself also acts as a capacitor by holding a certain amount of electrical charge until released. As a result of capacitor action, traps should maintain less than 5000 volts. Grids charged at more than 7500v will repel a fly.

Concern for Safety

U.L. Listed/Factory Mutual Approved

As a minimum requirement you should insist on an Underwriter's Laboratories (U.L.) Listed product. The engineers at both the Underwriter's and Factory Mutual laboratories make a thorough initial check into the safety of a product's design. Files are established, updated, and followed up with unannounced on-site investigations to the manufacturing facilities of each listed or approved product. Make certain the whole unit and not just the component parts are U.L. Listed. There are units on the market which are not U.L. Listed and do not meet the National Electric Code. Some practices such as using two ballasts to overdrive an attractant lamp are unsafe and the benefits only short lived.

If the Factory Mutual Approval label is present, you can be further assured that you are buying a quality product. The tests to acquire and retain the Factory Mutual Approval label are most stringent.

MAINTENANCE

Very little maintenance is required other than emptying and sanitizing the insect catch trays weekly (daily in production areas) and changing lamps once a year. With a well designed light trap, this takes a minimal amount of time. On the other hand, if a trap design requires 15 to 20 minutes with a screwdriver to complete these procedures with each of twenty or so traps, then you have quite a little job on your hands each week. You definitely want to consider how well a trap design allows for cleaning and re-lamping (see Fig. 12). Light trap maintenance services are provided by some pest control operators.

FIGURE 12. The easy-release latch and swinging guard door of this light trap makes cleaning and re-lamping a relatively simple process.

Change Lamps Yearly

Lamp phosphors will gradually lose their effectiveness over time. It's a good idea to replace your lamps (at least) once a year in the spring, so your traps are ready for the summer insect season.

Empty the Catch Trays

If the trays are not emptied on a routine basis, dead insect carcasses can be used as a breeding medium, permitting a tray to become a source of infestation. Periodic washing of the tray, lamps, and reflector is also recommended. If desired, a residual chemical (approved for the area in which the trap is placed) may also be applied to the catch tray.

Quality of Materials and Workmanship

Industrial light traps can be built to last. There are traps made over fifteen years ago that are still in operation today, some even using the original ballast and transformer. On the other hand, there are also light traps that do not last a month if they ever work at all. A good ballast or transformer should have at least a two year guarantee and an actual life beyond that. Checking on the quality of component parts and guarantees that a manufacturer offers is a good idea for anyone considering making a sizable investment in them. It will pay off down the road in lower maintenance costs. Find out what sort of tests are run at the factory to insure quality and performance, beware of guarantees that are too good to be true, and make sure the company will be around to give you service.

SUMMARY

Light traps are an effective integrated-pest-management tool with value as both control and monitoring devices. Though they often catch more insects than were previously thought to be present in a facility, they are not a magic wand and should not be expected to instantly zap away flying insect control problems. Light traps must be used in conjunction with good sanitation, proper lighting, and the necessary chemical procedures to really get the best from them.

Light traps should be installed strategically at and between the entries and the area(s) which need the most protection. Generally, light traps should be installed as low as possible to be the most effective. Only in special circumstances should the top of a trap be placed over five feet high. Taking into account the photoperiodic nature of many insect species, an effort should also be made to spread the traps evenly throughout a facility so that when such an insect does enter a responsive period there will be a trap nearby to take care of it.

How effective a flying insect control system will be is determined to a great extent by the amount of control achieved at the entrances. Light traps placed in double or triple-door entrance foyers are most effective. Bottlenecks, corners, and stairwells also make excellent trap locations. Consideration must be given as to whether one monitor trap is needed in a production area or more for control purposes.

The most important insect attracting features of a light trap are the size and type of both the ultraviolet lamp and the reflector surfaces, but consideration should also be given to the quality of component parts, materials and workmanship, the ease of maintenance, the quality and availability of service, and the safety of the traps. The U.L. Listed and Factory Mutual Approval labels are good indications of product safety.

LITERATURE CITED

Beck, S. D. 1968. Insect Photoperiodism. Academic Press, New York and London.

Cameron, J. W. M. 1938. The reactions of the housefly, Musca domestica L. to light of different wavelengths. Canad. J. Res. (D), 16:307-342.

Greenberg, B. 1973. Flies and Disease, Volume II. Princeton University Press, pp. 79-89.

Gilbert, D. E. 1967-1984. Unpublished research.

Hartsock, J. G. 1961. Relation of light intensity to insect response. USDA, ARS (ser.) 20-10, July:26-32.

Killough, R. A. 1961. The Relative Attractiveness of Electromagnetic Energy to Nocturnal Insects. University Microfilms, Inc., Ann Arbor, MI. pg. 71.

Kirkpatrick, R. L., D. L. Yancey, and F. O. Muzke. 1960. Effectiveness of green and ultraviolet light in attracting stored-product insects to traps. J. Econ. Entomol. 63:1853-1855.

Marzke, F. O., M. W. Street, M. A. Mullen, and T. L. McCray. 1973. Spectral responses of six species of stored product insects to visible light. J. Georgia Entomol. Soc., 8(3):195-200.

Mazokhin-Porshniakov, G. A. 1960. Colormetric study of the properties of colour vision of insects as exemplified by the house fly. [In English]. Biophysics 5(3):340-349.

Murvosh, C. M., and C. W. Thaggard. 1966. Ecological studies of the house fly. Ann. Entom. Soc. Amer., 59:533-547.

Pickens, L. G. 1983. Color responses of the face fly, Musca autumnalis. Exp. et. Appl. 32:229-234.

Pickens, L. G. 1984. Personal communication with author.

Pickens, L. G., N. O. Morgan, and R. W. Thimijan. 1969. House fly response to fluorescent lamps: Influenced by fly age and nutrition, air temperature, and position of lamps. J. Econ. Entomol. 62(3):536-539.

Pickens, L. G., R. W. Miller, and L. E. Campbell. 1975. Bait-light combinations evaluated as attractants for house flies and stable flies. J. Med. Entomol. 11:749-751.

Pimentel, D., and R. W. Fay. 1955. Dispersion of radioactively tagged Drosophila from pit privies. J. Econ. Entomol., 48:19-22.

Shura-Bura, B. L. 1952. Experimental study of migration of houseflies using radioactive indicators. Zool. Zhurn., 31:410-412.

Thimijan, R. W., and L. G. Pickens. 1973. A method for predicting house fly attraction of electromagnetic radiant energy. J. Econ. Entomol., 66(1):95-100.

Yerington, A. P., and R. M. Warner. 1961. Flight distances of Drosophila determined with radioactive phosphorus. J. Econ. Entomol., 54:425-428.

Zaidenov, A. M. 1960. On the study of the dispersal of houseflies (Diptera, Muscidae) by means of the luminescent method of marking in the city of Chita [in Russian]. Rev. Entom. U.R.S.S. 39:574-584.

ARTHROPOD PEST MANAGEMENT
WITH RESIDUAL INSECTICIDES

J. Larry Zettler and Leonard M. Redlinger

Editor's Comments

Residuals are a key tool, a key part of any
Integrated Pest Management system for insect control. In
their first table the authors list the EPA approved
residuals and the insect species on which they are
particularly effective.

Proper selection of a residual will depend not only
on the identified species requiring control but on the
conditions under which the residual needs to be used. The
chapter speaks to stability and cites a number of
stabilities of individual chemicals. You are advised to
become aware of your use variables and how they may
adversely affect the stability of the selected residuals.

Valuable discussions on the information on pesticide
labels and the resistance of insects to insecticides, which
grows steadily, are presented. Table 3 summarizes the
current information on resistance. This continues to grow,
as the authors state, "exponentially" and a likely result is
"the ability of virtually all pests to develop resistance to
all toxicants." Hence the importance of using all control
tools and to vary these including the insecticides used.

Brief summary remarks on the behavioral aspects of a
few of the common environmental insects of risk to industry
are listed.

The authors do not speak to the comparative toxicity
of residuals to the various insect stages (eggs vs immatures
vs adults) because such information is lacking.

Temperature was properly mentioned as an important
factor affecting the stability of insecticides. At lower
application temperatures the residuals will last longer and
be needed less because insect growth and reproduction are
slower. You might be interested in the results of an
"informal" poll which I took years ago of six insect control
experts, three of whom are contributors to this book. They
were asked, "Above about what temperature range should
residuals be applied for good insect control in food
warehouses whose temperatures will fluctuate with outside
temperatures?" The answers ranged from 35-40 to 55-60°F,
with the median being about 45°F. With species of concern
and other variables this is a rough "estimate," but it may
provide you with some guidance in your program.

F. J. Baur

109

ARTHROPOD PEST MANAGEMENT WITH RESIDUAL INSECTICIDES

J. Larry Zettler and Leonard M. Redlinger

Research Entomologists
USDA, ARS
Stored-Products Insects Research and
Development Laboratory
3401 Edwin St., P. O. Box 22909
Savannah, GA 31403

INTRODUCTION

Insecticides have been the main tool of the pest management specialist in his efforts at controlling arthropod pests for many years. Even though supplementary tools have developed through advancing technologies, it appears that insecticides will continue to remain at the heart of insect pest management technologies. One of the most efficient ways of utilizing an insecticide in a control program is via a residual treatment. A residual treatment is one in which the insecticide is applied to a surface to leave a deposit or residue which becomes available for subsequent pick-up and penetration into an insect pest when that pest walks on, crawls through, or eats the deposit.

Ideally, in terms of insect control, the most efficient residual treatment would be one in which every surface in a treated facility is evenly coated with residual deposits of the insecticide. However, the very nature of food handling and processing facilities precludes the indiscriminate spraying of insecticides and imposes restrictions on where insecticidal materials may be used. Thus, in terms of human safety and practicality, the use of residual insecticides in the food industry is limited to crack and crevice and spot treatments, i.e., treatment of the areas where pests live and hide such as behind baseboards, in storage areas, closets, around water pipes, doors, windows, under equipment, etc.

INSECTICIDES

There are a number of insecticides registered by the Environmental Protection Agency (EPA) and available for use as crack and crevice treatments (Table 1). Some of the materials exert their toxic effect through physical means, some through poisoning of the nervous system, and others through blocking the biochemical enzymes in the insect which are normally used by the insect to metabolize the insecticide to non-toxic metabolites. Effective pest management specialists make efficient use of these types of insecticides by knowing which to use under a given set of circumstances. Thus, the specialist must have a thorough understanding of the insecticides available for use and how these interact with the insect pest and with other insecticides.

Table 1. EPA – approved chemicals registered and labeled for use as residual sprays against pests occurring in the food handling and processing industry.

	Acephate	Bendiocarb	Boric Acid	Carbaryl	Chlorpyrifos	Diatomaceous earth	Diazinon	Dichlorvos	Fenthion	Malathion	Propetamphos	Propoxur	Pyrethrins (Synergized)	Resmethrin	Ronnel	Silica Gel	Trichlorfon
Angoumois grain moth	X								X	X			X	X			
Ants		X	X	X	X	X	X	X		X	X	X	X	X	X	X	
Black carpet beetles		X						X		X				X			
Cadelles		X							X							X	
Carpet beetles		X			X		X	X		X	X	X	X	X		X	
Centipedes		X			X		X	X				X	X	X			
Cereal moths					X								X				
Cigarette beetles					X		X	X		X	X		X				
Cockroach	X	X	X	X	X	X	X	X	X	X	X	X	X	X	X	X	X
Confused flour beetles		X			X		X	X	X	X			X	X			
Crickets	X	X		X	X	X	X	X	X	X	X	X	X	X	X	X	
Dark mealworms	X				X								X			X	
Dermestids	X	X						X					X				
Dried fruit beetles		X			X		X	X		X		X	X			X	
Drugstore beetles		X						X		X		X	X	X		X	
Drywood termites		X			X			X				X				X	
Earwigs	X	X		X			X	X			X	X	X	X			
Firebrats/Silver- fish	X	X	X	X	X	X	X	X	X	X	X	X	X	X	X	X	X
Fleas		X		X	X	X	X	X	X	X	X	X	X	X	X		X
Flies					X	X	X	X	X	X			X	X	X		X

Table 1. Continued.

Pest	Acephate	Bendiocarb	Boric Acid	Carbaryl	Chlorpyrifos	Diatomaceous earth	Diaxinon	Dichlorvos	Fenthion	Malathion	Propetamphos	Propoxur	Pyrethrins (Synergized)	Resmethrin	Ronnel	Silica Gel	Trichlorfon
Flour beetles		X			X		X	X		X		X	X	X			
Flying moths								X					X	X			
Fruit flies		X			X					X	X		X	X			
Grain beetles		X			X		X			X	X		X	X		X	
Grain mites										X			X	X			
Grain weevils					X					X			X			X	
Granary weevils		X					X			X		X	X	X			
House flies					X			X	X	X			X	X	X		
Indianmeal moths	X							X	X	X			X	X			
Lesser grain borers		X								X				X			
Mediterranean flour moths								X	X	X							
Millipedes		X		X	X		X	X			X	X	X	X		X	
Mosquitoes		X			X		X	X	X	X		X	X	X	X	X	
Red flour beetles		X			X		X	X		X			X	X			
Rice weevils		X			X					X			X	X			
Sawtoothed grain beetles								X	X	X	X	X	X	X			
Sowbugs/Pillbugs	X	X			X		X			X	X	X	X	X		X	
Spiders		X			X		X	X		X	X	X	X	X	X	X	
Tobacco moths		X					X	X		X			X	X		X	

113

Inorganic Insecticides

These insecticides are not as important now as they once were. However, several inorganic insecticides remain effective for insect pest control.

Borax

Borax is the trivial name for sodium tetraborate decahydrate (also known as sodium borate or sodium pyroborate). It is a white crystalline solid, odorless, and slightly soluble in water. The acute oral (AO) LD_{50} (Table 2) for rats is 2,660 - 5,190 mg/kg; the lethal dose to human infants is 5-6 g. It is highly phytotoxic. When borax or boric acid is ingested by a cockroach, death usually occurs within 24 hrs. However, death may occur within 2-10 days if cuticular absorption is the only mode of entry.

Boric acid

Also known as boracic acid and orthoboric acid, this material is a colorless, odorless, transparent crystal very similar in toxicity to borax.

Silica gel

Silica gel is the trivial name for silicic acid. It is a white, amorphous powder, insoluble in water, and non-toxic to humans and other warm-blooded animals. This insecticide kills by removing the waterproof layer of the exoskeleton through a continuous adsorption of its lipid elements or through abrasion, thus bringing about dessication. It is claimed that silica gel is relatively non-toxic; the applicator is nevertheless advised to wear a mask when applying because the dusts are irritating to the nose and mouth (=silicosis).

Control with this compound is slow, usually requiring several weeks or longer. In fact, during the first week after application, there may appear to be an increase in pests because the dust flushes the insects into the open. The dust has no effect on insect eggs and becomes ineffective on active life stages if it becomes wet.

Diatomaceous earth

This insecticide is a fine siliceous earth composed chiefly of the cell walls of diatoms, also known as fossil flour, celite, and super-cel. It is a white to light gray to pale buff powder which is insoluble in water. This, like silica gel, acts as a dessicant. The dust is capable of causing pulmonary fibrosis in man under certain conditions which include long exposure to high concentrations.

Organic Insecticides

This class of insecticides contains the more common chemicals used as residuals. Generally, they have a high acute toxicity to insects, being much less toxic to mammals. The oldest of the organic insecticides is pyrethrins, a crude mixture of pyrethroids. Yet, more than 150 years after its introduction, it still remains a viable insecticide, particularly because of its remarkably low toxicity for mammals. It is the most spectacularly effective insecticide (see Chapter 10) in that knock-down is instantaneous if the insect receives a direct hit. However, total recovery of the victim can occur under some conditions and consequently pyrethrins is usually formulated with synergists which prevent recovery. Pyrethrins exerts its toxic action by acting directly on the nervous system to disrupt transmissions of nervous impulses.

Table 2. Relative toxicities of insecticides registered and labeled for use as residual treatments in food handling and processing facilities.

Chemical Name	Type	Mammalian Toxicity (Rats)[1]		Manufacturer
		AO	AD	
Acephate	OP	866-945	>10,250	Chevron
Bendiocarb	Carbamate	40-64	555-600	Fisons
Borax	Inorganic	2660-5190	--	Kerr-McGee
Boric Acid	Inorganic	>2500	--	--
Carbaryl	Carbamate	307-986	>500 -> 4000	Union Carbide
Chlorpyrifos	OP	82-245	202	Dow
Diazinon	OP	66-600	379-1200	Ciba-Geigy
Dichlorvos	OP	25-170	59-900	Ciba-Geigy
Fenthion	OP	255-298	1680-2830	Chemagro/Mobay
Malathion	OP	885-2800	> 4000	Amer. Cyanamid
MGK-264	Synergist	2800	--	MGK
Piperonyl butoxide	Synergist	6150->7500	> 7500	FMC
Propetamphos	OP	119	2300	Sandoz
Propoxur	Carbamate	83-104	> 1000 -> 2400	Chemagro/Mobay
Pyrethrins	Botanical	200-2600	1800	Fairfield Amer. Corp, MGK
Resmethrin	Pyrethroid	2500	> 3400	Fairfield Amer. Corp., Rigo Co.
Ronnel	OP	906-3025	> 1000-2000	Dow
Silica Aerogel	Inorganic	--	--	W. R. Grace & Co.
Trichlorfon	OP	450-630	> 2000	Chemagro/Mobay

[1] LD_{50} dose expressed as milligrams/kilograms (mg/kg). AO = acute oral; AD - acute dermal.

Generally, the predominate types of insecticides which fall in this category are the organophosphorus (OP) and carbamate insecticides. Both types of insecticides exert their toxic action by inhibiting the enzyme cholinesterase, one of the naturally occurring substances in the nervous system which is essential for normal transmission of nerve impulses. Inhibition of cholinesterase causes a concomitant increase of the neurotransmitter acetylcholine which causes repetitive firing of the nerves, a condition which eventually causes paralysis.

Acephate (Orthene)

Acephate is an OP insecticide of moderate persistence with a residual activity of from 6-8 weeks at recommended use rates. The technical material is a white solid with a low vapor pressure and a cabbage-like odor. Acephate liquid concentrate may crystallize out at temperatures below 29°F. Product may be reconstituted by agitation and warming to 70°F. It is soluble in water (approximately 65%) and has low solubility in organic solvents. Water-based sprays are relatively stable.

Bendiocarb (Ficam)

This carbamate insecticide is moderately toxic but has little or no flushing action. The technical material is a white, odorless, crystalline solid. Spray solutions are odorless and usually stable in water. However, Bendiocarb is usually formulated as a wettable powder and consequently must be agitated to prevent settling out of solution. It is susceptible to degradation on alkaline surfaces, but, unlike many OP insecticides, is stable on stainless steel. It will not damage paints, plastics, or other surfaces where water alone causes no damage.

Carbaryl (Sevin)

This carbamate insecticide is toxic to many insect species and has moderate residual activity. The technical material is light tan to lavender, essentially odorless, and stable to UV light, heat, and pH under normal conditions of storage and use. However, it decomposes in alkaline environments. It is not corrosive to metals. Mammalian toxicity is rather low, particularly via the dermal route. Carbaryl can be synergized with certain other carbamate insecticides.

Chlorpyrifos (Dursban)

This OP insecticide is highly toxic to most insect pests and exhibits a rather long residual activity, usually up to 10 days - 2 weeks. The technical material is a white, granular crystal with a rather low vapor pressure and a mild mercaptan-like odor. It is stable for rather long periods in neutral or slightly acid conditions at room temperature. Because of the low vapor pressure, chlorpyrifos is often formulated as an aerosol.

Diazinon (Spectracide)

This OP insecticide is somewhat selectively toxic in that some insects can enzymatically transform it to a more toxic metabolite than can mammals. It is a cholinesterase inhibitor and its toxicity can, under some conditions, be synergized with the addition of certain other chemicals to the formulation. The technical material is a pale to dark brown liquid with a faint mercaptan-like odor. Diazinon tends to be unstable in acid conditions. Because of its widespread use, some insect pests, particularly the cockroaches, have developed a physiological resistance to it.

Dichlorvos (Vapona, DDVP)

Dichlorvos, or DDVP, is an OP insecticide which is quite toxic to most insect pests. Because of its relatively low vapor pressure, it is active as a vapor (c.f. Vapona[R] insecticide strip). It is a cholinesterase inhibitor and, like malathion, can be synergized by some compounds although this is not practical economically. The technical material is colorless to amber and has a mild chemical odor. Dichlorvos is stable only in the presence of hydrocarbon solvents. It undergoes hydrolysis (degrades) in the presence of water and is readily decomposed by strong acids and bases. It is corrosive to black iron and mild steel. Because it degrades in water, water-based spray applications should be avoided whenever possible. The effectiveness of water-based sprays is quite short, on the average of 1-3 days.

Dichlorvos is not highly selective in its toxicity to mammals and insects. Of the OP insecticides used in pest control, it is one of the most toxic to mammals, its AO toxicity being 25-170 mg/kg.

Fenthion (Baytex)

This OP insecticide, although requiring enzymatic transformation for increased toxicity, is relatively toxic to mammals compared to many other OP insecticides. Its mammalian toxicity is AO 255-298 mg/kg, well within the limits of safety, however. Residual activity is moderately persistent. A particular characteristic of this insecticide is its resistance to lime; sprays can be applied to fresh whitewash without degradation or without staining the whitewash. However, excess wetting of asphalt tile, rubber, and plastic materials should be avoided. Fenthion has seen frequent use against public health insects which are resistant to the chlorinated hydrocarbon insecticides.

Malathion (Cythion)

This is an OP insecticide which is quite selectively toxic to insects. For example, it is nearly 200 times more toxic to the cockroach as to the Guinea pig. The reason for the differential toxicity between insects and mammals is that malathion must be enzymatically converted by biological organisms to a more toxic form. Mammals generally lack the enzymes necessary for this conversion. Insects on the other hand are quite capable of making the conversion and thus become much more susceptible to malathion than do mammals.

Because of its selective toxicity, it is in widespread usage as an insecticide and is registered and labeled for a wide variety of insect pests. However, because of its long-term use, some insect pests, particularly stored-product insects, have developed a physiological resistance to it.

Malathion exerts its toxic action by inhibiting cholinesterase. Its toxic action can be synergized by certain compounds; however, in practice, this is not normally done. Technical malathion is a transparent brown to colorless liquid and exhibits a strong mercaptan-like odor. For this reason Premium Grade Malathion (more refined) is commonly used when odor is a concern. Malathion is stable when stored under the proper conditions; however, it decomposes at high temperature. Malathion attacks iron, steel, terne plate, tin plate, lead, and copper; the last two very seriously.

MGK-264 (Octacide 264)

This synergist, first introduced in 1944, is used to potentiate the toxicity of pyrethrins and some carbamates. It is usually incorporated in the formulation at the rate of 0.5 - 2%. It is a liquid with a bitter taste. Chemically, it is very stable, and is non-corrosive. This chemical is rather non-toxic to mammals via the oral route; interestingly, it is most toxic via the dermal route.

Piperonyl butoxide (PBO, Butacide)

This chemical was developed as a synergist for pyrethrins in 1947. It consists of a mixture of several compounds, the major component making up about 80% of the total. PBO is a pale yellow, odorless oil which is very stable chemically. It also is non-corrosive. When used with pyrethrins, it is incorporated at the ratios of 5:1 to 20:1, usually 10:1 by weight. Synergism of insecticides is generally believed to occur because PBO and related chemicals inhibit degradative enzymes which are responsible for metabolizing the insecticide.

Propetamphos (Safrotin)

Propetamphos is an OP insecticide with rather long (up to 2 weeks) residual activity against a wide range of insect pests. It is a reddish-brown oily liquid of rather high vapor pressure with a slight odor. It has been reported to be ineffective against certain malathion-resistant species of insects, however.

Propoxur (Baygon)

Baygon is a carbamate insecticide which gives rapid knock-down and rather long residual activity against most insect species. The technical material is a white crystalline powder with a faint, characteristic odor. It is relatively stable in water sprays but decomposes on alkaline surfaces.

Pyrethrins

The most common form in which pyrethroids are used is as an extract of pyrethrum, obtained by extracting the ground flowers to obtain a solution of pyrethroids along with waxes and pigments. The latter impurities are removed with chemical "clean-up" to leave the material called "pyrethrins." Pyrethrins is not a single compound but actually a mixture of 4 different compounds.

Pyrethrins has three peculiar attributes in relation to its toxicity. First, recovery from knockdown can occur due to the insect's ability to rapidly metabolize and degrade the molecule. Second, pyrethrins exhibits a negative temperature coefficient in relation to its toxicity. That is, it is usually more toxic at low temperatures. Third, pyrethrins has strong repellent activity to insects. As a result of this repellency, it serves as an excellent flushing agent for insects and is frequently used in conjunction with residual sprays.

The efficacy of pyrethrins can be improved by mixing with synergistic compounds. Synergists are compounds which, alone, are virtually non-toxic. However, when applied with certain insecticides, the toxicity of that insecticide is greatly improved. Typically, synergists act to block the metabolic breakdown of the actual toxicant by inhibiting key degradation enzymes. In the case of pyrethrins, the two most common synergists used in formulations designed for insect control are PBO and MGK-264.

The AO mammaliam toxicity of pyrethrins alone is 200-2600 mg/kg and the acute dermal (AD) toxicity is 1800 mg/kg. It is quite selectively toxic to insects. For example, pyrethrins is 1500 times more toxic to mosquitoes than it is to the Guinea pig.

Resmethrin (NRDC-104, SBP-1382, Synthrin)

Resmethrin is the common name for the synthetic pyrethroid originally designated as NRDC-104. It is somewhat more stable than the pyrethrins, but is decomposed fairly rapidly on exposure to air and sunlight. The technical material is a waxy solid, off-white to tan (amber) in color, with a characteristic odor. Its mammalian toxicity is AO 2,000 mg/kg. It is a contact insecticide with short residual activity. Synergists effective with pyrethrins do not work as well with the compound. It is formulated as an oil based concentrate or as an EC concentrate.

Ronnel (Korlan)

An OP, ronnel is a cholinesterase inhibitor and, like malathion, is enzymatically transformed to a more toxic metabolite by insects. Thus, it is rather selectively toxic to insects. The technical material is a white powder, has a relatively low vapor pressure (very active as a gas and is often formulated as an aerosol), and is unstable in alkaline and strongly acid media and upon prolonged exposure in aqueous preparation. Thus, water-based sprays are short-lived. Once used extensively in pest control, ronnel is presently not commonly formulated.

Trichlorfon (Dipterex)

This OP insecticide is closely related chemically to dichlorvos. In fact, it quite readily rearranges, with the loss of HCl, to the more toxic dichlorvos. Like dichlorvos, it is quite toxic to most insect pests but has a rather short-lived residual activity. Trichlorfon is particularly susceptible to hydrolysis under alkaline conditions. It is stable at room temperature but is decomposed by water at higher temperatures.

TOXICITY OF INSECTICIDES

All insecticides must be considered potentially toxic to man and animals (see Table 2). However, the degree of toxicity is one of several factors in the use of insecticides that determines the hazard. The primary hazard lies in failure to follow the precautions and directions for use indicated on the insecticide label. These precautions and directions depend not only on the degree of toxicity and the nature of toxicity of the insecticide but also on its stability. Some highly toxic insecticides that must be handled with great caution dissipate so rapidly upon exposure to certain surfaces and substrates that they create no serious residue problems.

Only the insecticides registered for use as crack and crevice or spot treatments must be used. A commodity that comes in contact with floors, walls, or machinery treated with an insecticide not registered for use in storage areas or food handling and processing facilities may become contaminated and be liable to seizure. Even when using insecticides which are registered for crack and crevice and spot treatments, care must be taken to avoid contamination of machinery and structures which might come in contact with a commodity. The insecticides should be applied in a pin-stream directly into cracks and crevices. Care must be taken to avoid misting which will lead to contamination. Spot treatments may be applied at moderate pressure (up to 35 psi) using a coarse fan tip to avoid atomizing or splashing of the spray. No individual spot will exceed two (2) square feet. Special attention should be given to the precautions on the label such as removing or covering exposed commodities before treatment.

Treatment for Poisoning

If a person becomes poisoned by an insecticide, call a physician, read antidote and first aid information on the insecticide container label, and give first aid immediately. If breathing has stopped, give artificial

respiration. If two persons are present, one should give first aid while the
other obtains the insecticide container and calls the physician. In general,
it is advisable to induce vomiting if the victim has swallowed a highly toxic
insecticide, is conscious, and the physician will be unavailable for at least
a half hour. A tablespoon of salt or baking soda in a glass of warm water
will help induce vomiting. Have the victim lie down and keep quiet until you
get advice from the physician. Keep the victim warm.

If a concentrate or oil solution has been spilled on the skin or clothing,
remove the contaminated clothing and wash skin with soap and water. If a
person feels sick while using an insecticide or shortly afterward, call a
physician immediately. In all cases, make available the insecticide container
and any attached labeling. Information provided by them is extremely valuable
to the physician. Inform the physician of all recent contacts or exposures to
insecticides; the one most obvious to you may not be the one to blame.

Disposal of Empty Containers and Surplus Insecticides

The careful disposal of empty insecticide containers and surplus
insecticides is an important part of safe insecticide use. The insecticide
label contains specific instructions for disposing of the insecticide
container. Generally, empty insecticide containers are buried in a sanitary
landfill after they have been crushed or punctured to prevent reuse.

Unless the label specifically states otherwise and if a suitable landfill
is not available, break or crush glass and metal containers (except
pressurized cans) and bury them in an isolated place where they will not
contaminate water supplies. Avoid mixing more insecticide than needed.
Following treatment pour any excess insecticides into a hole at least 18
inches deep dug in level ground in an isolated place where they will not
contaminate water supplies, and cover with dirt. Large drums that contained
insecticides should be sold to firms dealing in used drums and barrels. Do
not dump containers or leftover chemicals in gullies, ditches, streams, woods,
or trash heaps.

Residues

The pesticide that remains in or on raw farm products or processed foods
is called a residue. The Environmental Protection Agency (EPA) sets residue
tolerances under regulations authorized by the Federal Food, Drug, and
Cosmetic Act. The Federal Drug Administration (FDA) is responsible for
enforcing these regulations (See chapters 26 and 27). A tolerance is the
amount of residue that is judged safe for human use. Residues in processed
foods are considered to be food additives and are regulated as such. If too
much residue is found on a raw or processed product, the product may be seized
or condemned.

Labels and Labeling

Often, the meanings of these terms are confused or misunderstood.
Labeling is all the information that a pesticide manufacturer or his agent
supplies regarding its product, and includes such things as the label on the
product, brochures, flyers, and any information supplied by the pesticide
dealer.

The label is the information printed on or attached to the container of
pesticides. To the manufacturer, the label is a license to sell the product;
to the State or Federal Government, it is a means by which control is
maintained over the distribution, sale, storage, use, and disposal of the

products; to the buyer or user, it is the main source of facts which govern the correct and legal use of the product and any special safety measures needed.

Every pesticide label must contain specific information relating to the correct use of the product. There are 12 categories of specific information required by law to be printed on a pesticide label. These categories follow.

Brand Name
The brand name is the one used by the manufacturer in its promotions and advertisements and is the most identifiable name for the product.

Type of Formulation
Different types of pesticide formulations (i.e., liquids, emulsifiable concentrates, wettable powders, dusts, etc.) require different methods of handling. The label will tell what type of formulation the product contains.

Common Name
A common name is used to more easily identify pesticides which have complex chemical names. A chemical made by more than one company will be sold under several brand names, but the same common name or chemical name will be used on all of them.

Ingredient Statement
This statement lists the chemical make-up of the product. The amount of active ingredient is given as a percentage by weight or as pounds per gallon of concentrate and can be listed by either the chemical name or the common name. The inert ingredients need not be named but the label must show what percent of the contents they constitute.

Net Contents
This value can be expressed in gallons, pints, quarts, pounds, or other units of measure.

Name and Address of Manufacturer
The manufacturer or distributor of the pesticide must put the name and address of the company on the label so that the user will readily know who made or sold the product.

Registration and Establishment Numbers
The registration number shows that the product has been registered with the EPA. The establishment number tells what factory made the chemical. This number does not have to be on the label, but will be somewhere on each container.

Signal Words and Symbols
These are designed to provide the user with information regarding the human toxicity or hazards of a particular chemical. Specifically, the signal word tells how toxic the material is to people. These words are set by law and it is incumbent upon each manufacturer to use the correct one on every label.

Signal Words	Toxicity	Approximate Amount Need to Kill the Average Person
DANGER	Highly toxic	A taste to a teaspoonful
WARNING	Moderately toxic	A teaspoonful to a tablespoonful
CAUTION	Low toxicity or comparatively free from danger	An ounce to more than a pint

All products must bear the statement, "Keep out of reach of children."
The symbol of skull and crossbones is used on all highly toxic materials along with the signal words DANGER and POISON.

Precautionary Statement
This statement relates the ways in which the pesticide may be poisonous to man and animals. It also relates any special steps required to avoid poisoning (i.e., protective equipment needed). If the pesticide is highly toxic, it will inform physicians of the proper treatment for poisoning.

Statement of Practical Treatment
This relates the emergency first aid measures needed if swallowing, inhaling, or getting the pesticide on the skin or in the eyes would be harmful. It also tells what types of exposure require medical attention.

Statement of Use Classification
The label must state whether the pesticide is for general or restricted use. If the pesticide will harm the applicator or the environment very little or not at all when used exactly as directed, it will be labeled as a general use pesticide with the wording "General Classification." If the pesticide could cause some human injury or environmental damage even when used as directed, it will be classed as restricted use with the label reading, "Restricted use pesticides for retail sale to and application only by certified applicators or persons under their direct supervision." This statement must be at the top of the front panel of the label.

Directions for Use
This section of the label tells the correct way to apply the pesticide. It will tell specifically the pests the product is registered to control, the crop, animal, or other item the product can be used on, whether it is for general or restricted use, in what form the product should be applied, how much to use, and when and where the product should be applied. Other parts of this section include: the misuse statement (a reminder that it is a violation of Federal law to use the product in a manner inconsistent with the label); the reentry statement (if required, will indicate how much time must pass before a pesticide-treated area is safe for reentry by a person without protective equipment or clothing); the category of applicator (if required for the pesticide); and it will limit use to certain categories of commercial applicators and the storage and disposal directions.

ARTHROPOD PESTS

The list of insect and other pests which can be associated with the food industry is enormous. Some gain entry into the food plant via the raw materials used in the making and packaging of the final product; some gain access through cracks in the physical structure or around plumbing and electrical passages; still others enter through windows and doors. It is imperative that the pest management specialist be aware of these possible entry points into the food handling facility so that preventive measures can be taken to reduce the risk of infestation. In addition, the specialist should be aware of sanitation practices and their importance in preventing the establishment of insect infestations. These facts notwithstanding, insect infestations can and will occur in the food industry even under the best of conditions. Because of this, the pesticide specialist must become familiar with these pests, their habits and haunts, and the means by which they can be eliminated.

Many of the insect pests directly associated with agricultural commodities processed in the food industry are described in detail elsewhere (Chapter 5 and 6). However, the following list will provide additional information on the habits and haunts of other important anthropod pests.

Ants

Ants can be distinguished from termites in that their "waist" is constricted or pinched; termites have no such constriction. Also, the hind wings of the ant are considerably smaller than the front wings; those of the termite are very nearly the same size. Ants usually invade buildings from nests outside; however, some ants can nest in old, decaying wood of some buildings. Ants generally attack only food items, and carry these materials back to the nest. Good control of these insects requires finding the nest which may be done by tracing the ants' line of march from the food source. Proper insecticidal treatments can then be made.

Centipedes

While centipedes do not damage food supplies, they are a source of potential contamination. In many situations, they actually assist the pest management specialist in his control activities. These animals are long, flattened, and many-segmented predaceous arthropods with each segment bearing one pair of legs of which the foremost pair is modified into poison fangs. These fangs are used to kill insects and other small creatures for food. Centipedes normally nest outdoors and are migrants into buildings. Other occasional invaders like ground beetles, or those insects attracted to lights such as leaf hoppers, and moths are particularly a problem during the summer months.

Clothes Moths and Carpet Beetles

Although not close relatives, these two insects both feed on natural hair, feathers, wool, fur, horns, and hoofs. Larvae of these insects are attracted to such things as human sweat, urine, tomato juice, coffee, and beef gravy, presumably because of the salts they contain. They are particularly fond of dog foods and other types of food containing bone meal.

Cockroaches

There are several species of cockroaches known to infest food processing plants and associated buildings in the United States. These insects abound in warm, moist areas near hot water and steam pipes, sinks, electric motors behind refrigerators, and inside drains. Cockroaches are diurnal in that they forage and feed at night, hiding during the day. They are omnivorous, that is, they eat both plant and animal matter, and are especially fond of starches, cereals, sweets, book-binding paste, meat products, and can damage papers and fabrics. They feed on garbage as well as human food and thus may transmit human diseases. Cockroaches possess a characteristic unpleasant odor and fecal pellets. Some strains of the German cockroach have developed resistance to diazinon.

Crickets

Like centipedes, these insects nest outdoors and wander into buildings, seeking out moist, warm areas much like the cockroach. They are mostly nocturnal and omnivorous.

Earwigs

These animals usually nest outdoors and crawl into buildings seeking food. They are nocturnal, omnivorous, and actually feed on small insects and other pests.

Houseflies

Nearly all of the flies associated with the food industry are house flies. These insects breed in decaying organic matter such as manure, garbage, and food, and consequently are among the filthiest of pests. Although the adults can be controlled with residual sprays, the treatment of choice is aerosol sprays. The best method of control is elimination of the breeding site or suitable larvicidal treatment of these breeding areas.

Silverfish and Firebrats

Both these insects are nocturnal and feed on a variety of foods containing protein, sugar, and starch. They require a source of warmth and moisture, and will eat cereals, moist wheat flour, any paper on which there is paste or glue, starch in clothing, and rayon fabrics.

Spiders

For the most part, these animals are harmless and beneficial to man. They are predaceous creatures which either hunt or ensnare their prey by means of silk webs. Spiders feed on insects, small animals, and other spiders, and except for their potential to contaminate food products, are generally not regarded as pests. However, the presence of spider webs is offensive to FDA, and may be classified as filth or the existence of unsanitary conditions.

INSECTICIDE RESISTANCE

Resistance to insecticides is the development of an ability in a strain of insects to tolerate doses of toxicants which would prove lethal to the majority of individuals in a normal population of the same species. Resistance is a problem as old as insecticides. Soon after man began using insecticides to control pest insects, he became aware of this phenomenon when he discovered, after continued use of an insecticide, that he could no longer control the target pest with doses which once proved lethal.

The first reported case of insecticide resistance occurred in 1914, with the resistance of a scale insect to lime sulfur. In 1916, resistance of another scale insect to cyanide was reported. Prior to 1945, 13 insects or tick species were reported to be resistant to arsenicals, selenium, rotenone, cyanide, and other compounds. However, it was not until the advent in 1942 of DDT, the first synthetic insecticide, that insecticide resistance came into its own. By 1947, reports came in from all over the world that DDT was becoming less effective for insect control. Since that time, the number of authentic cases of insect pest resistance has grown exponentially, doubling every 6 years; at the present time, the number is over 400. It is estimated that by the turn of the century, nearly all pest insects will exhibit some degree of insecticide resistance. Those insect species occurring in the food processing and handling industry which have developed resistance to pesticides are listed in Table 3.

Insecticide resistance can be classified into two broad categories. Behavioral resistance describes the ability of an insect to avoid a lethal dose of an insecticide; physiological resistance describes the ability to degrade, metabolize, excrete, or otherwise render non-toxic, a lethal dose once it has been taken into the body. The greater majority of cases of insecticide resistance fall into the latter category.

Resistance is preadaptive and follows the principles of Darwinian selection. That is, the insecticide does not induce genetical changes in the insect. Rather, the genes for resistance are already present, although in low frequency. The insecticide merely selects out those few resistant genes and thus perpetuates the phenomenon of resistance.

An important aspect of resistance is the ability of one insecticide to induce resistance in an insect to a second insecticide without the insect ever having been exposed to the second insecticide. This phenomenon is referred to as cross-resistance and often times involves not just a single insecticide but whole classes of insecticides. Generally, OPs induce resistance slowly, much more slowly than do the chlorinated hydrocarbon insecticides, but cross resistance in this group can be extreme. For example, house flies selected 41 generations with parathion, an OP, developed a 7-fold resistance to the OP but developed cross-resistance of 3000-fold to DDT, 70-fold to methoxychlor, and 120-fold to lindane. When selection pressure was stopped, the flies lost their resistance to parathion but retained their resistance to DDT.

Likewise, carbamates may induce resistance to OPs and chlorinated hydrocarbons but the extent of cross-resistance is highly variable. Thus, cross-resistance, coupled with the diversity of insecticides available for use, is likely to result in the ability of virtually all pests to develop resistance to all toxicants.

Many attempts have been made to overcome the development of resistance in insects. Synergists, non-toxic compounds which potentiate the toxicity of certain insecticides, have been used to a limited extent, the formulation of PBO with pyrethrins notwithstanding. The synergists usually work by physiologically blocking the metabolism of the insecticide by the insect. These synergistic formulations are usually cost-prohibitive. However, the most reliable and economical rationale to overcome resistance is to rotate the use of available insecticides or treatments, switching from one to the other in an effort to reduce the duration of selection pressure. Resistance will usually decline when the selective agent is removed; however, subsequent use of the insecticide will normally result in more rapid buildup of resistance than during the initial selection.

The last line of defense against insect control failures, whether due to insecticide resistance or some other reason, is fumigation. Done properly, fumigation provides 100% control of the target pest. In recent years, however, some species of insects have developed resistance to some of the commonly used fumigants (see Table 3). Thus, the commonly held belief that insects cannot develop resistance to a fumigant is no more than just a myth.

Table 3. Insect pests which have been shown to exhibit physiological resistance to certain pesticides. 1/

Insect Pest	Pesticides					
	DDT	Cyclodienes	OP	Carbamate	Fumigant2/	Other
Almond Moth	X	X	X			Pyrethrins
Angoumois grain moth		X	X			
Ant		X				
Cockroach – German	X	X	X	X		Pyrethrins
Oriental	X	X				
Dermestids – Hide beetle		X	X			
Khapra beetle			X			
Flea – dog	X	X				
cat	X	X				
human	X	X				
Flour beetle – confused		X	X		MB, PH$_3$	
red	X	X	X	X	EDB, MB, PH$_3$	Pyrethrins, Bioresmethrin
Grain beetle – Merchant		X	X	X		EDB
Sawtoothed			X			
House flies	X	X	X	X		Pyrethroids
Indianmeal moth	X	X	X	X	PH$_3$	Pyrethrins
Lesser grain borer		X	X		PH$_3$	
Mosquitoes3/	X	X	X	X		Pyrethroids
Weevil – granary	X	X	X	X	EDB, MB, PH$_3$	Pyrethroids
maize	X	X	X	X		Pyrethroids
rice	X	X	X	X	PH$_3$, CPN	

1/ Data from Georghiou and Mellon, 1983 (See Literature Cited).

2/ Abbreviations for fumigants are as follows: EDB = ethylene dibromide, MB = methyl bromide, PH$_3$ = phosphine, CPN = chloropicrin.

3/ Virtually all species of mosquitoes tested are resistant to DDT and the cyclodienes; most are resistant to OPs, and some are resistant to carbamates.

126

APPLICATION METHODS AND EQUIPMENT

The efficiency of an insecticide dispenser lays in its ability to distribute the minimum amount of insecticide to kill the maximum number of pests in the least amount of time. In most cases, the amount of insecticide required to kill all the pests in a population is only a minute fraction of the amount actually applied because of uneven coverage or distribution resulting in areas of non-treatment. Thus, the efficiency of a dispenser depends on its ability to distribute the insecticidal material as evenly as possible. Adequate distribution for virtually all residual crack and crevice treatments can be obtained with any of the three methods of application: spraying, dusting, and brushing.

Spraying

The most popular and versatile sprayer for residual sprays is the compressed air or compression sprayer. In its simplest form, it is a pressure can; the can is charged with the insecticidal formulation and air is applied under pressure to fill the remaining space. The most familiar form of this class of sprayer is the 2-3 gal pneumatic sprayer which may be slung over the shoulder for portability. It may be obtained with a variety of nozzles including a pin-stream spray. This type of sprayer has the advantages of being economical, simple, and easy to use, clean, and store. A disadvantage of the hand sprayer is a lack of good agitation and screening for wettable powders so that the sprayer must be shaken often.

Dusting

All dusters operate on the principal of emitting a blast of air in which the dust particles are airborne. Hand dusters may consist of a squeeze bulb, bellows, tube or shaker, a sliding tube or a fan powered by a hand crank. Capacities of hand dusters range from 1 pint to 2 quarts. Some advantages and disadvantages of dust application over spraying are:

Advantages	Disadvantages
a. no mixing required;	a. takes longer to apply;
b. longer lasting;	b. drifts causing contamination;
c. penetrates well;	c. repells;
d. more stable;	d. ineffective when wet;
e. easily picked up by insects;	e. more expensive;
f. less hazardous to operator;	f. unsightly.
g. doesn't stain.	

Brushing

In some cases, it is advantageous to paint an insecticidal spray formulation on surfaces with a brush where spraying or dusting could contaminate adjacent food or equipment. This application allows very precise placement of the residue with no dust or mist drift. The paint brush may be cleaned and re-used but should be properly stored in a container specifically for that purpose.

The old adage which says that an ounce of prevention is worth a pound of cure is no more applicable than to pest management. The easiest way to avoid insect pests is to prevent their occurrence and spread. This can be done most effectively through sanitation. However, as stated previously, no matter how much effort is aimed at prevention, pest management failures can and will occur. An integral part of insect control with residuals includes the establishment of a treatment schedule or regimen whereby the premises are regularly inspected and/or treated. Inspection and reporting are excellent tools for policing and analyzing any pest control program (Chapter 4). Inspections must be made by qualified personnel who understand the production operation and who have a keen interest in the total field of sanitation. In addition, the specialist must be familiar with the habits of all pests which may invade the plant operation. A regular schedule is a must, and it must be tailored to the type of structure and the type(s) of activities carried out in the structure, however. Therefore, a particular regimen for one building may not be appropriate for another.

Factors Which Affect the Treatment Schedule

Many variables must be considered when establishing a schedule. The variables that affect the toxicity of a chemical include temperature, relative humidity, clean-up operations, chemical characteristics of the insecticide, accessibility of the residue, substrate upon which the residue is applied, and rotation of insecticides.

Temperature
Generally, the integrity of an insecticide is inversely related to temperature, i.e., as the temperature increases, the chemical activity decreases due to breakdown of the chemical. Consequently, the residual activity of insecticides will decrease much quicker during the summer months in warehouse situations which are subject to ambient temperatures. Consequently, more frequent spraying may be necessary in this situation than in an air conditioned building.

Relative humidity
The effects of relative humidity are most evident on dust formulations. Activity of a dust formulation is maintained only if the dust remains dry. Consequently, dusts can be rendered inactive if they are subject to high humidity. Also, exceptionally high relative humidity maintained over a long period of time will reduce the residual activity of certain spray formulations.

Clean-up operations
Various clean-up operations (mopping, stripping and waxing, vacuuming, etc.) dictate different treatment schedules. If a wet-processing facility is required to hose down the premises daily, a long-lasting residual chemical will be useless in the schedule. This situation may require a daily treatment of a short-lived residual like synergized pyrethrins. Vacuuming can remove active dust formulations and mopping can remove residues of spray formulations. A good treatment schedule must go hand in hand with the clean-up schedule of the particular structure of facility.

Chemical characteristics of the insecticide

As indicated previously, each chemical has its own inherent degree of activity and different insecticides differ in this particular trait. For example, under identical conditions of application, synergized pyrethrins may be active only a day or two while malathion may remain active for many months.

Availability of residue

Many times, particularly in warehouse situations or other storage areas, an insecticide residue may become inaccessible to the insect pest because dust, debris, or other materials accumulate and obscure the residual deposit. This situation is particularly evident in the handling of bulk commodities like grains, oilseeds, flour, etc., which produce large quantities of dust. In addition, facilities, equipment, and buildings associated with the handling of these commodities may be subject to infrequent cleaning.

Substrates

The type of surface upon which an insecticidal residue is deposited greatly influences the longevity and thus the activity of that residue. Due to its alkaline content, concrete causes rapid breakdown of the OP and carbamate insecticides so their activities decrease quickly on concrete. Absorbent substrates like wall board and certain types of wood make the residue inaccessible to the pest. However, some residues actually "bleed out" of absorbent substrates at a slow rate so that the longevity of these residues is increased. This concept has been fully exploited in "encapsulation chemistry" where the insecticide is actually encapsulated in slow-release formulations. The slow-release formulations are advantageous in that they reduce the frequency of applications.

Repetition

Continual use of a single insecticide to control a specific pest is not recommended because of the potential for the development of insecticide resistance. Periodic changes in the selection of an insecticide will provide longer lasting control.

SUMMARY

Insect pest management with residual insecticides has been the main tool of the specialist in his efforts at controlling arthropod pests for many years. A residual insecticide is one which is applied to a surface to leave a deposit or residue which becomes available for subsequent pick-up and penetration into an insect pest when that pest walks on, crawls through, or ingests the deposit. Residual treatments in the food processing and handling industry are limited to "crack and crevice" and "spot" treatments. These treatments can be made only in areas where pests live and hide such as behind baseboards, areas where plumbing enters or leaves the room, under sinks and electrical machinery, in flooring cracks, etc., and cannot be made to equipment, food products, or surfaces subject to food handling or processing, or to any other place which could lead to the accumulation of insecticidal residues and thus the subsequent contamination of food products. Treatments are usually made as water- or oil-based sprays or brush-ons, or as dusts, the main objective being to distribute the minimum amount of insecticide to kill the maximum number of insects in the least amount of time.

A number of insecticides and insecticidal formulations are presently registered and labeled for use as residuals. Each is different and proper use requires a thorough knowledge of the insecticide material as well as the habits and behavior of the particular target pest. All insecticides are toxic and potentially hazardous to man and animals, but generally, by definition, they are selectively toxic, being much more toxic to insects than to man and other animals. In order to safely utilize insecticides, one must follow the precautions and directions for use indicated on the insecticide label.

The phenomenon known as insecticide resistance is evident in some insect species and can cause control failures to occur. Possibly more important than resistance is the development of cross-resistance by some pest species whereby these pests become resistant to certain insecticides by virtue of their resistance to other insecticides. In view of the fact that some pests have developed resistance to fumigants, our last defense against insect pests, both of these physiological processes are becoming increasingly important and must be considered in any pest management program.

The easiest way to avoid insect pests is to prevent their occurrence and spread. This goal can be aided considerably through the use of sound sanitation procedures. However, even under the best of conditions, control failures can and will occur. Therefore, an integral part of any insect pest management system must be routine inspection and treatment. This pest management schedule must be tailored to the particular type of food handling and processing facility.

LITERATURE CITED

Anon. 1970. Control of pantry pests. USDA, ARS, MQRD Correspondence Aid, CA-51-3. 3 pp.

Anon. 1976. Controlling household pests. USDA, Home and Garden Bulletin No. 96. 30 pp.

Brown, A. W. A. 1951. Insect control by chemicals. John Wiley and Sons, Inc. New York. 817 pp.

Fonk, W. Don. 1970. Chemical control. In Fundamentals of Applied Entomology. [Ed] R. E. Phadt. The Macmillan Co., New York. 668 pp.

Georghiou, G. P., and R. B. Mellon. 1983. Pesticide resistance in time and space. In Pest resistance to pesticides, [Ed.] Georghiou, G. P., and T. Saito. Plenum Press, New York, 809 pp.

Herms, W. B., and M. T. James. 1961. Cockroaches and beetles. In Medical Entomology. The Macmillan Co., New York. 616 pp.

Laughlin, Paul E. 1954. Sanitation and pest control in the food industry. Modern Sanitation. May. p. 27-29, 61.

Mallis, Arnold. 1964. Handbook of pest control. Mac Nair-Dorland Co., New York. 1148 pp.

Martin, Hubert [Ed]. 1972. Pesticide Manual. British Crop Protection Council. England. 535 pp.

O'Brien, R. D. 1967. Insecticides. Academic Press, New York. 332 pp.

Rowe, John S. 1947. Pest control in food plants. Assoc. Food and Drug Officials, Annual Conference, Carlsbad, NM.

Sanderson, D. W. 1959. Pest control in the food plant. Canadian Food Industries, Dec. pp. 25-27.

Scott, H. G. 1962. How to control insects in stored foods. Pest Control. Dec. pp. 18, 20, 22, 24, 26, 28.

SPACE TREATMENTS IN FOOD PLANTS

J. H. Rutledge

Editor's Comments

This common use of insecticides is a supplementary approach which kills only exposed insects. The frequency of use is about monthly to interrupt the life cycles of the stored-product insects. In addition to consideration of the species of concern the timing can also be delayed as a function of environmental conditions, primarily temperature.

Space treatment does not require as sealed a space as does fumigation.

F. J. Baur

SPACE TREATMENTS IN FOOD PLANTS

J. H. Rutledge

Vice President-Food Protection Services
Lauhoff Grain Company
Danville, Illinois 61832

THE CONCEPT OF SPACE TREATMENT

The term "space treatment" may at times be confused with "space fumigation." It is essential to separate those terms because they refer to dramatically different methods of insect control. Space fumigation is a major project which requires that a building be sealed extensively or covered with a tarpaulin to render it as gas-tight as possible. This is followed by introducing a deadly, penetrating gas into the atmosphere and maintaining a killing concentration for anywhere from one to several days. Fumigants typically used would be methyl bromide or phosphine. Space fumigation is expensive, requires considerable time, and is undoubtedly the most hazardous method of controlling insects. However, successful space fumigation may render a structure virtually free of living insects, including the immature life stages.

Space treatment refers to fogging or misting-type applications. Through the use of specialized application equipment, liquid insecticides are broken down into minute droplets and dispersed into the atmosphere in a structure. Space treatment is generally more rapid and much less expensive than fumigation. Space treatment can usually be accomplished in a matter of hours rather than days. Buildings need not be sealed as extensively, although it is necessary to minimize air loss or air movement during the period of treatment. It must be realized, however, that space treatment basically kills only exposed stages of insects. In food plants, contact insecticides are used for such treatments. These chemicals do not impart a long-lasting residue for extended kill. Also, the insecticide does not penetrate to anywhere near the degree of a fumigant. Therefore, deeply hidden infestations or some immature stages of insects are not likely to be killed. For the most part such treatments will kill exposed adults and possibly larvae. In the Federal Register of August 10, 1973, the Environmental Protection Agency defined space treatment as "---the dispersal of insecticides into the air by foggers, misters, aerosol devices or vapor dispensers for control of flying insects and exposed crawling insects."

VARIOUS APPROACHES TO SPACE TREATMENT

Space treatment has been used extensively for many years in food plants as an insect control technique. The available types of application equipment have shown varying degrees of effectiveness. A "buyer's guide" to such equipment may be found in Pest Control magazine, January, 1984.

One style of equipment is referred to as "thermal fogger" or "smoke generator." This equipment works on a principle whereby liquid insecticide is dripped onto a heating element, thus creating a very dense insecticidal smoke or fog. This type of application is generally effective against flying insects such as adult moths, flies, and mosquitoes. It has been the author's experience that such treatments have not been particularly effective against stored products insect pests such as flour beetles, dermestids, etc. Some references (Brett, 1974) attribute this reduced effectiveness to the size of individual droplets of insecticide claiming that insecticidal smokes contain a high percentage of droplets which are less than one micron in size. This theory postulates that droplets that small will not impinge on insects to the degree necessary to be taken into their bodies in sufficient quantity to effect a kill. Others (Mallis, 1982) have theorized that the heat employed to generate the insecticidal smoke may reduce the effectiveness of contact insecticides such as pyrethrins. Regardless of these reservations, such treatments have been widely used in the past in food operations and are still employed by some today.

Other fogging devices operate by compressed air. The general principle is that of creating pressure inside a container of insecticide which forces the liquid out through restricted nozzle openings. The liquid under pressure passing through the constricted nozzle openings creates a mist or fog. Some units may consist of a single nozzle while others contain multiple nozzles, sometimes arranged in manifold fashion. Since nozzle orifices must be small in order to create a satisfactory fog, clogging of the nozzles frequently occurs. This can be caused by impurities in the compressed air supply or in the insecticide. If the nozzle openings are large enough to avoid clogging, this frequently results in creating a "wet fog" in which individual droplets of insecticide may range upward in size to 300 to 400 microns. Although the fog or mist thus created may be quite dense, a couple of problems may occur. The larger particles of insecticide will rapidly settle out of the atmosphere. This results in "oil slicks" on the floor a short distance from the point of discharge or in an oily film on processing equipment. This is very undesirable in food plants. Another problem caused by the rapid settling is that potential contact of the insecticide with the insect is reduced. The basic concept of space treatment is to deliver maximum dispersion of insecticide droplets which will be carried by normal air movement into inaccessible areas, thus maximizing the potential of contacting exposed insects. Whereas compressed air foggers may have certain drawbacks when particle size is critical, such units may have application when using an insecticide formulation such as 5% Vapona (DDVP, dichlorvos) for which particle size is less significant. This will be discussed later in more detail.

A third method of conducting space treatments is the use of pressurized aerosol containers. These pressurized containers of insecticide are available in sizes ranging from a few ounces to several pounds. A pressurized aerosol container that delivers a high percentage of droplets of insecticide in the 10 to 20 micron particle size can be quite effective for control of stored products insects in food operations. Insecticide cost may be higher for pressurized aerosol containers, but this may be offset by saving the cost of expensive application equipment and the labor required for its operation. Comparative cost analysis should be performed before a technician selects the technique to be used.

A concept of space treatment that has gained a great deal of popularity in food plants during the past ten years is generally referred to as ULV or ULD. Those initials stand for Ultra Low Volume and Ultra Low Dosage, respectively. Such terms may not be entirely appropriate. The key element in space treatments, especially when applying pyrethrins formulations, has to do with particle size of insecticide droplets. For this reason, the author would prefer to use the term "controlled particle size" when referring to this technique.

For example, previously mentioned pyrethrins are commonly used for space treatments in food plants. Invariably they are used in conjunction with piperonyl butoxide as a synergist, with some formulations incorporating a second synergist, N-octyl bicycloheptene dicarboximide (commonly called MGK-264). A very popular formulation is referred to as 3610, which contains 3.0% pyrethrins, 6.0% piperonyl butoxide, and 10.0% MGK-264. Total active ingredients are 19.0%.

Label recommendations for space treatments with undiluted material range from ½ to 1 ounce per 1,000 cubic feet. Another formulation similarly labeled for space treatments in food plants contains 0.5% pyrethrins and 5.0% piperonyl butoxide for a total of 5.5% active ingredients. Label recommendations for that product call for a dosage rate of 1 pint per 6,000 to 10,000 cubic feet. Let us assume the treatment of 500,000 cubic feet of space using the two above mentioned formulations at the lowest recommended dosage. Application of the 3610 at ½ ounce per 1,000 cubic feet would require 250 ounces which would contain 7.5 ounces pyrethrins and 47.5 ounces total active ingredients. Use of the second formulation mentioned at one pint per 10,000 cubic feet would require 800 ounces, but would only result in 4.0 ounces pyrethrins and 44.0 ounces total active ingredients, yet the 3610 product is commonly referred to as being a ULD formulation. However, since ULD has virtually become a generic term for a certain type of space treatment, that term will be used in this discussion.

Manufacturers of ULD equipment state that if the equipment is properly adjusted, the size of insecticide droplets produced will generally be 1 to 30 microns with a high percentage of droplets falling in the 10 to 20 micron range (Micro-Gen Equipment Corp., 1977). It is further stated (Brett, 1974) that this range of droplet size is considered ideal for impinging on the setae of insects, thus agglomerating to be taken into their bodies to effect a kill.

There are a number of manufacturers of ULD application equipment. Units are available in a variety of sizes ranging from single-nozzle, portable units to larger, dual-nozzle machines on wheels to multiple-nozzle units mounted on truck beds for out-of-doors applications. Equipment may be powered by electric motors or gasoline or propane engines. Newer systems incorporate in-place dispensers which are operated from centralized controls. Selection of equipment will depend upon such factors as size of the space to be routinely treated, ability to maneuver equipment, initial investment cost, and potential fire or explosion hazard. In food processing operations technicians will usually select electrically powered equipment in order to avoid carbon monoxide or possible spark from a gasoline engine. This is particularly significant in potentially dusty operations such as cereal mills.

The basic elements of operation are fairly simple. A typical unit may consist of an insecticide reservoir, a drive motor, a high-speed blower, and the nozzle(s) which are baffled. The drive motor powers the high-speed blower which is connected to the nozzle arrangement. The liquid insecticide is syphoned up to and through the nozzle. As the blower operates, air is forced past the baffles of the nozzle(s), thus creating a tremendous amount of air turbulence in the nozzle chamber. That air turbulence physically breaks the liquid into tiny droplets and blows it into the atmosphere. Theoretically, proper adjustment of the insecticide flow will result in droplets within the 1 to 30 micron size range. In addition to their ability to impinge on the insects, droplets of that size will tend to stay suspended in the air and be carried about by normal air currents. Even though a structure is closed up sufficiently to eliminate major air movement, there will still be some degree of natural air currents inside. In essence, normal air currents are used as the carrier to distribute the micron-size particles of insecticide into inaccessible places, thus enhancing the likelihood of contacting exposed insects and effecting a kill.

CHEMICALS USED FOR SPACE TREATMENTS IN FOOD OPERATIONS

Space treatments may only be conducted when the plant is not in operation and all people are out of the buildings. Quite often, this is other than daytime when insect activity is greater. Insecticides registered for space treatments in food plants are pyrethrins, Vapona (2-2-dichlorovinyl dimethyl phosphate), and a few of the synthetic pyrethroids. It is important to closely study insecticide labels to determine if they may be used for that purpose in your type operation.

As previously mentioned, a popular pyrethrin formulation for space treatments is the 3610 material applied with ULD equipment. With pyrethrins size of the insecticide droplets is critical to accomplishing maximum kill of stored products insects. This necessitates proper adjustment and calibration of application equipment. Equipment operating manuals and manufacturers technical personnel can properly advise in this regard. Many companies choose to use only pyrethrins because of allowable food residue tolerances and relative safety to personnel. Pyrethrins are considered to be rather weak insecticides, thus the objective is to make them more effective through improved applications. A microencapsulated pyrethrin product is also registered for space treatments in food plants. Microencapsulation generates a gradual release of pyrethrins, thus creating a residual effect from what would otherwise be considered a non-residual insecticide.

Another commonly used insecticide for space treatment in food plants is Vapona. Where its use is permitted, it has proven to be an excellent tool for the control of stored products insects. Vapona is somewhat unique as an insecticide in that it is quite volatile. Not only does it kill by contacting insects, but it is also effective as a vapor and in that regard acts similar to a fumigant. For space treatments in food plants, Vapona is formulated at various levels in carriers such as 1,1,1-trichloroethane or mineral oil. These are usually ready-to-use formulations.

A popular product for ULD-type treatments is 5% Vapona with 1,1,1-trichloroethane. Because of the volatility of Vapona and the carrier, particle size of this formulation is not nearly so critical. It will evaporate almost immediately upon release from application equipment and kill insects by vapor action. In fact, many experienced technicians prefer to dispense this formulation in a manner that allows the total dosage for an area to be released as rapidly as possible, claiming that this enhances its effectiveness. The author shares this feeling. Because it does convert to a vapor so rapidly, it is advisable to take greater care in sealing structures for Vapona treatment in order to minimize leakage. In general, the tighter the structure, the more effective the treatment. It is important to remember that Vapona is a contact insecticide, thus there is essentially no residual killing effect following space treatment.

Since particle size is not critical, technicians will often apply the above mentioned 5% Vapona formulation using compressed air fogging devices rather than ULD equipment. One popular device is the Insect-O-Jet four-nozzle sprayer (Spraying Systems Co.) which is available in ½ gallon or 1 gallon models. Use of this equipment enables one to arrange a system of treatment whereby exposure of application personnel is minimized. For example, one approach is to use a whole series of pre-filled units which in combination contain the required amount of insecticide to deliver the proper dosage for the space being treated. The units can then be activated by turning on the central compressed air system for a predetermined length of time to insure complete dispensing of the Vapona. This approach is especially suited for treating multiple-story buildings. Such a technique can even be tied into a computer so that the period of treatment is totally automated. In this instance the only personnel exposure would be that of filling the sprayer units. In addition, there are sophisticated, totally automated application systems that completely eliminate personnel exposure, and newer systems are continually being developed with that same objective. Use of such techniques is highly advisable because Vapona is considerably more toxic than pyrethrins, and as such is more hazardous to personnel making applications. It is extremely important to protect people against exposure to Vapona, and the best way to do so is by avoiding contact. With adequate planning and financial investment, this can be accomplished.

SAFETY CONSIDERATIONS

Product

Insecticides registered for space treatments in food operations will have statements on their labels indicating that food should be removed or covered during treatment and that all food processing surfaces should be covered during treatment or thoroughly cleaned before use. Obviously, the purpose of such precautions is to prevent the contamination of foods with illegal residues. In view of the fact that space treatment involves filling a structure with airborne particles of insecticide, there may be a possibility of some minute amount of contamination even when all label instructions are followed. This possibility would seem very remote because the insecticides used are short-lived. However, because that remote possibility exists, many technicians feel it is prudent to select an insecticide that has some form of residue tolerance in food. For example, if using pyrethrin formulations around certain processed foods, there are residue tolerances of 1.0 ppm for pyrethrins, 10 ppm for piperonyl butoxide, and 10 ppm for MGK-264 (21 Code of Federal Regulations 193.320, 193.360, 193.390). For Vapona, there is a maximum allowable residue of 0.5 ppm as the result of application on packaged or bagged nonperishable food (21 CFR 193.140). Also in the Federal Register, Vol. 48, No. 155, Wednesday, August 10, 1983, the Environmental Protection Agency granted a food additive tolerance of 3.0 ppm for the synthetic pyrethroid resmethrin, also known as SBP-1382 (21 CFR 193.464). It must be emphasized that the existence of residue tolerances in food certainly does not allow for careless applications which violate insecticide label instructions.

Personnel

As with any pesticide application, safety of personnel during space treatments is of paramount importance. As previously mentioned, such treatments may only be conducted when food plants are not in operation and all workers are out of the buildings.

Pyrethrin formulations are considered to be very safe insecticides which are only slightly toxic to humans. They may, however, be very irritating to the eyes, and may cause a burning sensation in the nose and throat if breathed as a mist. Some people may also demonstrate allergic reactions to pyrethrins such as skin rashes. If automated systems are not being used and it is necessary for persons to operate fogging or ULD equipment during space treatments, they must use personal protective equipment. This includes wearing protective clothing such as a coverall garment. It is also necessary to wear a proper respirator to protect the nose and throat and to wear eye protection. As a minimum, persons should use a cartridge type respirator with built-in eye protection or separate goggles. A full-face mask with appropriate pesticide canister will give more complete eye, nose, and throat protection.

Since Vapona is considerably more toxic, protection of personnel is even more critical. Again, automated or at least semi-automated systems are preferred. However, if circumstances are such that people must operate dispensing equipment during Vapona space treatments, personal protection should be as complete as possible. A water-proof coverall garment should be worn, preferably with an attached hood for head protection. There are light weight, relatively inexpensive, spunbonded olefin garments that are suitable for this purpose. The Vapona label states that rubber gloves must be worn while filling, running, or emptying application equipment. Gloves should be unlined rubber or neoprene and should be worn inside the shirt sleeves. Rubber boots are also advisable. For respiratory protection, the author feels that persons exposed to Vapona should wear a full-face mask with the appropriate canister. It is also advisable for operators to work in pairs so that help is readily available in case of emergency.

Necessary steps must be taken to prevent unprotected workers from entering an area under treatment. For Vapona, it is required that all entrances to the area under treatment be locked, and warning signs must be placed on such entrances. The Vapona label will also specify the required treatment time and give aeration instructions.

SPACE TREATMENT RESEARCH PROJECTS

In August and September, 1972, twelve tests were conducted at the Lauhoff Grain Company pilot corn mill using ULD application equipment (Micro-Gen SIW-5, electric) to perform space treatments. The objective of the tests was to determine the effect of ULD space treatments using various insecticides on the types of insect pests that may infest cereal mills or cereal products. The pilot corn mill is a three-level building (second floor, first floor, and basement) containing 42,894 cubic feet of space. It is an old building of brick construction with no windows. One large air duct through the roof was covered with polyethylene sheeting. No additional sealing was done other than closing doors and turning off air conditioning equipment during each test.

Test insects were exposed in ten different locations during each test. Five of these locations were inaccessible to varying degrees and five were openly exposed at different levels above the floor. Test insects were supplied and placed by personnel of the U.S.D.A. Stored Products Insects Research Laboratory, Manhattan, Kansas. They also determined insect mortalities. Insects used were red flour beetle adults (*Tribolium castaneum*) (25 per location/250 per test) and dermestid larvae (*Trogoderma inclusum*) (15 per location/150 per test). Test insects were placed in ½ pint, open-mouthed jars.

The procedure was essentially the same for each test. After placement of the test insects, the air conditioning was turned off and doors to the outside were closed. Inside doors between the different levels were left open. Each insecticide application began on the second floor and finished in the basement with approximately one-third of the total dosage dispensed at each level. Following the exposure period, air conditioning was turned on, outside doors were opened, and fans were used to aid in aeration. Test insects were then recovered for later mortality determination by U.S.D.A. personnel.

Since the pilot corn mill contains 42,894 cubic feet of space, the structure is small enough to conveniently do this type of testing, yet it is large enough to reasonably extrapolate the results to what one might expect in a larger commercial food plant. Basic information about each test and the overall test insect mortalities are given in Table 1.

For test #12, certain explanations are in order. Treatment was begun on the second floor with 20 ounces of 5% Vapona in the machine. Because of the volatility of the mixture, dispensing was much more rapid than anticipated. Shortly after moving to the first floor, the machine ran empty. Calculations based on dispensing time showed that approximately 17 ounces of 5% Vapona were applied on the second floor and 3 ounces on the first floor. The decision was made to extend the exposure period to two hours. The 100% mortality of test insects included two complete sets located in the basement where no insecticide was applied. The only opening between the first floor and the basement was a stairway with a standard passenger door left open. It is felt that this demonstrated the movement of Vapona vapors and a tendency to equalize in a total space.

It must be noted that the work reported herein was strictly on an experimental basis and does not imply that all the insecticide formulations used are permitted for space treatments in food plants. Individual insecticide product labels must be examined closely to determine if the treatment you wish to perform is permitted.

138

Table 1

Test	Insecticide Used	Amount	Dosage/ 1000 cu.ft.	Exposure Time	Insect Mortality Red flour beetle Adults	Dermestid Larvae
#1	2.5% pyrethrins 20% piperonyl butoxide in Klearol	18 oz.	0.42 oz.	1 hr.	36%	70%
#2	SBP 1382 3% in Klearol	18 oz.	0.42 oz.	1 hr.	68%	83%
#3	Malathion 15% in vegetable oil	36 oz.	0.84 oz.	1½ hr.	100%	91%
#4	SBP 1382 2% Bioallethrin 1% in deodorized kerosene	22 oz.	0.51 oz.	1 hr.	19%	31%
#5	SBP 1382 5% in deodorized kerosene	18 oz.	0.42 oz.	1 hr.	94%	89%
#6	SBP 1382 3.3% Bioallethrin 1.6% in deodorized kerosene	18 oz.	0.42 oz.	1 hr.	63%	89%
#7	SBP 1382 3% in Klearol	18 oz.	0.42 oz.	1 hr.	74%	81%
#8	SBP 1382 2% Bioallethrin 1% in Klearol	18 oz.	0.42 oz.	1 hr.	62%	91%
#9	2.5% pyrethrins 20% piperonyl butoxide in deodorized kerosene	26 oz.	0.60 oz.	Approx. 14 hr.	69%	72%
#10	2.5% pyrethrins 20% piperonyl butoxide in deodorized kerosene	18 oz.	0.42 oz.	1 hr.	57%	65%
#11	7.5% pyrethrins 75% piperonyl butoxide in deodorized kerosene	6 oz.	0.14 oz.	1 hr.	58%	67%
#12	Vapona 5% in Chlorothene NU	20 oz.	0.47 oz.	2 hr.	100%	100%

In February, 1975, six additional tests were conducted at Lauhoff Grain Company. Test insects were supplied and placed by and mortalities determined by personnel at the U.S.D.A. Stored Products Insects Research Laboratory, Savannah, Georgia. Two of the tests were performed in the pilot corn mill using the same general protocol as the previously described tests with some modifications.

Table 2 gives basic information about the two pilot plant tests followed by clarifications.

Table 2

	1st Test	2nd Test
Location	Pilot Plant	Pilot Plant
Formulation	PV-5244[1]	PV-5244
Dosage	1 oz./1000 cu.ft.[2]	0.5 oz./1000 cu.ft.
Equipment	RS-1 w/remote nozzle[3]	RS-1
Exposure Time	1 hour	1 hour

Test Insect Mortalities

		1st Test	2nd Test
Confused flour beetle adults	dish	27%[4]	13%
	cage	10%	6%
Black carpet beetle larvae	dish	3%	15%
	cage	2%	7%

[1] The insecticide PV-5244 contains as active ingredients 0.5% Vapona, 2.0% pyrethrins, 4.0% piperonyl butoxide, and 4.0% MGK-264.

[2] One difference between the two tests was the dosage rate.

[3] A second difference was the application equipment. The RS-1 with remote nozzle allowed for directing the mist into inaccessible areas whereas the unit without the remote nozzle could merely be wheeled through the area being treated.

[4] Test insects were placed in both open petri dishes and small wire cages. It was felt that perhaps the insecticide would impinge on the fine mesh wire to some degree and not reach the insects within. Mortality was less in the wire cages. Overall insect mortalities were generally poor.

Two additional tests were conducted in a maintenance warehouse containing 226,000 cubic feet. The intent was to simulate food warehousing conditions. Table 3 shows basic information about those tests followed by clarifications.

Table 3

	1st Test	2nd Test
Location	Maintenance Warehouse	Maintenance Warehouse
Formulation	V-500[1]	5% Vapona in methyl chloroform
Dosage	0.5 oz./1000 cu.ft.	0.5 oz./1000 cu.ft.
Equipment	SPACE I[2]	MS-2
Exposure Time	2 hr. 40 min.[3]	6 hr.

Test Insect Mortalities

		1st Test	2nd Test
Confused flour beetle adults	dish	+99%	+99%
	cage	94%	+99%
Black carpet beetle larvae	dish	15%	63%
	cage	13%	60%

[1] The insecticide in both tests was 5% Vapona. Insecticide used in the first test appeared to be off-quality. It had been premixed and shipped to the test site in tin cans. The chemical had changed color indicating possible reaction with the metal cans. In the second test, Vapona concentrate was freshly diluted to 5% with methyl chloroform. This did show improved efficacy, especially against black carpet beetle larvae.

[2] The SPACE I unit used in the first test was equipment designed to dispense the insecticide from one location with the use of a built-in fan. The MS-2 unit used in the second test was a dual-nozzle unit that had to be wheeled throughout the area being treated.

[3] Note that there was a difference in test insect exposure time.

The final two tests were conducted in a commercial warehouse facility containing 800,000 cubic feet. Table 4 lists basic information about those tests followed by clarifications.

Table 4

	1st Test	2nd Test
Location	Commercial Warehouse	Commercial Warehouse
Formulation	V-500[1]	5% Vapona in methyl chloroform
Dosage	0.5 oz./1000 cu.ft.	0.5 oz./1000 cu.ft.
Equipment	MS-2[2]	SPACE I
Exposure Time	6 hrs.[3]	6 hrs. 15 min.

Test Insect Mortalities

Confused flour				
beetle adults	dish	90%		97%
	cage	78%		87%
Black carpet				
beetle larvae	dish	12%		63%
	cage	12%		54%

[1]The same problem existed with the insecticide in these two tests as described in clarifications for Table 3.

[2]Note the difference in style of application equipment as described in the clarifications for Table 3.

[3]Note there is only a slight difference in exposure time.

CONCLUDING REMARKS

Space treatment in food plants is a method that has been utilized for many years to aid in the control of insect pests. Improved application equipment and insecticides have increased the effectiveness of space treatments, thus enhancing the popularity of this approach among technicians.

It must be emphasized that space treatment is only one tool. It is not a panacea which will effectively eliminate all insect problems. Certainly space treatment does not replace the many other efforts that are part of pest management. By far the most effective means of controlling insect pests is good cleaning, sanitation, and housekeeping. The use of insecticides must be viewed as strictly an aid in control of insects that cannot be effectively dealt with through other means. Insecticides must never be used in place of good sanitation.

Unfortunately, the use of some insecticides is usually required to maintain adequate control of insects in most food plants. The mere nature of food processing is highly attractive to a variety of insects. Many food plants contain structural or equipment defects that offer excellent harborage for these pests. Inbound ingredients or other supplies may also serve as a source of insects which may go undetected.

Routine use of space treatments in conjunction with other good pest management practices can be a valuable aid to technicians charged with the responsibility of insect control. Scheduling such treatments at approximately monthly intervals will help disrupt the life-cycles of many of the insect pests that infest food plants. During warm weather periods when insect pressure is generally greater, increasing the frequency to every two or three weeks may be necessary. No schedule can be recommended that will be applicable to all operations at all times. However, thorough inspection of food plant facilities by trained personnel can readily determine if the existing program is adequate.

SELECTED REFERENCES

Brett, R. L., "Micron Generation: Droplet Sizes, Their Measurement and Effectiveness", Pyrethrum Post, April, 1974.

Mallis, A., Handbook of Pest Control, 1982; Katz, H. L., "Equipment".

Micro-Gen Equipment Corp., "Micron Generation: Optimum Pest Control Through Optimum Sized Insecticide Droplets", 1977.

Walter, Vernon E., The Industrial Fumigant Company, personal communication.

Wilbur, Donald A. Jr., The Industrial Fumigant Company, personal communication.

FUMIGATION OF RAW AND PROCESSED COMMODITIES

E. J. Bond

Editor's Comments

Every food plant and warehouse will need to consider on varying occasions having a fumigation performed of a commodity, a building, the insides of equipment, or a conveyance. This chapter and the three following speak to these four possible needs.

Fumigation can affect a complete kill (not easily but it can be done) of all stages of the insects because of the penetrating capabilities of the fumigants. To be sure of a complete kill, in addition to the proper level and time, it would be well to place samples of the insect, including eggs, appropriately in the space to be fumigated (see the previous chapter for examples of the approach).

For commodities, be aware of the need to avoid possible illegal residues and off qualities depending on what is used on what (methyl bromide is the main concern).

F. J. Baur

FUMIGATION OF RAW AND PROCESSED COMMODITIES

E. J. Bond

Research Centre, Agriculture Canada
University of Western Ontario
London, Ontario, Canada   N6A 5B7

INTRODUCTION

Fumigation is a widely used procedure that has served an important role in pest control programs for many years. Fumigants are particularly useful for the control of stored product insects because they diffuse and penetrate into places where other forms of control are impractical or impossible. By definition a fumigant is a chemical which, at a required temperature and pressure, can exist in the gaseous state in sufficient concentration to be lethal to a given pest in a prescribed period of time, and fumigation is the process of applying the gas under appropriate conditions to control the target organisms.

The number of materials that has suitable characteristics for fumigation is limited to a very few chemicals, and these vary widely in chemical composition and properties. Consequently, the types of formulations, methods of handling, methods of analysis, and the purposes for which they are used vary considerably. The choice of fumigant depends largely on its ability to give effective economical control of insects without adversely affecting the commodity.

All of the fumigants used to control pest organisms are toxic to human beings, and they also may have other adverse properties - they may be highly flammable or corrosive; they may produce offensive odours; they may be phytotoxic; or they may leave harmful residues in food materials. Usually adverse effects can be obviated by choosing the most suitable fumigant for the particular treatment in question and by applying proper methods of handling and use. A few of the fumigants are known to have or are suspected of having the potential for producing long term chronic effects on human health and some are listed as carcinogens. Appropriate precautions should be taken to avoid exposure to all fumigants and additional precautions should be taken to prevent any contact with compounds that are carcinogenic.

In applying a fumigant to a commodity it is particularly important to carry out the operation in such a way that the insects are controlled without damaging the commodity and without creating any hazard for personnel in the area near the treatment. The use of effective methods of detection and analysis of the fumigant is especially important. Where any possibility of exposure of personnel exists atmospheres should be monitored with appropriate equipment and threshold limit values (TLV), such as those recommended by the American Conference of Governmental Industrial Hygienists (1983-84), should be strictly adhered to. Misuse or abuse of fumigants can lead to accidents that endanger human life or property, and consequently may give adverse publicity to the practice of applying chemicals to food commodities for insect control.

FUMIGANTS FOR COMMODITY TREATMENTS

Several of the fumigants that have been used for the control of insects on food commodities are listed in Table 1. Some of these materials, especially methyl bromide and phosphine, are widely used for many different kinds of treatments, while others have more limited application. The choice of fumigant is

Table 1. Principal fumigants used on stored products.

Name	Molecular weight	Boiling Point (C)	Flammability by vol in air (%)	Remarks
Methyl bromide	94.9	3.6	Nonflammable	General fumigant
Phosphine	34.0	-87.4	1.79	General fumigant; good penetrating ability
Ethylene oxide	44.1	10.7	3-80	Controls microorganisms as well as insects
Hydrogen cyanide	27.0	26.0	6-41	General fumigant; poor penetrating ability
Ethylene dibromide	187.9	131.0	Nonflammable	Mainly spot fumigant admixed with other fumigants
Carbon tetrachloride	153.8	77.0	Nonflammable	Used chiefly in mixtures with flammable compounds to reduce fire hazard and aid distribution
Carbon disulphide	76.1	46.3	1.25-44	Grain fumigant; Usually mixed with nonflammable compounds
Chloropicrin	164.4	112.0	Nonflammable	Powerful tear gas; has been used as grain and spot fumigant
Ethylene dichloride	98.9	83.9	6-16	Usually mixed with carbon tetrachloride
Dichlorvos	221.0	120 at 0.14 mm	Nonflammable	Good only in free spaces; very toxic to most stored product insects.

Source: Adapted from Monro (1969).

determined largely by the type of problem encountered and the properties of the material concerned. Low boiling point compounds, e.g. phosphine and methyl bromide, are used where penetration into the commodity is important. The higher boiling point compounds such as the liquid fumigants (e.g. ethylene dibromide, ethylene dichloride, carbon tetrachloride, and carbon disulphide), are used for spot treatments or in other situations where gas-tight conditions cannot be achieved or maintained. Usually, particular formulations are designed by manufacturers for specific types of treatments. Some of the features of a fumigant that may determine its suitability for a specific treatment are given in Table 2. In applying fumigants to food materials, the dosages and other recommendations given on the label should be closely followed.

Methyl Bromide

This fumigant is very effective against all stages of stored product pests, although the pupal stages of insects and the egg and hypopal stages of mites are usually hardest to kill. To control mites a dose 50% higher than that used for insects is required.

Methyl bromide penetrates relatively quickly into commodities and the vapours desorb and dissipate rapidly during the aeration period at the end of the treatment. For bulk commodities, such as grain, where distribution throughout the commodity may be a problem, methods have been developed using forced air circulation systems to aid in penetration of this fumigant. Information on such procedures may be obtained from the manufacturer or from publications such as Monro (1969), Storey (1967, 1971a, 1971b). Procedures employing carbon dioxide as a carrier in promoting distribution have also been used (Calderon and Carmi, 1973).

Methyl bromide is effective against insects over a wide range of temperatures and has been used successfully in cold climates at -5 to -10°C (23-14°F). Treatments for a wide variety of commodities are recommended for temperatures ranging down to the boiling point (3.6°C or 38.5°F) of the fumigant (see Monro, 1969; USDA, 1976) and in some instances successful results may be obtained below this temperature. However the dosages required to kill pest organisms are greatly increased at low temperatures (Bond, 1975; Bond and Buckland, 1976).

Methyl bromide is supplied by manufacturers as a liquid in steel cylinders up to 816 kg (1800 lb) capacity or in 454 g (1 lb) and 680 g (1.5 lb) cans. When applying this fumigant, the liquid should not come in direct contact with food materials, but should be vapourized so that the gas will diffuse through the spaces and into the commodity. Heat is often required to vapourize the liquid, and several devices have been developed for this purpose. Usually a coil of copper tubing immersed in heated water is used and, as the liquid is passed through this coil, sufficient heat is transferred to produce the vapour at the orifice. For amounts of methyl bromide greater than 2.3 Kg (5 lb), copper tubing 15.3 m (50 ft) in length and 1.3 cm ID (1/2 in), coiled in a large container of water heated to 65°C or above, should be used. For smaller amounts of methyl bromide a 7.6 m (25 ft) coil of 1 cm ID (3/8 in) copper tubing should be ample. For some treatments, the liquid is dispersed through a spray nozzle and the tiny droplets are vapourized by absorbing heat from the atmosphere. Tubing or fittings of aluminium should never be used where they may come in contact with liquid methyl bromide. A comprehensive review of the properties and uses of methyl bromide as a fumigant is given by Thompson (1966).

Methyl bromide can be applied economically and effectively, under a wide range of conditions, without producing excessive residues in food material. However, repeated treatments on the same material may leave appreciable residue of inorganic bromide. Some precautions should be taken with this fumigant to avoid the production of undesirable odours on certain foodstuffs and other materials. Odours sometimes develop in bread made from wheat flour that has been treated with methyl bromide, and materials containing sulphur are particularly prone to odour

Table 2.   Some important characteristics of fumigants used for

	Methyl bromide	Phosphine[2] (hydrogen phosphide)	Ethylene dibromide	Ethylene dichloride
Commercial formulation	liquid in cans & cyl- inders	pellets tablets sachets	liquid	liquid
Solubility in water (g/100 ml)	1.3 at 25°C	slight	0.43 at 30°C	0.87 at 20°C
Penetra- tion	very good	excell- ent	poor	poor
Sorption	some	negligible	high	high
Desorption & aeration	fairly rapid	rapid	slow	slow
Toxicity to insects	high	high	high	low
Rate of kill	rapid	slow	slow	slow
Effect on germina- tion	variable	almost negligible	variable	little
Adverse reactions	combines with sulphur to create odours	corrodes copper		

1 More detailed information may be obtained from Monro (1969).
2 Phosphine is generated from aluminium or magnesium phosphide.

problems. A list of some of the materials that are adversely affected by methyl bromide is given in Table 3.

Good methods of detection and analysis have been developed for methyl bromide and these will permit both effective and safe use. Thermal conductivity analysers are particularly useful for monitoring fumigant concentrations under practical operating conditions. Interference refractometers, which measure the refractive index of fumigants, are also available. Detector tubes, containing chemicals which react with methyl bromide, are supplied by a number of manufactur- ers. They have a range of measurement from 2-200 ppm and are useful for health safety purposes. The halide leak detector is very useful for detecting leaks of the fumigant from treated areas, however, it should not be used for health safety purposes as it is not sensitive at the presently accepted TLV level of 5 ppm.

commodity fumigation.[1]

Carbon tetrachloride	Carbon disulphide	Ethylene Oxide	Hydrogen cyanide	Dichlorvos
liquid	liquid	liquid in cylinders with CO_2 or a hydrocarbon	liquid in cylinders, calcium cyanide granules	liquid, plastic strips
0.8 at 20°C	0.22 at 22°C	very sol.	very sol.	slight
poor	poor	fair-poor	fair-poor	negligible
high	high	some	some	
slow	slow	intermediate	intermediate	high
low	low	intermediate	intermediate	high
slow	slow	slow	rapid	slow
little	little	severe	little	
		may form ethylene chlorohydrin		

Phosphine

This fumigant diffuses rapidly and penetrates deeply into large bulks of grain, stacks of bagged materials, and other tightly packed commodities to give effective and economical control of insects. Exposure periods of 3-5 days or longer at temperatures above 5°C (41°F) are recommended. Phosphine has no appreciable adverse effect on food materials and it leaves no significant residue. It does react with some metals, particularly copper and its alloys, under certain conditions, as high humidity, and may damage electrical or other equipment with copper components (Bond et al., 1984).

In a fumigation treatment, phosphine gas is produced on site from tablets, pellets, or sachets of aluminium phosphide (magnesium phosphide formulations are available in some countries). On exposure to atmospheric moisture, the formulation releases the phosphine at a controlled rate and the gas diffuses throughout the space to kill the insects. In warehouses, the formulations are spread out on sheets of paper, and in transportation facilities, such as railway

Table 3. Materials which should not be exposed to methyl bromide.[1]

Foodstuffs
- iodized salt stabilized with sodium hyposulfite
- full fat soya flour
- certain baking sodas, cattle licks (i.e. salt blocks), or other foodstuffs containing reactive sulfur compounds
- fresh fruits and vegetables[2]

Seeds, bulbs, and plants
- seeds and bulbs to be used for planting[2]
- nursery stock and other living plants[2]

Pets
- all pets, including fish and birds

Rubber goods
- sponge rubber
- foam rubber, as in rug padding, pillows, cushions, mattresses, and some car seats
- rubber stamps and other similar forms of re- claimed rubber

Furs, woolens, horsehair, feathers
- especially in feather pillows

Leather goods
- particularly white kid or other leather goods tanned with sulfur processes

Viscose rayon
- those rayons processed or manufactured by a pro- cess in which carbon disulfide is used

Vinyl, Paper
- silver polishing papers
- certain writing and other papers cured by sulfide processes
- photographic prints and blueprints stored in quantity
- "carbonless" carbon paper

Cellophane

Photographic chemicals
- "darkroom" chemicals, but not cameras or film

Rug padding
- foam rubber, felt, etc.

Cinder blocks, mixed con- crete, mortar
- occasionally pick up odours

Charcoal

1 If uncertain about adverse reactions, conduct a trial fumigation on a small quantity of the material.
2 For specific information contact Dow Chemical, U.S.A.

boxcars or freight containers, they may be placed as a "prepac" of pellets or in envelopes on the inside of the door. For treatment of bulk grain, tablets, pellets, or sachets of the formulation may be inserted in the grain stream as it flows into a storage, or they may be inserted into the grain mass with probes. Methods of applying and dispersing phosphine through low flow rate recirculation systems in grain are being tested (Cook, 1980). On grain-carrying ships, an "in transit" procedure for treating grain in the holds is being developed (Leesch et al., 1978; Redlinger et al., 1979; Zettler et al., 1982). Rigorous safety procedures should be followed if this treatment is to be employed (IMCO, 1981).

Analysis of concentrations of phosphine can be made most conveniently with glass detector tubes. These are supplied for concentrations ranging from 0.1 to 1000 ppm. Infrared gas analysers and portable gas chromatographs are available for more precise analysis.

Liquid Fumigants

The principal liquid fumigants used for commodity fumigation include ethylene dichloride, ethylene dibromide, carbon tetrachloride, carbon disulphide, and occasionally chloropicrin. Some of these may be used individually but usually they are applied in mixtures of various proportions. Mixtures are formulated to combine or complement the effective properties of two or more compounds; for example, carbon tetrachloride is usually admixed with carbon disulphide to reduce the fire hazard of the inflammable vapour of carbon disulphide.

The main application of the liquid fumigants is for treatment of grain in bulk. Distribution of the fumigant and control of the insects are often obtained by adding the liquid to the grain stream as it enters the storage, or in flat storages, liquid fumigants can be sprayed over the surface of the grain to diffuse and penetrate by gravity. Forced air recirculation systems have been used to prevent stratification of components of mixtures and to improve distribution.

Liquid fumigants may remain as residues of unchanged compounds in food materials for many days or weeks after application. Residues of carbon tetrachloride have been found in fumigated wheat and maize 3 months after treatment and a substantial proportion of this accumulates in milled fractions, especially the bran (Jagielski et al., 1978). Ethylene dibromide may remain for similar periods of time and has been found in baked goods made from treated wheat (Rains and Holder, 1981). For analysis of the liquid fumigants in air, glass detector tubes may be used.

In general, it may be said that the liquid fumigants have been used successfully for many treatments of grain and grain products, but they are being replaced by fumigants, such as methyl bromide and phosphine, which are more effective and less hazardous to use.

Other Fumigants

Dichlorvos (DDVP, Vapona) is an excellent insecticide for rapid disinfestation of empty storage spaces. It is effective in the vapour phase against a wide variety of stored product insects but it does not penetrate into materials. Various formulations and methods of application have been used for a number of situations. Dichlorvos-impregnated polyvinyl chloride strips, which slowly emit the vapour over a period of time, have been found to give good control of moths in warehouses containing bagged goods. Aerosols from pressure cylinders and oil mist sprays have also been used effectively. For some pest control programs, dichlorvos is used with considerable success in conjunction with periodic general fumigation treatments.

Ethylene oxide is used extensively in certain treatments of foodstuffs, particularly spices and some processed foods to prevent microbial spoilage. It is normally mixed with a non-flammable carrier as carbon dioxide or a freon-type hydrocarbon to reduce flammability.

Hydrogen cyanide has been widely used for fumigating many foodstuffs, grain, flour mills, and warehouses. Although it is very toxic to insects and leaves little or no permanent residue, it is strongly sorbed and does not penetrate well into many materials. It has been largely replaced by other fumigants with more favourable properties.

FACILITIES AND PROCEDURES USED IN FUMIGATION

Fumigants are regularly applied in many different kinds of enclosures, from fumigation chambers to warehouses, other storage facilities, transportation facilities, and under gas-proof sheets. Generally, any space that can be enclosed and made gas-tight is suitable for fumigation. The dosages, recommendations for application and, regulatory requirements for the treatment of food commodities are outlined on the labels accompanying containers of fumigant. Recommendations should be carefully followed to ensure safety in use and to comply with legislative requirements.

Fumigation Chambers

(a) Atmospheric Chambers. For many types of bagged and packaged food materials, fumigation is most conveniently and efficiently done in specially designed fumigation chambers. Chambers designed for use at atmospheric pressure are adequate for most treatments, however, vacuum fumigation is recommended for certain densely packed or absorbent materials. Fumigation chambers are usually located in or nearby storage or processing facilities where goods can be treated and easily moved to uninfested areas. Possible locations for fumigation chambers in relation to storage facilities, as suggested by the UK Ministry of Agr. (1973), are shown in Fig. 1.

In choosing the site for a fumigation chamber, due allowance should be made for adequate ventilation so that personnel are not exposed to fumigant concentrations above the TLV levels. Also, properly ventilated space for treated commodity should be provided as fumigant will continue to desorb and dissipate for many hours after the end of a treatment. An effective fumigation chamber must be:

1. soundly constructed so it is gas-tight

2. provided with an efficient system for applying and distributing fumigant

3. provided with an efficient system for removing fumigant at the end of treatment

4. located so that infested goods can be handled conveniently

5. located and operated to present no hazard to personnel working near the chamber

The maximum economic size for a fumigation chamber is probably of the order of 100 m^3 (3500 ft^3). The dimensions are best decided by the responsible manager, but as a general guide for efficient and even distribution of fumigant, a chamber should be approximately twice as long as it is wide with a height of 2-3 m (7-10 ft).

Fig. 1 Possible positions for fumigation chamber in relation to storage
facilities (U.K. Ministry of Agr. 1973).

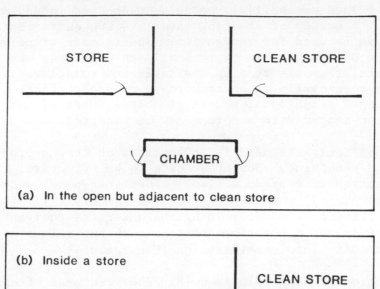

(a) In the open but adjacent to clean store

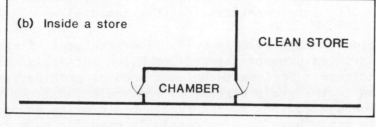

(b) Inside a store

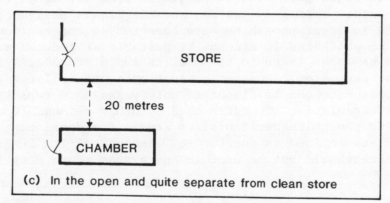

(c) In the open and quite separate from clean store

The most satisfactory type of chamber and the one which is likely to give
the minimum of trouble from leakage is one with a floor of concrete, walls of
brick, poured dense concrete or other similar solid building material and a flat
roof of reinforced concrete. Chambers made of timber framing covered with sheets
of plywood or other material are likely to produce initial difficulties in sealing
and continuing difficulties due to deterioration or damage. Anyone contemplating
the construction of an atmospheric fumigation chamber may refer to publications
such as the UK Ministry of Agr. (1973), USDA (1976), Monro (1969).

(b) Transportation Facilities. (see also Chapter No. 14) Several
carriers, such as railway cars and freight containers, can be made sufficiently
gas-tight to serve as fumigation chambers. When railway cars are fumigated with
methyl bromide, they are often isolated on separate sidings where the fumigant
will not be a hazard to human beings. Some "in-transit" treatments, where the
fumigant is applied and the railway car is permitted to proceed to destination,
are done with phosphine. Also grain cargoes are treated in the holds of ships for
"in-transit" fumigations. Appropriate warning signs with the name of the fumigant
and date of treatment should be posted on any vehicle under fumigation.

(c) Gas-proof Sheets. The availability of gas-tight sheeting materials has greatly facilitated fumigation treatments by making it possible to treat infested goods in situ without moving them from storage. The goods most often fumigated under sheets are bagged commodities, cereals, processed foods, and dried fruit in cartons or boxes. A number of sheeting materials are sufficiently impervious to gases that they can be used for fumigation. Sheets made of polyethylene or polyvinyl chloride 0.1 to 0.15 mm (4 to 6 mil) in thickness are suitable for many types of treatments; they are readily available and relatively inexpensive; however, they may tear easily on sharp corners or projections. For large scale operations in exposed situations, sheets of coated fabrics, such as nylon, terylene, or cotton coated with neoprene may be required.

The principal fumigants used for treatments under gas-proof sheets are methyl bromide and phosphine. For treatments to be effective, it is necessary to obtain a concentration-time product (i.e., a designated concentration level for a specified exposure period) which will give satisfactory kill of all stages of insects or mites present. In determining the dosage of fumigant, especially of methyl bromide, it is often desirable to take both the volume of space and the weight of the commodity into consideration (See Table 4).

(d) Fumigation of Materials in Bulk. The treatment of commodities such as grain in large quantities presents special problems for application and distribution of fumigant. Various techniques such as probing fumigant into the commodity, addition to the grain stream, or recirculation techniques have been used. Recirculation systems for moving fumigant-air mixtures and displacing the interstitial air in grain have been successfully used for many years for fumigants such as methyl bromide, hydrogen cyanide, and the liquid fumigants. Recirculation procedures for the fumigant phosphine have been tested and applied only recently. Because mixtures of phosphine in air may be unstable at reduced pressures, considerable care has been taken in the past to avoid procedures that might produce a flame or explosion. However, test treatments in large grain storages have shown that phosphine can be dispersed safely and more rapidly and uniformly by a "low-flow" recirculation procedure than by other methods (Cook, 1980). The procedures used for phosphine recirculation are in the developing stages and may be further varied and refined in the future. Because of its instability at low pressures, phosphine should not be used in any system where very low pressures are produced.

(e) Vacuum Fumigation Chambers. The primary objective of vacuum fumigation is to hasten and improve penetration of the fumigant. Vacuum fumigation is used in some food manufacturing industries for treatment of packaged cereals, spices, and prepared foods as well as for plant quarantine purposes and for tobacco fumigation.

A vacuum fumigation installation is considerably more expensive than an atmospheric chamber and is usually employed for specialized treatments where the gas cannot easily penetrate into the commodity. Vacuum fumigation chambers are made commercially and are available from several manufacturers.

The fumigants used in vacuum chambers are mainly methyl bromide and ethylene oxide. For some treatments, the ethylene oxide may be supplied in mixtures with a non-flammable halogenated hydrocarbon or carbon dioxide. Phosphine is unstable at low pressures (Dalton and Hinshelwood, 1929) and should not be used in vacuum facilities.

APPLICATION OF FUMIGANTS

For a fumigant to be effective, it is necessary to maintain a certain

154

concentration of the gas in a confined space for sufficient time to control all stages of the pest organism. Dosages of fumigant are sometimes expressed in terms of concentration-time products (c x t products) to meet this requirement. The concentrations and times needed to control a variety of stored product insects have been determined by experimentation. For example, it is known that for the cadelle, (Tenebroides mauritanicus) 99% of the insects will be killed by a concentration of 33 mg/l (33 oz/1000 ft^3) methyl bromide in a 5 hr exposure at 20°C (68°F). The product of concentration and time (i.e., 33 x 5 = 165 mg/l x h) is termed the c x t product and is the amount needed to kill 99% of the insects. Other combinations of concentration and time (within reasonable limits), as for example 16.5 mg/l and 10 hours that give the same c x t product of 165 mg/l x h will produce the same level of control.

For commercial treatments the dosage schedules are designed to give complete control of the insects and also to compensate for factors, such as sorption of fumigant by the commodity. By determining concentrations of the gas during the exposure period, the c x t product can be calculated and used to ensure that the insects are exposed to sufficient fumigant for effective control. Further information on the c x t product may be found in the manual by Monro (1969). Some dosage schedules for the fumigant methyl bromide specify the concentrations that must be maintained for the prescribed period of time or they may provide specific dosage rates for commodities with different absorptive properties. A dosage schedule designed to give satisfactory control of insects (excluding the khapra beetle, Trogoderma granarium) in different commodities over a range of temperatures is given in Table 4.

The use of the c x t product method of applying a fumigant like methyl bromide does much to ensure the success of a treatment. It allows the applicator to compensate for improper application of fumigant, loss through leakage of gas from the space, or absorption by the commodity, by adding more gas or extending the exposure time. By careful monitoring of concentrations during the course of the treatment and making coincident adjustments for insufficient fumigant, the success of the operation can be assured.

While the c x t method is useful for most fumigants it cannot be easily employed with phosphine. Although concentration and exposure time are still the main factors that determine toxicity of this fumigant, the length of the exposure time is of great importance. Phosphine is a slow acting poison that is absorbed slowly by some insects even at high concentrations. Therefore, high concentrations may not increase toxicity, in fact, they may cause insects to go into a protective narcosis that will allow them to survive. In phosphine fumigation certain minimum concentrations are required and therefore gas analysis should be carried out to ensure the presence of sufficient gas. For most treatments, the manufacturers' directions will provide adequate treatment if no excessive loss through leakage or sorption occurs.

Temperature. The most important environmental factor influencing the action of fumigants on insects is temperature. The effectiveness of fumigants normally declines as the temperature falls; most treatments are carried out at temperatures ranging from 10 to 35°C (50-95°F). Successful treatments with methyl bromide can be done at low temperatures, even below freezing, however, the amount of the fumigant required is greatly increased at the lower temperatures and penetration into materials is impeded. Fumigation with phosphine is not recommended below 5°C (40°F).

ANALYSIS OF FUMIGANT CONCENTRATIONS

As the success of fumigation treatments is dependent on the concentrations

Table 4. Dosages of methyl bromide recommended for control of stored product insects in cereals and other foodstuffs (Thompson 1970).

Commodity	Exposure period (hours)	Space dosage (g/m³)[1]			Commodity dosage (g/tonne)[2]		
		below 10°C	10–20°C	above 20°C	below 10°C	10–20°C	above 20°C
Rice,barley,cocoa beans,currants,sultanas, raisins	24	25	15	10			
Wheat, maize,oats, lentils	24	25	15	10	40(1.4)	30(1.1)	20(0.7)
Sorghum, nuts,dates, figs	24	25	15	10	80(2.8)	60(2.1)	40(1.4)
Flour,pollards,rice, bran,groundnuts,oil seeds,empty sacks	48	25	15	10	80(2.8)	60(2.1)	40(1.4)
Oilseed cakes and meals,fish meal	48	25	15	10	160(5.4)	120(4.2)	80(2.8)

1 g/m³ is equivalent to oz/1000 ft³.
2 values in brackets are oz/short-ton.

of gas in the atmosphere during the treatment analysis of these concentrations is of considerable importance. For fumigations involving methyl bromide, thermal conductivity meters or interference refractometers that will indicate concentrations in g per m³ or oz per 1000 ft³ are most useful. Samples of the gas can be drawn by sampling tubes from critical points in the commodity and the space being treated, to give good indication of distribution of fumigant. If the concentration of gas is deficient, more can be added or the length of the exposure time increased. The halide leak detector is widely used with methyl bromide and other halogenated fumigants to detect leakage of fumigant and also to give some indication of the presence of fumigant in the working atmosphere. These detectors can give discernable indication of methyl bromide down to about 15 ppm in air, however, they should not be relied on for testing atmospheres at low concentrations around the TLV (5 ppm).

With phosphine, glass detector tubes are used for measuring concentrations of the gas. Infrared gas analysers and gas chromatographs are also available for fumigant analysis. A portable gas chromatograph that is suitable for analysing the high concentration used for insect control as well as the low concentration around the threshold limit value for human health is now available (Bond and Dumas, 1982).

Note: It should be pointed out that three methods of expressing gas concentrations are in common use: weight per volume, parts by volume, and per cent by volume. For some purposes it may be desirable to convert one set of values to another, as for example, parts per million (ppm) to g/m^3 or $oz/1000 ft^3$. To make these conversions the reader should refer to the calculations given by Monro (1969).

SAFETY PRECAUTIONS

When handling and applying fumigants it is essential to know the hazards posed by each of the compounds, to know how to detect and measure concentrations of them in the atmosphere, and to know the precautions necessary to avoid the hazards. (see also Chapter No. 22) Threshold limit values are established so that safe conditions can be maintained in areas where personnel are working. Appropriate detection equipment should be used to ensure that no one is exposed above the established TLV levels. Some of the properties of fumigants related to health hazards are given in Table 5.

In planning a fumigation, particularly large scale treatments such as general mill, plant, or warehouse fumigations, all details should be carefully planned beforehand, with the necessary equipment available and with personnel involved in the operation carefully instructed on the procedures to be used. Safety equipment, such as gas masks and gas detection equipment, must be in good working order and appropriate arrangements should be made for both first aid treatments and emergency hospital care to deal with any accidental exposure to the fumigant. Even in small scale routine fumigations, strict safety precautions should be followed. Warning placards that indicate the fumigant being used and the date of application should be posted at the site of the treatment and they should be removed after aeration of the space and the commodity is complete.

SUMMARY

In this chapter the fumigants that are most used for insect control in raw and processed commodities are described, along with properties and procedures for their use. Some information is also given on methods for analyzing the gases and on safety precautions that should be taken to protect workers from their toxic effects.

When pest control programs are planned for the protection of food commodities, the matter of choosing the procedures that are appropriate to the problems involved is of considerable importance. Fumigation is just one of a number of methods that is used for the control of pests in food commodities. Several preventative measures that will exclude pest organisms or prevent infestations from developing may be carried out. Location and design of facilities, comprehensive inspection procedures for all areas where food is stored, transported, or processed, and good housekeeping and sanitation practices can do much to avoid infestation of insects and mites.

For situations where infestations do develop, several control procedures are available. These may include such techniques as low temperatures (freeze outs, refrigeration), high temperatures (as in heating of mills to temperatures lethal to insects), aeration and cooling, the use of controlled atmospheres with carbon

Table 5. Threshold limit values (TLV) for the principal fumigants
and properties associated with health hazards.

Fumigant	TLV[1] (ppm)	Approx. odour threshold (ppm)	Skin absorption	Chronic effects
Methyl bromide	5	none[3]	slight	affects nervous system
Phosphine	0.3	1[4]	negligible	not known
Ethylene oxide	(5)[2]	300–1500	some	possible carcinogen
Hydrogen cyanide	10	1–5	rapid	not known
Ethylene dibromide	( )	25	some	possible carcinogen
Ethylene dichloride	10	50	some	possible carcinogen
Carbon disulphide	10	1	some	affects nervous system
Carbon tetrachloride	5	60–70	some	possible carcinogen
Chloropicrin	0.1	1–3	some	not known
Dichlorvos	0.1		some	not known

1 ACGIH (1983-84).
2 Parenthesis indicate proposed values (if available).
3 At high concentrations methyl bromide has a sweet musty odour.
4 The carbide-garlic odour associated with phosphine is due to impurities, which can be separated from phosphine under some conditions.

dioxide or nitrogen, and residual sprays as well as fumigants. The choice of treatment will be determined by effectiveness of the procedure in controlling the pests, its safety for the personnel involved and the economic aspects of the entire operation.

LITERATURE CITED

American Conference of Governmental Industrial Hygienists (ACGIH) 1983-84. Threshold limit values for chemical substances and physical agents in the work environment. ACGIH, 6500 Glenway Ave. Bldg. D5, Cincinnati, Ohio, 45211. 91 pp.

Bond, E. J. 1975. Control of insects with fumigants at low temperatures: response to methyl bromide over the range 25° to -6.7°C. J. Econ Entomol. 68:539-542.

Bond, E. J., and C. T. Buckland. 1976. Control of insects with fumigants at low temperatures: toxicity of mixtures of methyl bromide and acrylonitrile to three species of insects. J. Econ Entomol. 69:725-727.

Bond, E. J., and T. Dumas. 1982. A portable gas chromatograph for macro- and microdetermination of fumigants in the field. J. Agr. Fd. Chem. 30:986-988.

Bond, E. J. , T. Dumas, and S. Hobbs. 1984. Corrosion of metals by the fumigant phosphine. J. Stored Prod. Res. 20:57-63.

Calderon, M., and Y. Carmi. 1973. Fumigation trials with a mixture of methyl bromide and carbon dioxide in vertical bins. J. Stored Prod. Res. 8:315-321.

Cook, J. S. 1980. Use of controlled air with fumigants in mass grain storages. Report of Degesch Technical Meeting, Sept. 1980, 79-85.

Dalton, R. H., and C. N. Hinshelwood. 1929. The oxidation of phosphine at low pressures. Roy. Soc. Proc. A, 125:294-308.

IMCO. 1981. Recommendations of the safe use of pesticides on ships. Intergovernmental Maritime Consultative Organization 101-104 Piccadilly St. London, U.K. MSC Circ. 298:24pp.

Jagielski, J., K. A. Scudamore, and S. G. Heuser. 1978. Residues of carbon tetrachloride and 1,2 dibromoethane in cereals and processed foods after liquid fumigant grain treatments for pest control. Pestic. Sci. 9:117-126.

Leesch, J. G. et al. 1978. An in-transit shipboard fumigation of corn. J. Econ. Entomol. 71:928-935.

Monro, H. A. U. 1969. Manual of Fumigation for Insect Control. FAO Agricultural Studies No. 79, FAO, Rome

Rains, D. M., and J. W. Holder. 1981. Ethylene dibromide residues in biscuits and commercial flour. J. Assoc. Off. Anal. Chem. 64:1252-1254.

Redlinger, L. M., et al. 1979. In-transit shipboard fumigation of wheat. J. Econ. Entomol. 72:642-647.

Storey, C. L. 1967. Comparative study of methods of distributing methyl bromide in flat storages of wheat: gravity penetration, single-pass and closed recirculation. USDA Marketing Research Report No. 794, Washington, D.C. 16 pp.

Storey, C. L. 1971a. Distribution of grain fumigants in silo type elevator tanks by aeration systems. USDA Marketing Research Report No. 915, Washington, D.C. 17 pp.

Storey, C. L. 1971b. Three methods for distributing methyl bromide in farm type bins of wheat and corn. USDA Marketing Research Report No. 929, Washington, D.C. 7 pp.

Thompson, R. H. 1970. Specifications recommended by the United Kingdom Ministry of Agr. Fisheries and Food for the fumigation of cereals and other foodstuffs against pests of stored products. OEPP/EPPO, Ser. D, No. 15, 9-25.

U. K. Ministry of Agr. 1973. The design, construction and operation of fumigation chambers for use at atmospheric pressure. Ministry of Agr. (Publications) Tolcarne Dr., Pinner, Middlesex HA52DT, UK. 12 pp.

USDA. 1976. Plant protection and quarantine treatment manual. USDA Animal and
    Plant Health Inspection Services, Federal Center Bldg., Hyattsville, Maryland
    20782, U.S.A.
Zettler, J. L., et al 1982. In-transit shipboard fumigation of corn on tanker
    vessel. J. Econ. Entomol. 75:804-808.

FUMIGATION OF STRUCTURES

Robert Davis and Phillip K. Harein

Editor's Comments

The authors present an objective, detailed, and progressive overview of how to conduct a fumigation of buildings such as plants or warehouses. You should not need to do this often, but when you do, it must be done correctly to accomplish the objectives of eliminating the insects without poisoning people or damaging products.

F. J. Baur

FUMIGATION OF STRUCTURES

Robert Davis and Phillip K. Harein

Stored-Product Insects Research and Development Laboratory
USDA, Agricultural Research Service
Savannah, GA   31403
and
Department of Entomology
Hodson Hall, 1980 Folwell Avenue
University of Minnesota
St. Paul, MN   55108

INTRODUCTION

Fumigants are chemical pesticides that kill as gases and are toxic to most forms of life including man. For this reason fumigants must be rigidly controlled in their use.

However, fumigants are a necessary and important part of our arsenal in our war against insects that contaminate and infest our raw and processed agricultural commodities. As gases the effectiveness of fumigants lies both in their ability to penetrate into cracks and crevices and into the commodity being treated, and to diffuse throughout the confined area to be fumigated.

In this chapter we will not consider the fumigation of particular commodities as it is treated elsewhere in this book (see Chapter 11). Rather our concern is to present the principles and logic for their safe and effective use in the fumigation of the structures that are used to store and process commodities.

The selection of an appropriate fumigant is of utmost importance. Special consideration must be given to many factors such as toxicity to the pest, volatility, penetrability, corrosiveness, flammability, residue tolerances, offensive odors, application methodology, and economics. For the fumigation of structures we have limited our presentation to four fumigants – hydrogen phosphide, methyl bromide, hydrogen cyanide, and sulfuryl fluoride. The reader is referred to Table 1 for common uses and main characteristics of these fumigants.

The preparation for a fumigation and the actual fumigation should follow a logical procedure. The remainder of this chapter will be a step by step discussion of this procedure.

PREMISE INSPECTION

When it has been decided that a fumigation may be required it is essential that a thorough and complete inspection of the structure be made. Each structure will be unique and will require its own specific plan.

However, the following are some questions which will usually require answers before attempting a fumigation of any typical structure.

- If the structure itself is not infested, could the infested commodity be moved from the building and fumigated elsewhere?
- Assuming that removal of the infested commodity from the building is not practical, can you fumigate the commodity in place?
- What is the amount of space occupied (cubic feet) by the commodity?
- What is the size (cubic feet) of the building?
- Can the structure be made reasonably airtight, or will it be necessary to tarp the entire building?
- From what construction materials is the structure built? (Fumigants will pass through cinder block with no difficulty and methyl bromide will react with them.)
- Are there broken or missing window panes that must be replaced?
- Are there cracks in the ceiling, walls, or floors that must be sealed?
- Are there floor drains, cable conduits, water pipes, windows, and doors or other openings that will require sealing?
- How are you going to seal air conditioning ducts and ventilation fans?
- Will interior partitions interfere with fumigation circulation?
- Are the interior partitions gas tight so that they can be relied upon to keep the fumigant from entering other parts of the structure?
- Are all parts of the building under your control?
- If not, can these other operations be shut down during the fumigation?
- What are the building contents - raw product, machinery, processed product?
- Can any of them be damaged by the fumigant?
- Can such items be removed before the fumigation?
- If they cannot be removed, can they be otherwise protected?
- Where are the pilot lights?
- Where do you shut the gas off?
- Where are the electrical outlets and main panels? Of what voltage are they?
- Will the circuits be live during fumigation?
- Can the outlets be used to operate your fumigant circulating fans?
- If you cover the entire structure with a tarpaulin, can you make a good, tight ground seal?
- Is there shrubbery next to the building that might be damaged either by the fumigant, or by your digging to make an airtight fumigation seal?
- Can this shrubbery be moved?
- How close is the nearest building?
- Does the adjacent building have air conditioning or other air intakes that could draw the fumigant inside -- particularly during aeration?
- How are you going to aerate your structure after fumigation?
- Are there exhaust fans, and where are the fan switches?
- Are there windows and doors that can be opened for cross ventilation?
- Does the building contain any high priority items that may have to be shipped out within a few hours notice?
- If so, can you make provisions for interrupting the fumigation and aerating the building within a certain time requirement?
- Is the structure to be fumigated so located that your operations may attract bystanders? If so, you should consider asking for police assistance to augment your own guards.
- Where is the nearest medical and fire facilities?
- What is the telephone number of the nearest poison control center?
- What is the availability of safety equipment?
- Are all your personnel properly trained? If not, what is the availability of training?

Once you are convinced that you have covered everything, prepare a checklist of things to do and of materials needed. Don't rely upon your memory.

Table 1. -- Selected Fumigants For Treating Structures[1]/

	Hydrogen Phosphide (Phosphine, PH_3)	Methyl Bromide (CH_3Br)	Hydrogen Cyanide (HCN)	Sulfuryl Fluoride (SO_2F_2)
Major Uses	Space, Grain Vehicles, Food and Feed, Structures	Space, Grain Vehicles (Static), Spot, Food and Feed, Structures	Space, Grain	Structures
Gas From	Tablets, Pellets Sachets, Plates	Cans, Cylinders	Cylinders	Cylinders
Speed of kill	Slow	Quick	Very Quick	Quick
Penetration	Excellent	Good	Fair	Good
Ease of aeration	Excellent	Good	Fair	Good
Mixed w/ other gases	No	Chloropicrin	No	Chloropicrin
Sorption	Negligible	Yes	Yes	Yes
Molecular weight	34.04	94.94	27.03	102.6
Sp. Gr. (air = 1)	1.214	3.27	0.9	2.88
Solubility in water	Very Slightly	Very Slightly	Infinite	Very Slightly
Latent heat of evaporation Cal/g.	102.6	61.52	210	20,034
Odor	Carbide, Garlic	None at low concentrations Sweet musty at high conc.	Almond	None
Boiling point	- 87.4°C	3.6°C	26°C	-55.2°C
Skin absorption	Negligible	Yes (Slow)	Yes (Rapid)	Negligible
Chronic poisoning	No	Yes	No	No

164

Table 1. -- Continued

	Hydrogen Phosphide (Phosphine, PH_3)	Methyl Bromide (CH_3Br)	Hydrogen Cyanide (HCN)	Sulfuryl Fluoride (SO_2F_2)
Threshhold limit value	0.3 ppm	5 ppm	10 ppm	5 ppm
Skin blistering	No	Yes	No	No
Flammability	Self-combustible above 1.79%	No	Yes 6-41%	No
Reacts w/	Copper, Silver, Gold	Sulfur containing compounds and aluminum when in liquid form.		Non-reactive
Gas mask	Organic vapor	Organic vapor	Acid gas	Acid gas
Canister	Yellow w/grey stripe	Black	White w/ green stripe	White w/ green stripe

1/ Taken in part form chart made available by The Industrial Fumigant Company, Olathe, KS.

SEALING

In the fumigation of structures one of two approaches may be used. The first approach is where the building itself is used to contain the fumigant. In this approach the walls must be relatively gas tight and the building openings closable and/or sealable. The other approach is where the entire structure is covered with a tarpaulin. This latter approach must be used if the pest problem (such as woodborers or drywood termites) involves an infestation of the structural materials. It is most important to achieve a well sealed structure prior to fumigating and the following areas may require special attention.

Roof Areas

It is important to seal all roof vents and other openings except roof drains. In the initial inspection of the roof all vents and other openings will need to be located. An assessment of whether or not the roof is gas tight will also need to be made. Flat roofs composed of layers of roofing felt will usually not offer a problem. However, corrugated metal roofs may not be gas tight and can pose a sealing problem at the roof-wall juncture.

The traditional round vents can be sealed quickly and easily with heavy duty plastic bags placed over the vents and taped at the base. Other equipment such as air conditioners are most easily covered with plastic and sealed to the roof with elongated rope-like sand bags (sand snakes). Dust collectors or similar equipment which may extend to greater heights above the roof may require complete covering with plastic sheeting or tarpaulin.

Doorways and Windows

Most windows, except on the most modern of buildings, will require some sealing. The older wood window frames and sashes will usually require complete covering with polyethylene sheeting. Other types of windows may be adequately sealed with tape or strips of plastic and tape.

Doors and doorways offer more of a challenge because of their great variability in size and types of construction material. The simple walk through metal door and metal door jam offer little problem and are easily sealed with tape. Most other doors, particularly of an over head type, should be covered completely, and sealed with glue or tape or both.

Miscellaneous Openings

Small openings around pipes and electrical conduits also need to be sealed. Use a permanent material to prevent future entrance of pests. Small ventilator fan openings are easily sealed with tape and polyethylene sheeting. Floor drains will need to be sealed if they do not possess water traps. Floor cracks and expansion joints may either require taping or permanent repairing to prevent gas losses. Concrete and cinder blocks are often not gas tight and will require sealing especially if they form a portion of the outside walls. If the building is constructed totally of concrete or cinder blocks a complete tarping of the structure may be necessary.

SELECTION OF FUMIGANT AND ITS DOSAGE

The proper selection of a fumigant and its dosage is mandatory. Most fumigants are restricted by law and must be applied by or under the direction of a certified applicator. The application must also be in accordance with the directions on the manufacturer's Environmental Protection Agency (EPA) approved label. Each fumigant, however, has its own particular physical and chemical properties and will perform reliably under similar circumstances. Thus when using fumigants experience is of great value. Therefore, in the selection of one of the four fumigants from Table 1, we must never underestimate the importance of their physical and chemical properties. Dosage (i.e.; dose or quantity of fumigant and time of exposure) is also an important consideration and is, therefore, presented in detail on the label as to the minimum and maximum allowable dosages and exposure times. For a more detailed discussion of the properties of these four fumigants and many other fumigants the reader is referred to Monro (1969).

Hydrogen phosphide is approved for use on most raw and processed foods. Methyl bromide and hydrogen cyanide are approved for use on a lesser number of commodities. Sulfuryl fluoride is not approved for use where it will come in contact with any raw or processed commodity, but is used for fumigating structures for wood infesting insects such as termites and wood borers. Again the importance of reading the manufacturer's label and other literature cannot be over emphasized.

EVACUATION OF STRUCTURE AND OTHER PRELIMINARIES

In the evacuation of personnel from structures to be fumigated it is most important to work closely with the other management so to assure that the evacuation is complete. This will necessitate the use of an employee roster so each employee can be accounted for before releasing the gas. It will also be necessary to accomplish a complete walk through of the entire premise. This inspection of the premise will not only be a visual check, but one in which you should use a loud voice to alert anyone who may have otherwise not been noticed. When this walk through has been completed guards should be posted and the building locked to prevent reentry. It is also recommended that notification be given at this time to the local fire and police departments and any private security company of your intent to fumigate, the fumigant to be used, proposed date of fumigation, safety equipment required for reentry, and fire hazard rating. The same notification should be given to pertinent medical organizations of your intent and provide them with copies of all available literature (labeling materials) from the fumigant manufacturer. At this time prepare warning signs, make final arrangements for security, and establish your two person team(s) that will release the fumigant and perform initial post fumigation activities.

REHEARSAL AND PLACEMENT OF FUMIGANT

The value of a rehearsal for the fumigant release and subsequent procedures cannot be overemphasized. Each member of the two man release team(s) will need to know exactly where each cylinder of gas is located or where to place the solid fumigant formulation and how long it will take to complete the release of the fumigant. Cylinder valves need to be quickly opened and closed to be sure they are in working order and canisters of aluminum phosphide should be placed at exposure locations. Fans, if auxillary air movement is required, need to be tested before releasing the fumigant. Open gas flames and any electrical equipment that will produce a high temperature must be turned off at this time. Participants conducting and supervising the fumigation should also be briefed on the availability of medical and other emergency arrangements and facilities. Warning signs should be posted at this time.

FUMIGATION

Wherever possible, when fumigating with cylinders of gas, the release should be made from outside the building. Where outside applications are made, care must be taken that all gas lines are in good condition and no leaks will occur.

When fumigants must be released from the inside the structure, the route must be planned that will take the two person team(s) away from the gas and towards a safe exit. There should not be any need to return to an area being fumigated. One 2 person team is normally used but other teams may be added if necessary to reduce the release time or the chance of exposure of personnel.

Hydrogen phosphide producing fumigant formulations provide safety by having a delayed release of the gas but some forms require considerable time to open the required number of hydrogen phosphide producing units. The following suggestions will help:

* Have all flasks or similar hydrogen phosphide formulation containers at point of discharge before the first one is opened.
* If pellets are used, have large pieces of heavy brown paper (minimum 3' x 3') on which to dispense the pellets.
* Equip each team with retainers to prevent the pellets from rolling off the paper. Packages are often difficult to open, so team should have a knife or jar opener available.

Cylinders of gas should be released carefully and in succession. It is usually better to have all cylinders opened by one man while his partner double checks to be sure that none are missed. Steady the cylinder with one hand while the valve is turned open with the other hand. Open the cylinders all the way to avoid nozzles from freezing shut.

AERATING THE BUILDING

Once the exposure period is complete, aeration should be started by opening the windows and doors on the ground floor that can be opened without entering the building. Attempt to provide cross ventilation. Where ventilators are accessible from the outside they should also be opened at this time. The ground floor should be allowed to aerate until the fumigant detector shows that the fumigant concentration has diminished to the point where it is safe to enter the structure while wearing an approved gas mask and recommended protective clothing (Mackison et al., 1978). At this time two people (or teams of two people each) should begin opening windows, starting at the bottom and working upward. These technicians should not try to open all windows on any single floor the first time through but should open only those windows that are necessary for cross ventilation and return to the outside as soon as possible. They should not remain inside the building for prolonged periods (not more than 15 minutes). The fans should be turned on and allowed to run when aeration begins and continued until aeration is complete. After the building has been partially aerated, the technicians, again wearing gas masks, should open as many of the remaining windows as needed to complete the aeration. No one should be allowed inside the building without a gas mask until all parts of the building have been checked with an appropriate fumigant detector that shows it is safe. Once the aeration has been completed, usually in two or three hours, the building can be returned to management for normal operations. Remember, it is advisable to leave in place such sealing that will not hinder aeration and operations, so that this sealing does not have to be replaced for future fumigations.

PERSONAL PROTECTION AND GAS DETECTION

Personal Protection

Fumigant gases can enter the body in two ways: (1) through the skin, and (2) through the lungs. Of these two modes of entry, the human respiratory system presents the quickest and most direct avenue of entry because of its intimate association with the circulatory system. There are two additional respiratory hazards to be considered when fumigating:

1. Oxygen deficient air.
2. Air laden with fumigants.

The normal content of oxygen in the air is 20.9 per cent by volume. Oxygen concentrations below 16 per cent will not support combustion and are generally considered unsafe for human exposure because of harmful effects on bodily functions, mental processes, and coordination. At very low oxygen concentrations, collapse can be immediate, without warning, and death can ensue within minutes. Occupational Health and Safety legislation requires that the oxygen percentage in a working place be not less than 19.5. In assessing exposure conditions, it is important to remember that oxygen deficiency can occur in confined spaces by displacement of air by other gases and vapors.

Air contaminants include gaseous material in the form of a true gas or vapor, such as a fumigant. Respiratory protective devices vary in design, application, and protective capability. The user must, therefore, assess the inhalation hazard and understand the specific use and limitations of available equipment to assure the proper selection. Respiratory protective devices are tested and approved by National Institute of Occupational Safety and Health/Mine Safety and Health Administration (NIOSH/MSHA) for protection against a wide range of inhalation hazards, including oxygen-deficient and/or highly toxic atmospheres. While not all types of respiratory protective devices are covered under the current NIOSH approval requirements, it is desirable to select NIOSH/MSHA approved equipment whenever possible.

Respiratory protective devices fall into two classes: air-purifying, and self-contained or supplied-air breathing apparatus. Gas masks with air-purifying canisters provide respiratory protection against certain specific gases and vapors in concentrations up to 2 percent (20,000 ppm) by volume or as specified on the canister label. Specific gas masks are approved by NIOSH/MSHA 30 CFR Part 11, Subpart 1.

Self-contained and supplied-air breathing apparatus provide complete breathing protection for various periods of time based on the amount of air or oxygen supplied and the breathing demand of the wearer. The basic types of self-contained breathing apparatus are:

(1) Oxygen Cylinder Rebreathing Type,
(2) Chemical Oxygen Rebreathing (self-generating) Type, and
(3) Demand and Pressure-demand Types.

Self-contained breathing apparatus are approved under NIOSH/MSHA 30 CFR Part 11, Subpart H.

Gas Detection

Gas detection equipment for use with fumigants is readily available from most distributors of fumigants. The importance of having such equipment available to determine the fumigant concentration during the fumigation and assessing the safety of a structure during aeration cannot be overemphasized. Table 2 presents a few currently accepted types of detection equipment.

Table 2. Currently accepted types of detection equipment and devices.

Fumigants	Detectors
Phosphine	Detector tubes
Methyl bromide	Detector tubes Halide gas detector Thermal conductivity analyzer Interference refractometer
Hydrogen cyanide	Detector tubes Color indicator papers
Sulfuryl fluoride	Thermal conductivity analyzer

SUMMARY

The handling and use of fumigants to control pests in structures is an endeavor that is not to be taken lightly. Carelessness or ignorance can result in the death of the fumigator or of innocent bystanders, destruction of the usefulness of the product being treated or a simple failure to control the pest. Fumigants in most instances will be labeled as restricted pesticides

and training and certification will be required before they may be purchased
and used. Consideration of the recommendations presented herein and strict
adherence to the manufacturer's Environmental Protection Agency's (EPA)
approved label will in most situations assure a safe and effective fumigation.

Notice - For precise information on fumigant selection and use, the
reader always should refer to a current fumigant label.

LITERATURE CITED

Mackison, F. W., R. S. Stricaff, and L. J. Partridge. 1978. Pocket guide to
     chemical hazards. NIOSH/OSHA, Cincinnati, 191 pg.
Monro, H. A. U. 1969. Manual of fumigation for insect control. FAO
     Agricultural Studies. Rome No. 79 2nd Ed (Rev.) 381 pg.

SPOT FUMIGATION

James C. Dawson

Editor's Comments

Spot fumigation has been a very valuable tool for the food industry's efforts avoiding adding insects to the product stream via infestations internal to processing equipment. With the recent suspension of ethylene dibromide and the likely similar action against the less effective and more toxic carbon tetrachloride-ethylene dichloride, industry is in desperate need of suitable replacements. Improved cleaning and cleanout are always desirable and possible. Use of insect traps in equipment is an excellent monitoring device but hardly suitable for the necessary control. The growing use of plant fumigations with methyl bromide (also suspect safety-wise) is expensive and time consuming. An adjunct/alternate is more extensive fumigation of incoming commodities. The redesign of the equipment to permit more efficient cleaning is a long-range option.

The future of spot fumigants and spot fumigation is in flux.

F. J. Baur

SPOT FUMIGATION

James C. Dawson

President, Ferguson Fumigants, Inc.
93 Ford Lane
Hazelwood, Missouri 63042

THE SPOT FUMIGATION CONCEPT

Modern standards of sanitation require the prevention of insect infestation and insect propagation in equipment that manufactures and handles dry food commodities. Grain millers, cereal food processors, prepared mix manufacturers, dog food manufacturers, and brewers are just a few of the industries that process or handle dry food commodities. The product as well as the equipment used to produce or handle dry foods must be kept free of insect infestations.

Insect infestations normally result from inbound infested product (see chapters 11 and 14). Product residue or static stock inside the equipment becomes a breeding ground for insects. The red flour beetle, Tribolium castaneum (Herbst), is able to fly and may contaminate food processing equipment by attraction to a food source. This insect is more prevalent in the southern United States and is probably the most predominant insect contaminant in the cereal processing industry. The confused flour beetle, Tribolium confusum (Jacq duVal), is the most frequently identified "bran bug" (flour, sawtoothed grain, flat grain, and others) as well as being the most widely distributed insect in the cereal food processing industry. It is this insect, its life cycle, and its habitat that the usual spot fumigation program is designed to control.

Federal Food and Drug investigators will be satisfied with nothing less than insect-free food processing equipment. The manufacturing facility must also be free of infestations. The finding of an infestation in the equipment is considered sufficient proof that the finished product could also be contaminated. Chapter 4, Sections 402(a)3 and 4 of the Federal Food, Drug, and Cosmetic Act provide the legal requirement of the food processors to insure that processing equipment is kept free of insect contamination. To quote the law, "Adulterated Food, Section 402 - a food shall be deemed to be adulterated (a)3 if it consists in whole or in part of any filthy, putrid or decomposed substance or if it is otherwise unfit for food; or (a)4 if it has been prepared, packed or held under insanitary conditions whereby it may have become contaminated with filth or whereby it may have been rendered injurious to health." It is therefore imperative that possible infestations be eliminated on a regular basis during the preparation of food products. Close attention must be given to prevention of recontamination.

A preventive spot fumigation program is designed to eliminate insects introduced as part of the material flow and the propagation of infestation in static stock inside the food processing equipment. A regular insecticidal program (see chapters 9 and 12) including fogging must be initiated to control insect infestation outside the equipment. Most plants are currently using (see chapter 10) a pyrethrin base insecticide or DDVP (2,2-dichlorovinyl dimethyl phosphate) for fogging.

The conditions inside a food processing plant are usually ideal for insect propagation. Under optimum conditions, most "bran bugs" will propagate every 21 to 30 days. Because of this life cycle, a regular preventive spot fumigation can be scheduled. A spot fumigation performed every three to four weeks will prevent insect propagation in food processing equipment.

THE SURVEY

For most effective results, it is important that the spot fumigant be well distributed throughout the interior of the equipment. Proper distribution will allow the development of fumigant concentrations that are lethal to all stages of insect life over an 18 to 24 hour exposure period. Application points must be selected strategically to achieve distribution of the fumigant.

Special attention must be given to areas where static stock is likely to develop during food processing. Static stock that contains a shot of spot fumigant will release lethal gas during the exposure period, helping to maintain deadly gas concentrations. The following is a basic list of application points for a cereal processing system:

EQUIPMENT	APPLICATION, LOCATION
Bucket Elevator (Head)	Each side, just below the head
Bucket Elevator (Boot)	Center of the boot
Screw Conveyors	
4" Diameter	Every 20 feet
9" Diameter	Every 12-15 feet
12" Diameter	Every 10-12 feet
Sifters	Directly into the top of each section
Roll Stand	Each side of feeder housings
Suction Fan	Eye and feed-in spouts
Purifiers	
Conveyor bottom purifier	Each end of each conveyor
Hopper bottom purifier	Shots fanned across the hoppers
Weighing hoppers, small storage bins, dump hoppers, and the like	Application to be made in such a manner that vaporization takes place in the upper portion of the space. A cellulose sponge may be used to assist in holding the fumigant

To assist in proper distribution of the fumigant, it will be necessary to drill application holes in the equipment. The holes must be large enough to accept the tip of an application gun. To eliminate dusting of the food product during processing, we recommend that standard snap cap oiler fittings be permanently installed in these holes.

To maintain consistent application by different spot fumigation crews, it is important that the application points be well marked. For best results, the application points on each floor or in each section of the food processing area should be marked and numbered in the order in which they will be used. A checkoff list is an effective tool for the fumigation crew supervisor.

APPLICATION EQUIPMENT

For the most cost-effective application, it is best to utilize spot application equipment that will automatically measure the amount of fumigant being applied. This is accomplished in one of several ways.

The LITTLE SQUIRT TOTER spot fumigant applicator (see Fig. 1) is designed for application of one or more gallons of spot fumigant. This equipment is all stainless steel and brass to prevent corrosion and to allow for easy maintenance. It will automatically meter a two-ounce shot of fumigant when the trigger of the applicator gun is pulled. It has an automatic shutoff design for safety and will operate with or without being connected to a compressed air source.

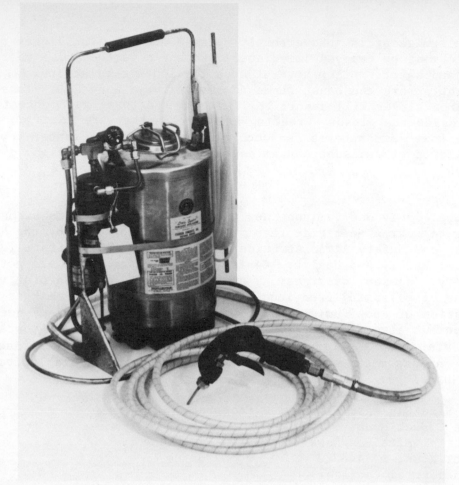

Fig. 1. LITTLE SQUIRT TOTER spot fumigant applicator. This handy piece of
equipment can make spot fumigant application both safe and efficient.

For manufacturing facilities, primarily flour mills, that require the
consumption of 10 or more gallons of fumigant for each spot fumigation, larger
equipment is available. For those facilities with a freight elevator, the
LITTLE SQUIRT Portable may be a better choice. Standpipe systems are also
customer designed for individual needs by fumigant suppliers.

In some cases, a manifold system may be considered for application to all or
part of the equipment requiring spot fumigation. Manifolds can be operated from
a central location and are normally used to apply spot fumigant to areas that
are hard to reach or physically impossible to get to. The cost of installation
must be weighed against the speed and convenience of application as well as the
additional safety afforded the application crew.

The label on the fumigant contains the application instructions which must be
followed.

SAFETY EQUIPMENT

The most important factor in understanding the use of safety equipment (see
also chapter 22) when utilizing the spot fumigant is to understand the hazards
the fumigant presents to the human body. Inhalation presents the greatest
risk. Physical contact with the liquid spot fumigant can create a problem if the
liquid is held close to the skin. The gas may also create problems if confined
in high concentrations next to the skin.

The liquid spot fumigant when held close to the skin will cause chemical
burns similar to third-degree heat burn. In more severe cases, a group of water
blisters or a single large blister will develop followed by swelling of the
extremity. In these cases, medical attention is required. Burns can develop if
the liquid spot fumigant is allowed to get under a ring or a wristwatch.

If the fumigator is inadvertently sprayed and the liquid gets on his clothing it must be removed immediately and the affected areas must be washed with soap and water. Allow several days for leather to aerate.

Fumigant vapors can cause burns in several ways when held close to the skin. Rubber gloves will absorb the fumigant and toxic gas concentrations will build up inside the gloves, creating atmospheres that are toxic to the skin. These same toxic atmospheres can occur inside leather or rubber boots should an accidental drop of fumigant contaminate them during application.

The Gas Mask

The label (only one fumigant now has EPA registration) requires the applicator to wear a pesticide respirator jointly approved by the Mining Enforcement and Safety Administration and the National Institute for Occupational Safety and Health. Canister style respirators are normally used in our industry. Nausea, dizziness, a staggering gait, and blurring vision are early signs of poisoning from inhalation. A full-face gas mask is the standard recommendation of spot fumigant suppliers. Not only is there a need to protect respiration but also to protect the eyes from liquid fumigant contamination.

Mine Safety and Appliances makes the most widely used full-face gas mask equipment and gas mask canisters. These products are available from most fumigant suppliers:

    Super Size Gas Mask/Ultra View Facepiece
      Type C Back Mount - #457095
    Industrial Size Gas Mask/Ultra View Facepiece
      Front Mount - #448923
    GMC Industrial Canister
      2% Organic Vapor - #449888
    GMC - S Super Size Canister
      2% Organic Vapor - #448965

The GMC-SS-1 #77713 is not recommended for use during spot fumigation. Although it is approved for 2% organic vapors (the maximum allowable concentration), the total amount of protection does not equal those canisters built specifically for organic vapor protection.

Detection Equipment

Upon completion of the exposure period, gas detection equipment is used to insure that aeration of the building is complete and the atmosphere is safe for production employees. Two types of equipment are available for testing the atmosphere.

The halide detector requires an open flame for gas detection. The intensity of color in the flame, ranging from a faint tinge of green to a strong blue, will indicate gas concentrations ranging from 25 to 25,000 ppm. This detector will not detect gas in the atmosphere below 25 ppm and therefore is not suited for determining that the atmosphere is clear for worker safety.

The Draeger or Auer (brand names) pumps, together with the correct gas detection tubes, will measure fumigant gases between .5 and 50 ppm. A single tube may be reused in several areas until gas is detected, making the pump detector a useful, economical, atmosphere-monitoring device.

THE SPOT FUMIGATION

Long before the actual spot fumigation begins, the entire spot fumigation crew should thoroughly review the following:
  • Have a complete knowledge of the spot fumigant application points as well as the sequence of application.

- The spot fumigant label should be read by the crew paying particular attention to safety precautions and first aid in case of an accident.
- The gas masks must be checked and fitted by each individual crew member.
- Make absolutely sure that the canisters have not been used and that they are dated well within the expiration date.
- The application equipment must be checked for proper operation.
- The processing equipment must be run as dry as possible to eliminate excessive stock. Static stock is desirable and will be of benefit to the fumigation.
- Insure that drafts have been eliminated within the processing equipment.
- Make sure that all spout lids are in place, elevator heads and boot slides are closed, and vents to the exterior of the building have been sealed.
- Check all doors and windows to make sure they are closed, helping to eliminate drafts.

In order to help monitor the effectiveness of a spot fumigation, "test cages" can be used to identify areas of low concentration. A test cage normally contains a number of live adult insects as well as larvae. Test cages are placed in strategic areas inside the equipment. It is important to locate the cages so that they do not receive a direct shot of spot fumigant during application. Live insects, after a spot fumigation, may indicate a need to adjust a few application points or take a closer look at stopping internal drafts that dilute the fumigant concentration.

Warning signs must be posted before application begins. The start time, exposure period, and a telephone number must be on the warning signs should problems arise.

The liquid spot fumigant is a heavy gas. For this reason it is best to start the spot fumigation of a multilevel food plant in the basement or lowest level. As the fumigation crew moves up through the structure, they will be moving away from the gas that has been applied. Upon completion of the spot fumigant application, the crew will spend only a very short time in toxic atmospheres as they go down and out of the building.

When the required exposure period is expired, the equipment is aerated, usually using the existent dust collection or pneumatic system.

All spot fumigants are classed by the Environmental Protection Agency (EPA) as "RESTRICTED USE PESTICIDES." In order to purchase and use these materials, the spot fumigation crew or a member of the crew must be certified by the state. The crew may also operate under the direct supervision of a certified applicator. Check with your local state agency for certification requirements.

Pesticide products containing ethylene dibromide were canceled by the EPA on February 3, 1984. This cancellation affected several widely used spot fumigant formulations. The only liquid spot fumigant remaining on the market is a formulation of 75% carbon tetrachloride and 25% ethylene dichloride. It is expected that EPA will issue a position document against carbon tetrachloride and its use as a fumigant sometime in 1984. Cancellation of this last liquid spot fumigant formulation is inevitable.

Degesch America, Inc. is currently working with state agricultural agencies to expand the use of Magtoxin® pellet - prepak for spot fumigation. Application and retrieval of the individual packets will be by hand with special attention to accurate count control. The current proposed label calls for a minimum 36-hour exposure. Data are currently being accumulated for Federal EPA approval.

SUMMARY

Spot fumigation was developed as a preventive sanitation procedure. The products that have been available in the past have been effective in maintaining control of the development of insect populations inside dry foods processing equipment. With the loss of ethylene dibromide, we must find an effective product and application procedure in an effort to continue to use a procedure that will help eliminate insect development inside dry foods processing equipment without the expense of a regularly scheduled general fumigation.

The need for and application of spot fumigants to prevent and eliminate insect infestation in food processing equipment has been discussed.

FUMIGATION OF FOOD SHIPMENTS IN-TRANSIT

Ernest A.R. Liscombe

Editor's Comments

This chapter deals with avoidance rather than control. In-transit fumigation is frequently necessary to protect shipments from insect infestation. The author covers his subject completely.

Whereas the sanitation and physical integrity of boxcars have improved, they still offer problems. Industry needs to continue its move to air-slide type cars.

F. J. Baur

FUMIGATION OF FOOD SHIPMENTS IN-TRANSIT

Ernest A.R. Liscombe

Soil Chemicals Corporation
P.O. Box 531
Morgan Hill, CA 95037

INTRODUCTION

If the consumer is to receive a clean, unadulterated product, it is essential that at each step in the process, from the shipping of the raw ingredients to the distribution of the finished product through retail outlets, extreme care must be taken to see that no contamination occurs (Freeman, 1958).

The Food, Drug, and Cosmetic Act became law in 1938. As a result, food plant sanitation received more emphasis and, over the past 25 or 30 years, great progress has been made in grain milling and processing techniques to preclude insect contamination (Cotton, 1960). Unfortunately, the same can not be said for the condition of many of the nation's railcars.

Railcars have long been known to harbor insects (Cotton, 1962). Although railcars are generally cleaned and/or sprayed with an insecticide before loading, the areas of the car where insects can hide and breed are generally inaccessible to the cleaning and spraying operation.

It has been well documented that, at times, substantial quantities of the nation's grain and peanuts go into storage containing varying numbers of stored-products insects. As these infested commodities are transported by rail from place to place, there is a great potential for spreading the infestation because of the structure of many of the boxcars.

During World War II, a shortage of boxcars in Canada necessitated the building of many large temporary grain storage facilities. From time to time serious insect infestations occurred, and due to congested conditions making fumigation difficult, large numbers of boxcars were diverted to these trouble spots so grain could be moved to terminals and treated there. Investigations at that time showed that live insects were still present in the grain residues left in 71 per cent of the cars after unloading (Liscombe, 1966).

In 1965, it was found that 12 per cent of 120 randomly selected empty boxcars sampled in June contained important stored-products pests. When the same number of randomly selected boxcars were sampled in August of that year, the infestation level was up to 33 per cent (Liscombe, 1966).

The flour milling industry relies on the railroads as its major means of transportation both for the movement of bulk grain and for finished flours and bakery goods (Smith, 1982). In the past 15 years the number of hopper cars has been gradually increasing. This helps to reduce the spread of insects because hopper cars do not have areas where product residue can collect and remain to become infested. It does nothing, however, for reducing infestation present in the grain before it is loaded.

The shipment of bagged flour in boxcars to bakeries around the country is rare today. This method of shipment has been replaced by bulk unit-trains resulting in increased transportation efficiency and lower freight rates, and by pressure differential trucks that deliver customized bulk flour and other products to the bakery doors. This method of movement has been speeded up by the fact that manufacturers have tried to move processing plants closer to the customer. It is always cheaper to ship a raw commodity than a finished product. In the very competitive marketplace we have today, all such efficiencies are important.

Damage free (D.F.) cars and rebuilt refrigerator cars (reefers) have also come into use for transporting flour, bakery products, and other processed cereals. Provided these cars are used solely for this class of merchandise, there is rarely a problem with infestations in-transit.

Although the food industry relies heavily on the railroads for transporting their goods, it also uses truck trailers, intermodal containers, unmanned barges, and ocean-going ships (Freeman, 1965; Freeman, 1968; Freeman, 1971; Hurlock, 1964). Each of these modes of transportation can be a potential source of infestation to the commodity being transported. It is possible to conduct an effective fumigation in each type of conveyance but, since each has its own peculiarities, they must be discussed separately.

CONVEYANCES USED FOR IN-TRANSIT FUMIGATION

Descriptions

Railcars

There are three basic types: (1) a steel frame boxcar with walls of a single layer of tongue and groove lumber, hence there are no pockets where grain or other residue can hang up. The end walls are double; (2) a steel boxcar with a regular wood lining and a grain trench at the base of the side walls. End walls are double; and (3) an all steel hopper car. Most are closed with slot-type or round hatch covers. Many can be unloaded pneumatically.

Truck Trailers or Vans

These units are common throughout the country today and carry a huge tonnage of food commodities each year. They are generally constructed of steel with a wooden floor, although some have a steel frame with a wooden skin.

Intermodal Containers

These are typically trailer vans without a chassis or wheels. They are normally constructed with a wooden floor and a steel frame and skin.

Unmanned Barges, Dry-Bulk Carriers, and Tankers

These are of welded steel construction so they are very tight. Unmanned barges carry no crew, while dry-bulk carriers and tankers do. Dry-bulk carriers and tankers must be constructed so that no cargo holds which are to be fumigated are directly below or attached to quarters occupied by the crew. In dry-bulk carriers, the holds are large and designed to carry only dry cargo in bulk. In tankers, the holds are generally much smaller in size and larger in number and are designed to carry either dry or liquid cargo in bulk. In many instances, a tanker will carry liquid crude oil in one direction and grain on the return trip. The holds must be very well cleaned each time they are emptied so there is no contamination of the next cargo.

PREPARATION FOR FUMIGATION

Ascertaining Tightness

Railcars

Boxcars are quite gas tight when new, but time and hard use can change this condition rather quickly. One of the best ways to determine leakage in any particular car is to fill it with smoke and close the doors. The escape of the smoke will pinpoint the location and amount of leakage. Should the fumigator not consider it feasible to seal large openings then the car should be rejected and replaced by one in better condition. A Dyna-Fog® machine filled with a solvent which produces smoke when burned makes a good smoke generator.

The steel construction of hopper cars makes them very suitable for fumigation. Care must be exercised, however, to see that the hopper lids are not warped and that they can be closed and sealed.

Truck Trailers or Vans
They can be examined with a smoke generator as above.

Intermodal Containers
They can be examined with a smoke generator as above.

If a smoke generator is not available, then boxcars, trailers, and intermodal containers can be examined for tightness by entering the conveyance on a bright day and closing the doors. A close examination of all interior surfaces and joints for the presence of daylight will tell the fumigator how tight the conveyance is and where any leaks are. Special attention should be given to inspection of containers with wooden (tongue and groove) floors. These floors are seldom tight enough to retain the fumigant gases. Extra care should be taken to see how the doors close. This is especially true in the case of boxcars where, in some cases, the lower door track is hard to seal because it is not completely attached to the boxcar and is open on the back side. This type of construction is hard to see from the outside when the door is closed.

Unmanned Barges, Dry-Bulk Carriers, and Tankers
Because of their welded steel construction these conveyances are very tight. Consequently they make excellent fumigation chambers. The fumigation of cargo on barges and ocean-going dry-bulk carriers and tankers is a very specialized field. When a fumigator undertakes such a job, he should always employ the services of a marine surveyor for the pre-fumigation inspection. It is this inspection that determines the integrity of the vessel and forms the basis for a 'go' or 'no go' decision on the fumigation. Under no circumstances should a fumigator undertake the in-transit fumigation of cargo on tween deckers (where there are large openings in each of the decks in the hold), or in the holds of any ship which are below or attached to areas of the ship occupied by the crew.

There are additional steps the fumigator must take in preparation for the fumigation of a dry-bulk carrier or tanker. As outlined in Code of Federal Regulations, Volume 46, Section 147A (Anonymous, 1981), he must meet with the person in charge of the vessel and outline in detail what is to be done, what fumigant will be used, how it will be used, and explain fully all hazards, safety precautions, and symptoms of poisoning. Two members of the crew must also be taught how to use gas detection and respiratory protective equipment, and at least one gas detector kit and two gas masks, fitted with canisters for the fumigant being used, must be on board the vessel when it leaves port. The fumigator must also have in his possession a copy of the referenced Regulations and he must notify the Captain of the Port, in advance, of specific information concerning the treatment.

Cleaning and Sealing

Railcars
There is very little work required to prepare a hopper car for fumigation. Make certain that the car has been properly cleaned and/or washed and be sure to seal the small holes or vents located in each end wall of the car. Failure to seal these vents will probably result in a fumigation failure.

The story is considerably different for boxcars. As previously outlined, boxcars often contain broken linings behind which insects and debris can remain. Over the years, the railroads have attempted to partially solve this problem by fitting fiberglas batts into the end walls. This has drastically reduced insect contamination (McSpadden, 1966).

A boxcar is best cleaned by banging on the wooden lining to dislodge any residue which might be behind it. Next, it is swept out and sprayed on the interior with an approved insecticide (Schesser, 1967b). Any cracks or crevices found during the inspection should be repaired before the interior walls are covered with kraft paper (coopered) which is always done when the product to be loaded is a finished food.

It is also possible to use a mixture of methyl bromide:ethylene dibromide (EDB) (70:30) to fumigate behind side and end linings while the car is empty (Schesser, 1967c). Although the technique is usually not completely effective in destroying insects, it is of considerable value in helping to reduce the insect population. There is presently a great deal of concern over the use of EDB and its registration may be cancelled in the near future.

The next step in preparing a boxcar is to seal the door that will not be used for loading. This can best be done by going inside the car and glueing a piece of polyethylene film over the entire opening. By using a paint brush and a liquid adhesive, the polyethylene film can be completely sealed to the door frame and floor. Take care to seal any other cracks or crevices found during the original inspection.

While the floor may be a possible source of leakage, tests have shown little or no advantage to be gained from covering it with polyethylene. Also, the use of such a floor covering can be a hazard, since it may slip or shift under the forklift used for loading (personal communication).

Trailer Vans

These are usually ready to accept product as they are received. Any areas of leakage detected during inspection should be repaired at this time.

Intermodal Containers

These units are similar to trailers and are generally relatively gas tight if they are in good condition.

Unmanned Barges, Dry-Bulk Carriers, and Tankers

The preparation of holds on barges and ocean-going ships requires sealing all openings leading from the space to be fumigated (except the hatches through which the loading of the cargo will be accomplished). Tape and polyethylene are generally used for this operation. Care must be taken to see that any other openings such as smoke detector tubes and stuffing boxes are sealed off or prepared in such a way that gas from the fumigated area will not be able to enter occupied areas of the vessel.

CHOICE OF FUMIGANT

The choice of a fumigant for in-transit fumigation is not complex but is of extreme importance. Before a given fumigant is applied to a conveyance loaded with grain, processed foods, or feeds, it must be ascertained that the label for the fumigant permits such treatment and it is considered to be most effective for that usage, as will be discussed later.

There are basically three types of fumigants available today which may be used for in-transit fumigation. These are (1) aluminum phosphide preparations, (2) liquid fumigant mixtures, and (3) methyl bromide. Each fumigant type has special characteristics which may render it more suitable for a specific use. For ease of understanding, each will be discussed separately.

Aluminum Phosphide Preparations

These can be used effectively to fumigate all kinds of stored products in all types of conveyances (Schesser, 1967a; Quereshi et al, 1967; Freeman, 1968; Leesch et al, 1978). Labeling has been approved for the treatment of railcars in-transit, but trailer vans and intermodal containers must remain static while they are under the gas. An exception to this rule allows containers under fumigation to be put on board a container ship providing special precautions are taken. These will be described in a later section.

Military specifications for the use of aluminum phosphide to treat commodities in railcars were drawn up in 1970 (Anonymous, 1971). These standards define the requirements for fumigation of all railcars containing certain infestible items purchased by the Department of Defense between May 1 and October 31 each year. In 1972, the Armed Forces Pest Control Board recommended aluminum phosphide as the fumigant of choice for the Department of Defense (Anonymous, 1974).

Under present regulations of the United States Federal Grain Inspection Service (FGIS) it is permissible to use aluminum phosphide to fumigate infested grain on certain types of ships after they have been loaded and are ready to sail (Anonymous, 1977). When the fumigant is applied under the supervision of an FGIS Inspector, a clear grade certificate can be issued. This procedure can only be used on dry-bulk carriers, tankers, and ocean-going barges. A recent addition to the regulations (Anonymous, 1982b) now permits the layering of aluminum phosphide tablets or pellets with the cargo. In other words, one-third of the total fumigant dosage for that hold is applied after one-third of the hold has been filled with grain. Then the second one-third of the grain for the hold is loaded and another one-third of the fumigant dosage is then applied. Finally, the last one-third of the grain is loaded and then the last one-third of the fumigant dosage is applied. This method is permitted only on dry-bulk carriers, and tankers. It is also permissible to use a combination method utilizing a liquid 80:20 formulation (carbon tetrachloride:carbon disulphide) and aluminum phosphide on bulk carriers.

For the use of aluminum phosphide on unmanned barges as well as in intermodal containers destined to be placed on board a ship, one must be aware of the specific requirements of the United States Coast Guard (USGC). This Agency has issued two Special Permits; one designated 2-75 for unmanned barges, and one designated 52-75 for intermodal containers.

Special Permit 2-75 has designated specific companies (fumigant suppliers) as parties to the Permit. The Permit modifies portions of 46 Code of Federal Regulations 147A and a copy of it must be in the possession of the fumigator when the treatment is made.

Special Permit 52-75 is issued to specific shippers for the movement of containers by ship while under fumigation. Any shipper requesting it can have his name added to the Permit by writing to the Coast Guard. Although the Permit is written specifically for aluminum phosphide preparations, other types of fumigants could be allowed if the Coast Guard is asked.

The substance of the Permit allows containers treated with aluminum phosphide to be placed on board a ship without aeration providing the container is (a) stored on deck, (b) placed at least 20 feet from vent intakes, crew quarters, or regularly occupied working areas, and (c) properly placarded. The shipper must also notify the recipient of the container as to the date of fumigation and the type and amount of fumigant used. A copy of the Permit must also be carried on the vessel during the voyage.

Some research has been done in which containers under fumigation with aluminum phosphide were stored on a container ship below deck. No hazard was observed during the ocean voyage because of the extensive air movement around the containers in the hold (Childs, 1972). Further study is needed, however, before the Coast Guard will consider changing their regulations to permit the random below deck storage of containers under fumigation.

Liquid Fumigants

There are a number of registered formulations which can be used to fumigate bulk grain as well as processed foods and feeds, in all types of conveyances. When used for the treatment of grain found to be infested at the time of loading into dry-bulk carriers, tankers, and ocean-going barges, the FGIS and USCG regulations must be followed as with the use of aluminum phosphide. Because of the dangers inherent in the use of carbon tetrachloride, proposed changes in the registration of certain liquid fumigant formulations could reduce or eliminate the use of these products.

Methyl Bromide

This fumigant can be effectively used on raw grain and processed foods in railcars, trailers, and containers, as well as on bagged commodities in the holds of ships (Thompson, 1966). This fumigant has not been approved for treatment of grain in-transit by ship under the FGIS Regulations, and when bagged food is fumigated in the holds of ships, the vessel must remain stationary at the dock until it is aerated.

DIRECTIONS FOR USE OF FUMIGANT

As stated in the introduction it is mandatory that for each fumigant all label directions and instructions be followed. There are a number of points needing elaboration regarding each of the fumigants labeled for various in-transit fumigation uses.

Aluminum Phosphide

Railcars

When tablets or pellets are used for the fumigation of bulk grain or bulk feeds in boxcars, the fumigant may be spread on the floor of the car or added as the commodity is loaded. For processed foods the tablets or pellets may be placed on paper trays or put in moisture permeable envelopes. This is necessary since it is not legal to permit the fumigant formulation to come into direct contact with processed foods. Although the fumigant formulation can come in direct contact with animal feeds, when the feed is bagged this method is still preferred. Packaged aluminum phosphide pellets and sachets containing granular material are also available. The packages of fumigant are held in the boxcar by taping them to the wall or onto a piece of cardboard above the load line, or placing them on top of the lading (Schesser, 1977).

When used in hopper cars, the tablets or pellets are mixed with bulk grain or feeds as they are loaded or, in the case of slot-top cars, they can be probed in after loading is completed. When bulk processed foods such as wheat flour are treated in hopper cars, there can be no contact between the fumigant formulation and the lading, so tablets or pellets are placed in envelopes or on trays which are taped to the top side of a cardboard disc taped over the hatch opening. Prepackaged pellets and sachets may also be taped to such cardboard discs.

Trailer Vans

Bulk grain and feeds can be treated by placing tablets or pellets on the floor before loading, or by adding them as loading progresses. Bagged foods can be fumigated by using moisture permeable envelopes, trays, sachets, or prepackaged pellets. It bears repeating that trailers can not be moved on the highways under fumigation.

Intermodal Containers

The fumigation of these conveyances is similar to that of trailers. Remember that containers under fumigation may be put on a container ship without aeration providing USCG specifications relating to such movement are met.

Unmanned Barges

These have to be dealt with separately from ocean-going ships (dry-bulk carriers and tankers) since the law differs with regard to the two types of vessels. Bulk grain and feeds on barges are treated by probing tablets or pellets into the product after loading is completed. Bulk and bagged processed foods are treated by using envelopes, trays, prepackaged pellets, or sachets attached to cardboard so there is no possibility of the fumigant formulation coming into contact with the food.

Dry-Bulk Carriers, Ocean-Going Barges, and Tankers
Bulk grain fumigated in the holds of ships under the Regulations of FGIS and USCG must comply with pertinent requirements. The Regulations permit the layering of tablets and pellets or probing after loading is completed on both dry-bulk carriers and tankers. On tankers, aluminum phosphide may also be used in conjunction with liquid fumigants consisting of 80 percent carbon tetrachloride:20 percent carbon disulphide (Anonymous, 1982a). There is no aluminum phosphide labeling which permits the fumigation of bagged foods or feeds in the holds of ships in-transit.

Liquid Fumigants

Railcars
Both boxcars and hopper cars loaded with bulk grain may be fumigated by pumping certain fumigant mixtures onto the grain surface when loading has been completed. For processed foods and feeds a different fumigant should be chosen.

Trailer Vans
They are treated the same as are railcars.

Intermodal Containers
They may be treated the same as are railcars, but subsequent shipping of the containers may well preclude the use of liquid fumigants.

Unmanned Barges
The fumigant may be pumped onto the grain surface after the loading has been completed. For processed foods and feeds one of the other types of fumigants should be chosen.

Dry-Bulk Carriers, Ocean-Going Barges, and Tankers
The FGIS and USCG Regulations specify how these fumigants are to be used on ships for the treatment of bulk grain found to be infested during loading. Liquid fumigant mixtures are not used for the treatment of bulk or bagged processed foods or feeds.

Methyl Bromide

Railcars
Boxcars and hopper cars of bulk grain, feeds, and processed foods (bagged or in bulk) may be treated by injecting the fumigant into the headspace over the load. When using methyl bromide in cans, the product must be released from outside the car. Cans are not to be punctured and tossed in on top of the load before sealing of the conveyance. Research conducted some 30 years ago (Phillips et al, 1953; Redlinger, 1957) indicated that methyl bromide did not penetrate very well through a car of bulk grain or flour. For boxcars of grain it is possible to use a portable probe system for recirculation, with the gas being placed in the headspace (Anonymous, 1963). Air is drawn from the floor to pull the gas down through the grain bulk. Although this system is effective, it precludes shipping the car until the fumigation is completed. Methyl bromide can be used effectively to fumigate boxcars of bagged processed foods and feeds, but for boxcars containing bulk feeds and for hopper cars of grain or flour, a different fumigant should be chosen.

Trailer Vans
They are treated the same as are railcars for bagged commodities. The van can not be moved until the fumigation is completed.

Intermodal Containers

They are treated the same as are railcars for bagged commodities. Treatment must be completed before the container may be moved.

Unmanned Barges, Dry-Bulk Carriers, and Tankers

Labeling is available permitting the use of this fumigant on barges and ships, but use is usually restricted to bagged commodities, and the vessel must remain static during the fumigation. The crew must also be removed from the ship during the exposure period. Methyl bromide is not used to fumigate dry-bulk carriers and tankers under Aux. 19 Regulations of FGIS (Anonymous, 1977).

POST APPLICATION PROCEDURES

Sealing

Railcars

On boxcars, the door through which the car was loaded is sealed with masking tape. On hopper cars, hatches are usually covered with a polyethylene film (over the cardboard disc when aluminum phosphide has been used), then the covers are closed and sealed with masking tape.

Trailer Vans

Doors are sealed with masking tape.

Intermodal Containers

Doors are sealed with masking tape.

Unmanned Barges, Dry-Bulk Carriers, and Tankers

The hatch covers on vessels (when properly closed and dogged) are designed to preclude the entrance of water. It is often necessary, however, to seal certain vents and some hatches with polyethylene and masking tape, as the canvas covers used to prevent the entry of water are not gas tight.

Placarding

The proper use of placards is a requirement of law and is certainly very necessary if in-transit fumigation is to be done safely. Make certain the placards are dated as to time of treatment, list the name of the chemical used, and contain the name and phone number of the fumigator or a person to contact in case of an emergency.

Railcars

Special requirements for certain railcar shipments are to be found in 49 Code of Federal Regulations Sections 173.426(a) and 173.426(b) (Anonymous, 1981). Section 173.426(a) specifies that railcars containing product fumigated with a flammable liquid or flammable gas must remain static for at least 48 hours after treatment, or until they have been ventilated so as to remove the danger of fire or explosion due to the presence of flammable vapors. Section 173.426(b) says that any railcar fumigated with a poisonous liquid, solid, or gas must be placarded on each door or near thereto with placards as described by this section. In the case of fumigated hopper cars, the placards must be placed on each hatch cover and on the wall of the car adjacent to each ladder.

Trailer Vans

Placarding must comply with the above referenced sections of 49 Code of Federal Regulations.

Intermodal Containers

They must also comply with the above referenced sections of 49 Code of

Federal Regulations.

It is important to remember that trailers and intermodal containers may not be moved over the highway under gas.

Unmanned Barges, Ocean-Going Barges, Dry-Bulk Carriers, and Tankers

These conveyances must comply with the placarding requirements set out in 46 Code of Federal Regulations 147A and USCG Special Permit 2-75.

Shipping

Once the fumigant application has been completed and placards have been put in place as required, the conveyance may be turned over to the appropriate person for shipping. It is the responsibility of the person or persons shipping the goods to notify the recipient that the conveyance has been treated, what fumigant was used and in what quantity, and that the shipment will arrive under fumigation. The shipper should also provide information on any special precautions which should be exercised and the best procedures to follow for aeration of the cargo.

RECEIPT OF CONVEYANCE UNDER GAS

Aeration

Railcars

Depending on transit time and the tightness of the car, there could be a relatively high concentration of gas present when the car is opened regardless of the fumigant used (Schesser, 1967c). Before the car is entered it should be allowed to aerate in a place where the escaping gas can diffuse harmlessly into the atmosphere. Should the aeration take place in a large warehouse complex, extra caution must be taken to see that the workplace remains free of toxic concentrations of fumigant. Should it be necessary to enter a car to unload it, the gas concentration must be determined using the appropriate detection device to be sure the concentration is at or below the MAC (maximum acceptable concentration) value for the fumigant (Anonymous, 1981).

When pressure differential type hopper cars have been treated with aluminum phosphide they may be unloaded directly upon receipt. Test work by the author has shown that by the time the bulk flour from such a car is carried into the storage bins of the receiving location, there is no gas remaining.

When cars of grain have been treated with liquid fumigants, extreme caution must be exercised if the car must be entered for sampling or unloading. Boxcars of grain so treated can take days or even weeks, in inclement weather, to reach a point where they can be entered safely. Hopper cars do not need to be entered for unloading, but care must be taken if they are sampled, and also when treated grain is taken into the elevator. Respiratory protection is recommended during sampling procedures.

Railcars of bagged commodities treated with methyl bromide can also contain toxic concentrations of the gas at the time they are unloaded. To prevent an accident, the concentration of gas should be determined first using a suitable detector. Remember that a halide leak detector will not register a flame color change below 60 ppm, and at such a concentration the gas would still be hazardous to health (Anonymous, 1968). This fumigant being heavier than air can require extended periods of time to dissipate, depending on atmospheric conditions.

The Environmental Protection Agency (EPA) has ruled that for railcars, although restricted use pesticides such as aluminum phosphide and methyl bromide must be used by or under the direct supervision of a certified applicator, a certified applicator need not be on hand to open a railcar that has been treated with one of these fumigants (personal communication from EPA).

Trailer Vans

These must be aerated before transit. Aeration would proceed as for similar commodities in railcars.

Intermodal Containers

The same can be said for these conveyances as for trailers in the sense that they are aerated before movement unless going on a container ship under the Coast Guard Special Permit 52-75. Aeration would proceed as for railcars.

Unmanned Barges, Ocean-Going Barges, Dry-Bulk Carriers, and Tankers

Unmanned barges are almost always unloaded mechanically, so that no one need enter the barge to work on the grain surface. Under these circumstances, no undue hazard should exist for personnel when aluminum phosphide has been used. Care must be exercised in the elevator when grain which has been treated with liquid fumigant mixtures is handled, since vapors present in the grain can escape into the air as the grain moves over the conveyor belt.

The hatches of ocean-going vessels are opened when the ship docks, and the headspace over the grain is allowed to aerate. It has been demonstrated that when aluminum phosphide was used, workmen could then enter the holds and stand on the grain surface to work the unloaders (Redlinger et al, 1979). In cases where liquid fumigants had been used, there was some gas given off from the grain surface as the cargo was carried into the elevator, and men had to be cautious both on the surface of the grain and in the elevator (Leesch et al, 1978).

REMOVAL OF FUMIGATION MATERIAL AND WARNING SIGNS

The physical removal of any remaining fumigation material is necessary only when the cargo has been treated with aluminum phosphide in envelopes, on trays, or as prepackaged pellets, or sachets. In these instances the residual dust is either removed from its container and stirred into a drum of water to which a small amount of detergent has been added or, in cases where prepackaged pellets have been used, the packages are placed in the water mixture and weighted down to keep them fully submerged. Following complete deactivation of the trace amounts of aluminum phosphide remaining in the dust, the slurry or prepackaged pellets are disposed of according to label directions.

All warning placards must be removed and destroyed before the fumigation is considered complete.

SAFETY

It is required by law that personal protective equipment be provided for persons handling fumigants and/or fumigated commodities, and that they be trained in their proper use. Code of Federal Regulations 29, Section 1918.102 defines what is required to comply with the law (Anonymous, 1981). Section 1918.106 of 29 Code of Federal Regulations defines requirements for the protection from drowning for persons working or walking on the decks of barges on the Mississippi River System and the Gulf Intercoastal Waterways.

Since the wording of various sections of all Codes of Federal Regulations changes from time to time, copies of present wording are not attached. They are referenced so that the reader will know what part of such a document to look at to find current wording.

Insecticides used to fumigate stored grain, processed foods, and animal feeds are of necessity, extremely toxic. For this reason every precaution must be exercised in their use. From experience it is known that on occasion, a railcar, trailer, or container has been shipped without benefit of placards, even though it was fumigated. Should someone have unknowingly gone inside to sample the commodity or start to unload it, poisoning could have resulted.

There have also been reported incidents where one of the fumigants has been used to treat grain being trucked to a receiving elevator. The trucker mistakenly believed the fumigant would kill the insects in several hours but not be dangerous either to himself or to personnel at the elevator. This was not the case, and it presented a serious hazard that could easily have been avoided. Because unloading personnel at an elevator never really know whether grain has been fumigated just prior to receipt, a gas reading should always be taken with gas detection equipment if there is any unusual odor present on the grain or there is any question as to whether the grain has been fumigated.

SUMMARY

Insect infestation is a problem in agriculture at each step in the chain from producer to consumer. The food manufacturer who sells a finished product to the consuming public has the added responsibility to see that his unadulterated finished product gets to the consumer in a similar condition. It is very important to the consumer that there be no insect infestation present (either live or dead) in the food he buys, hence the reputation of the manufacturer goes "on the line" each time one of their products is shipped. It is for this reason that the in-transit fumigation of foods and feeds is a very critical step in the marketing chain.

Great progress has been made in boxcar sanitation over the years, but due to the fact that most conveyances carry both raw and finished products indiscriminantly, the need for an effective in-transit fumigation program will remain of the utmost importance.

LITERATURE CITED

Anonymous, 1963. A portable recirculation system for fumigating bulk products in freight cars. U.S.D.A. Bull. 24, 4 pp.

Anonymous, 1968. Methyl bromide torch, refrigerator leak detector, or halide torch. Cal. Safety News, Sept., 1 pp.

Anonymous, 1971. Military standard in-transit fumigation of freight cars. Mil Std-1486B, Nov. 24, FSC 6840, 12 pp.

Anonymous, 1974. The Armed Forces Pest Control Board Technical Information Manual No. 11, 40 pp.

Anonymous, 1977. Shiphold fumigation. U.S.D.A. FGIS, Grain Inspection Manual, GR Instruction 918-6, Aux. 19, Rev. 2, Sept. 12, 9 pp.

Anonymous, 1979. TLV's. American Conference of Governmental and Industrial Hygienists, Cincinnati, Ohio, 94 pp.

Anonymous, 1981. 29 Code of Federal Regulations. Superintendent of Documents, Washington, D.C.

Anonymous, 1981. 46 Code of Federal Regulations. Superintendent of Documents, Washington, D.C.

Anonymous, 1981. 49 Code of Federal Regulations. Superintendent of Documents, Washington, D.C.

Anonymous, 1982a. In-transit fumigation of grain loaded aboard tanker type vessels. U.S.D.A., FGIS, FGIS Instruction 919-1, April 21, 4 pp.

Anonymous, 1982b. Interim policy-fumigation of insect infested grain aboard bulk carriers and tankers. U.S.D.A., FGIS, Supplement to FGIS letter dated June 23, 6 pp.

Childs, D.P. 1972. Tobacco consignments fumigated in containers while in-transit. Container News, October, 2 pp.

Cotton, R.T. 1960. Pests in stored grain and grain products. Burgess Publishing Co.

Cotton, R.T. 1962. How to prevent in-transit infestations. Northwestern Miller, Vol. 267(12):32-34.

Freeman, J.A. 1958. The control of infestation in stored products moving in international trade. Proc. 10th Int. Congress Ent., Vol. 4:5-16.

Freeman, J.A. 1965. On the infestation of rice and rice products imported into Britain. Proc. 12th Int. Congress Ent., London, pp. 632-634.

Freeman, J.A. 1968. Problems in the carriage of infested commodities in freight containers. Proc. Int. Container Symp., 2 pp.

Freeman, J.A. 1971. Problems of transport. Prevention and control of infestation in conventional cargo ships. Third Brit. Pest Control Conf., Paper No. 18, 9 pp.

Hurlock, E.T. 1964. Infestation of foodstuffs from the United States of America inspected in the United Kingdom between 1953 and 1961. Bull. Ent. Res., Vol. 55(1):173-192.

Leesch, J.G. et al. 1978. An in-transit shipboard fumigation of corn. Jour. Econ. Ent., Vol. 71:(6):928-935.

Liscombe, E.A.R. 1966. Insect and rodent contamination of boxcar origin. Northwestern Miller, March, pp. 32, 33, 34.

McSpadden, P. 1966. Infestation and rodent contamination of boxcar origin. Grain and Cereal Products Sanitation Conference, K.S.U., pp. 115-119.

Phillips, G. L., and R. Latta. 1953. The value of forced circulation in the fumigation of bulk commodities in freight cars. Down to Earth, Vol. 8(4):13-15.

Quereshi, A. H., and J. Riley. 1967. Studies on the fumigation of bagged groundnuts with PHOSTOXIN® in railway waggons. Rep. Nigerian Stored Product Research Inst. Tech. Rep. No. 10, pp. 99-103.

Redlinger, L.M. 1957. Studies in the fumigation of bulk rice in freight cars. Rice Jour., Vol. 60(3):10, 12, 14.

Redlinger, L.M. et al. 1979. In-transit shipboard fumigation of wheat. Jour. Econ. Ent., Vol. 72(4):642-647.

Schesser, J.H. 1967a. A comparison of two fumigant mixtures for disinfesting empty railway freight cars. Northwestern Miller, July, 1 pp.

Schesser, J.H. 1967b. A dichlorvos-malathion mixture for insect control in empty railcars. American Miller and Processor, Sept., 4 pp.

Schesser, J.H. 1967c. Official report/phosphine fumigation of processed cereal products in railcars. American Miller and Processor, Jan., 6 pp.

Schesser, J.H. 1977. Fumigation of cereal grains and processed products in transport vehicles with phosphine from Detia Ex-B®. Jour. Econ. Ent., Vol. 70(2):199-201.

Smith, M.J. 1982. Comments on transportation as related to the milling industry. Assoc. of Operative Millers Bulletin, Sept., pp. 3962-3964.

Thompson, R.H. 1966. A review of the properties and usage of methyl bromide as a fumigant. Jour. Stored Prod. Res., Vol. 1:353-376.

HEAT STERILIZATION (SUPERHEATING)
AS A CONTROL FOR STORED-GRAIN PESTS IN A FOOD PLANT

Kenneth O. Sheppard

Editor's Comments

The use of heat is a viable alternative to fumigation
of plants for insect control. A key factor favoring
selection of this approach is that sealing of the building
is less critical for successful application. Safety, to
plants and personnel, and lower costs are the other main
advantages.

This means of control has been around since about 1915
but has been little used by industry. That is now changing
since it is finally getting the deserved publicity. The use
of heat also merits consideration in the pre-planning or
design of new plants or facilities so that appropriate steps
are taken to maximize its effectiveness while minimizing
risks to equipment such as packaging.

F. J. Baur

HEAT STERILIZATION (SUPERHEATING) AS A CONTROL
FOR STORED-GRAIN PESTS IN A FOOD PLANT

Kenneth O. Sheppard

The Quaker Oats Company
2811 South Eleventh Street
St. Joseph, Missouri 64502

ABSTRACT

Many large mills and food plants are periodically fumigated with toxic gases to get a general kill of insect life. The involved hazards are great as the toxic gases are quite lethal, and the procedure is effective only at much expense in materials, time, and preparation.

From the standpoint of safety, effectiveness, and utility, heat sterilization (the control of insects by superheating) has become a time-proven method. The effect is that insects are killed by raising temperatures to 130°F (54°C) using unit heaters having good circulation capability that will maintain the required temperature for about 30 hours. Since insects do not have the ability to cool themselves but come to the temperature of their surroundings, the time/temperature guide will produce the desired results in facilities that lend themselves to this procedure.

Heat sterilization has long been recognized by The Quaker Oats Company as one of its principle measures in the plant's overall master plan for sanitation. This technique is extremely effective when properly balanced with sanitation/housekeeping, insecticides, and competent inspections. Alone, it is not sufficient to offer a modern food plant the needed protection in light of today's regulatory activity, not to mention properly serving a better-informed, quality-conscious consumer.

Although much is to be said for heat sterilization, realistically, it is not a panacea for insect control as it is not without some after effects. These can, however, be minimized with proper guidance, forethought, planning, and practical experience.

INTRODUCTION

The purpose of this chapter is to provide a straightforward and honest approach to one method of insect control in a food manufacturing facility. As the title implies, the temperature of a room or enclosed area is raised to and maintained at a level that becomes too hot to sustain insect life that is exposed or may be hiding in cracks or crevices. With the appropriate tools and preparation, a facility that has the capability and lends itself to heat sterilization can attain an effective level of insect control. From the onset of this discussion, it is necessary to understand that heat sterilization is not a panacea for insect control but must be worked into a mill or food plant's master plan of sanitation control along with housekeeping, competent inspections, and insecticides. If these are not properly balanced, the program will perform much the same way an automobile

in need of a tune-up. For those already practicing heat sterilization,
little of what is written here will be new. Hopefully, this information will
offer insight to or enlighten any troublesome area that would inhibit a
strong program. For those who would seriously consider implementing a heat
sterilization program, it is important that all alternatives be weighed,
processes thought through, and advantages and disadvantages considered and
discussed before any final decisions are made. The sections on methods,
materials, and preparation for heat sterilization will serve as a guide to
planning and preparation. Then realization of safety factors, practicality,
and economics of using superheated air versus a toxic gas will be realized.
There is much that needs to be said at this point concerning the role of
every individual having responsibility for sanitation or quality assurance in
a mill or food plant. The success or failure of a heat sterilization will
depend not only on equipment and people functioning properly, but management
that is educated to and understands the value, necessity, and overall
rational of this procedure. Support must come from the top and follow
through to front line supervision. Without this support, something less than
mediocrity is all that can be expected. Much time, effort, and money will
have been wasted. An introduction to the topic of heat sterilization would
be incomplete without brief mention that there are a few inherent
disadvantages with regard to long term effects of high temperature on
equipment and various parts of machinery. These seem to concentrate mainly
in packaging departments of food plants where high speed, sophisticated
electrical equipment can be effected. Details of this aspect are covered
later with suggestions on alleviating or lessening the affects which, in
time, become an almost established part of the heat sterilization routine.
If properly planned for, any drawbacks mentioned in this work can be prepared
for in advance, making Monday morning start-ups less of a problem.

REVIEW OF PREVIOUS WORK

The use of heat in one form or another has long been recognized as one
of the most effective methods of destroying insect life. The heat of the sun
was utilized in early times by primitive people for killing insects in grain,
and it is probable that these observations of the effectiveness of solar heat
led to the later use of artificial methods of sterilizing grain and cereal
products (Cotton, 1950).

Early entomology in East Asia records that the government of the Shang
Kingdom in China appointed anti-locust officers who used bonfires to dispose
of caught insect pests in 1200 B.C. (Masayasu and Yosiaki, 1973).

According to Goodwin (1912), heat was used for the general sterilization
of a flour mill in this country as early as 1901 by a miller who fitted up
his mill for heating after he had made the observation that flour-infesting
insects were killed by the heat in the vicinity of steam pipes leading to a
corn drier. However, it remained for Dean (1911) to investigate the
effectiveness of heat in flour mills on a scientific basis and to call
attention to its practicality. As a result of investigations conducted over
the period 1910 to 1913, he developed a method that was so successful that it
was adopted by mills in many parts of the country. This method, Dean states,
was thoroughly tested by workers in the Bureau of Entomology of the United
States Department of Agriculture and by various state entomologists and found
to be highly effective, inexpensive, and free from hazard to the workmen.
The latter feature appealed to many millers who hesitated to use fumigants
that were dangerous to life (Cotton, 1950).

On reviewing the available literature, Dean (1911) wrote that the
French long ago knew the value of heat and devised contrivances, called

insect mills, for the heating of infested grain. Experiments made by
Professor F. M. Webster in 1883 to ascertain the amount of heat required to
destroy the Angoumois grain moth gave these results: a temperature of 140°F
continued for nine hours literally cooks the larvae or pupae; a temperature
of 130°F for five hours is fatal, as is also 120°F for four hours; while
110°F applied for six hours was only partially effective. It was also found
in the Webster experiments that wheat could be subjected to a temperature of
150°F for eight hours without impairing its germinating properties (Dean,
1911). Dean stated nearly all the experiments of this nature were made
relative to the discovery of a method to destroy Angoumois grain moth, and
from the results of these experiments, some of the experimenters and other
writers have assumed that many of the grain insects could probably be
destroyed in the same manner but that it would require a higher temperature
to destroy the adults than the larvae or pupae (Dean, 1911).

     It was written that superheating was superior to either fumigation or
freezing, two of the most popular methods for insect control in Montana flour
mills. This method, by which the insects are killed by raising the
temperature to about 120° to 130°F, has been known and used in other parts of
the country for some twenty years, but the recent improvements in units
heaters now give it several very important advantages....the new method of
heating mills by means of unit heaters which require no elaborate piping
system and which cause the heat to be evenly distributed overcomes many of
the former disadvantages of heat sterilization as an insect control measure
(Pepper & Strand, 1935). Pepper & Strand (1935) wrote that Goodwin, in 1922,
determined the relative susceptibility of different stages of insects to heat
and found that even though differences as great as 8° to 10°F were required
to effect the destruction of different species, most of the common insect
pests succumb readily to temperatures of 120° to 130°F with practically no
injury to cereal products (Pepper & Strand, 1935). Pepper & Strand further
stressed, by keeping the air in circulation, it is possible without using
excessive heat to raise the floor surface temperature high enough to kill all
species of flour-infesting insects. The most significant results dealing
with floor surface temperature show that at a quarter of an inch above the
surface a temperature of 120°F was reached in four hours. By laying a
mercury thermometer on the floor, the temperature recorded will be that at
approximately one-quarter inch above the surface. But the actual surface was
not heated to 120°F until the eighth hour. The confused flour beetle, the
most important mill insect in most localities and the one most likely to
escape a heat treatment by migrating to the coolest place, is scarcely more
than one-fiftieth of an inch off the surface. These insects will be at a
temperature from 5-10 degrees lower than the temperature indicated by a
mercury thermometer which was a half inch above the floor. This
stratification of the heated air just next to the floor makes it necessary to
continue the heating several hours after thermometers on the floor indicate
that killing temperatures have been reached (Pepper & Strand, 1935). They
also pointed out, for employees going through a mill during a heat
sterilization, that it was much more convenient to read thermometers which
are hanging at about shoulder height....records so obtained are, at most, of
only relative value....due to a temperature gradient in the first half inch
above the floor. From this it may be concluded that temperatures taken at
shoulder height are of very doubtful value so far as indicating the places
where insects could escape the effects of high temperatures (Pepper & Strand,
1935).

METHODS AND MATERIALS FOR HEAT STERILIZATION (SUPERHEATING)

This section deals mainly in the areas of reasoning for and planning of the various tools, equipment, and aspects of superheating a modern food plant. Long range planning or master planning for incorporating heat sterilizations into a mill or food plant's manufacturing schedule is as critical as performing an actual sterilization itself. Some past practices and previous mistakes made regarding scheduling were to plan these exercises to fit into the convenience of a holiday routine. While that type of thinking was giving a type of insect control, too much time was elapsing between sterilizations allowing succeeding life cycles of insects to occur. It was becoming obvious that increasing the number of sterilizations to coincide with insect life cycles was necessary. Based upon the total needs of our sanitation program, a representative scheduling is presented here as an example of how to perform six heat sterilizations per calendar year.   Two of the exercises would fall in the colder months and four in the warmer months. Actual examples of scheduling dates are as follows:

1. January   22, 23, and 24
2. April   8, 9, and 10
3. May   27, 28, and 29
4. July   1, 2, and 3
5. August   27, 28, and 29
6. October   15, 16, and 17

Factors and variables affecting any type of scheduling are numerous, therefore, flexibility can and should be exercised when needed to arrange for heat sterilization scheduling to be moved up or back a week.

The choice of energy sources used to produce energy in the form of heat is determined by the local situation. Since many facilities use boilers to produce steam, that is the logical choice. The steam, piped to each floor throughout a food plant, offers a ready source of heat to the places where insect control is desired. At each floor level, heating and ventilating units (usually two to a floor depending on size) are set up and connected. These can be floor mounted or suspended to meet any space and application requirement.

The various species of cereal-infesting insects differ very little in their susceptibilities to high temperatures. Exposures to temperatures ranging from 120° to 145°F for 24-30 hours will generally destroy all stages of insects which commonly infest mills. The success of using heat to control mill insects depends on a number of factors. Large temperature gradients occur at points where convection currents are strong or where there is inadequate air circulation at ceilings and floors. If not properly opened up, pockets of static air will develop in machinery and machinery housings. Double wall spaces, crevices in walls and floors, deposits in bins, behind ledges and beams, and cracks or openings around doors and windows present still more locations that need to be checked. Obviously, the conditions necessary for successful heat sterilization vary from plant to plant.

To further enhance the heat sterilization, other appropriate preparations will be necessary. Equipment should be allowed to run empty, elevator boots opened and dragged out, all sweepings and sacks of product, portable tanks of product, and so forth removed. It is especially not advisable to leave sacks against an outer, unheated wall. Insects seeking to escape the desiccating effects of the heat can use them to find protection. Dead stock should be removed from conveyors under reels, etc., so that the insects will not be able to find any place where they can escape the heat and remain insulated from it. The temperatures prescribed earlier will kill all forms of insect life, but it must be remembered that the escape of only a few eggs or a few insects will provide a starting point for reinfestation.

Ahead of heating, an approved residual spray should be applied at floor-wall junctures, across doorway thresholds to unheated buildings, and where the insects have been found by observation to go when the heat is turned on. The floor-wall junctures of an unheated space next to heated areas should be treated also.

Sealing of the facility is one aspect of traditional fumigation type insect control that is, in most cases, intensive and very costly. The fact is that exhaust fans at open windows, openings around manlifts, or areas of walls where conveyors pass through from a heated area to one that is not heated are about the only instances where sealing is necessary. This can be accomplished with polysheeting and bondmaster glue or duct tape. It is not critical to get an airtight seal, just one to contain the heat and not allow very much heat loss. The question of what kind of sealing, how much, and its extent becomes not as critical as a regular fumigation. This can be done at the end of the preparation prior to turning the heat on.

Pre-sterilization cleanup consists of three areas.

Overheads should be cleaned first, removing any or all dust or product accumulations from pipes, spouting, and light fixtures.

Equipment will then be cleaned off and emptied out, i.e., conveyors, sifter boxes opened and emptied, covers and access doors opened on all equipment.

Floors - all materials and debris on floors will need to be swept or dry mopped. Use of compressed air is not advised as it rescatters dust and other accumulations back up on equipment and overheads previously cleaned. It is extremely important that no accumulation be left that can give insects harborage or insulation from the heat.

Second in importance to cleanup, prior to heating, is the removal of all heat-sensitive items, i.e., equipment, tools, ingredients, paper supplies, and fire extinguishers. Many of these items can be left in unheated areas with no problem. Many of them should be taken to fumigation vaults or chambers and gassed as a further added measure to prevent reinfestation from occurring once heating is completed. Assuming that all preparations prior to sterilization are satisfactory, then the turning up of the heat and maintenance of the prescribed temperature of 130°F for about 30 hours can be carried out. In order to attain killing temperatures at the floor surface level and inside certain types of machinery, it is necessary to maintain eye level temperatures at 130° to 145°F for 24 to 30 hours. Provided there is adequate circulation of air to decrease temperature gradients between ceiling and floor, heat stratifying in an area will not result. If steam jets are used to introduce moisture at the outlets of the unit heaters, the jets should be turned off 2 to 3 hours before the heat is turned off. This will avoid condensation of moisture on metal parts and on the outer wall surfaces as the building cools.

Because of large differences in building design and construction and in types of machinery, it is important that the adequacy as well as the efficiency of a particular sterilization exercise be established by actual experiment. This can be done with appropriately placed insect test cages containing live insects, thermometers at various locations, and chart recording temperature thermometers for tracking on temperature behavior during the course of the prescribed treatment.

Once the heat has been turned on and is up to the prescribed temperature, temperatures are taken on each floor and recorded every 2 hours around the clock for the next 30 hours. During the "rounds" on which temperatures are recorded, any adjustments of unit heaters up or down can be made. Leaks which occur can be repaired. Steam traps and header valves are checked to see that they are clean, do not leak, and are staying operational. Sprinkler heads should be rated at a temperature that will prevent discharge yet allow for variances up to 155°F. Care should be taken with any wooden milling machinery to make sure that these areas keep a required level of humidity, thus preventing damage to equipment such as sifters. Fire extinguishers as well as all aerosol containers must be removed from any heated area for the obvious safety hazard involved. Movement in and out of any heated area requires no special safety equipment nor is time lost in resealing doors that opened or shut repeatedly. They must fit well, but a seal similar to one used to contain toxic gases is not required.

At the conclusion of the exercise, the heat is promptly turned off; doors and screened windows are then opened. Ventilating fans are left on to continue air movement aiding in the cooling of various rooms. Educating plant employees and management as to what to expect will be discussed in further detail under the heading of Effects of Superheating on Facilities, Equipment, Products.

EFFECTS OF SUPERHEATING ON FACILITIES, EQUIPMENT, PRODUCTS

Through years of experience and experimenting, it has been learned what equipment is adversely affected and what is not during the last heat sterilization. Also what follow-up is needed as to where insects are found and how the outside temperature and relative humidity affect the performance in buildings of considerable size during the period of heat sterilization. This follow-up is important so that problems and the same mistakes will not be made in the future, possibly hindering the heat's ability to do the required job. The effects of high temperature for long duration have shown no real damaging effects on equipment used specifically for milling and mixing operations. The same cannot be said of sophisticated, high speed, electronic packaging equipment that has been introduced into the working environment. In an effort to keep insect pests from migrating from the heated mill areas into the cooler packaging areas, these areas are also heated. This can be done safely if necessary precautions are taken by physically removing all heat-sensitive electrical components or by protecting them from the heat by building insulation around them during their installation. Another method that can be used is to request firms producing electrical components to engineer in certain heat resistant capabilities which would insure the life of whatever piece of equipment was exposed to high temperatures of long duration. Where removal is practically not possible, insulated rooms can be constructed, complete with air conditioning depending on exactly what is necessary. No matter what steps are taken, there will still be some problems with "Monday morning" start-ups. It will be easy to put the total blame for poor packaging line performance on heat sterilization, for instance. But after careful checking, those in front line supervision did not make sure all glue was cleared from lines going to sealers. These situations can be not only frustrating but tend to discourage Management's attitude toward this particular tool of insect control.

It is important that all involved be educated as to what to expect before, during, and after a heat sterilization. Other items that should be noted prior to planning a heat sterilization are lubricant loss from machinery bearings; manlift belts may need retightening; plastics, rubber, and other synthetic materials should be checked for vulnerability. Some floor coatings such as epoxies might develop cracks after repeated heatings. Expansion joints, i.e., floor, floor to wall, and overhead should be observed for any changes. Certain caulks and paints tend to deteriorate at an accelerated rate. All weigh belts should be loosened. Scales should be recalibrated.

When weighing all the above items against the cost and safety hazards involved with the use of toxic fumigants, the final analysis may not swing in favor of heat sterilization. If not, those are decisions best handled by competent sanitarians, engineers, and operations people. If the decision is in favor of heat sterilization as an alternate pest control method, approach the exercise with an open mind and attitude. It has to be remembered that heat sterilization is not a panacea for insect control. Rather it is a tool to be utilized along with housekeeping, sanitation, and the use of insecticides. By themselves, no one aspect can effect a safe and thorough program of pest control. But, with the right blend and management commitment, a proper balance can be achieved.

SUMMARY

The summation of this work would best be stated by giving the simple definition of heat sterilization, that is, superheating a room or specific area for a prescribed time at a specific temperature. It offers many positive aspects, yet it is not without some negative features.

When one considers that an extra lifetime could be spent trying to insect-proof food plants, especially those of considerable size and age, every available tool and resource should be considered. The ways insects gain entry to mills and food plants are innumerable. Either there are problems with the building construction or there are inaccessible void spaces within product systems allowing for insect harborage and breeding.

Heat sterilization will kill insects, but better results will be obtained by using this tool as a preventive measure rather than as a reactive measure. The most simple approach to preventive insect control has to start with upper management commitment to sanitation and GMP's. This has to be effected through plant employees from plant management to the sweeper with the lowest seniority. The key here is with the front-line supervisors.

The success or failure of a heat sterilization will not always be mechanical. What could be labeled as a pre-determined failure would be the lack of conscious effort and perseverance on the part of some supervisory person to keep equipment operating and in good repair.

The survival ability of mill insects or stored-product pests is at times baffling. The education responsibility of the sanitarian to the employees is necessary to avoid the continuation of acts or procedures which can allow future occurrence of insect activity. All then claim a share of the responsibility in making heat sterilization work.

LITERATURE CITED

Cotton, Richart T. 1950. Pests of Stored Grain and Grain Products, Book 1.
Dean, George A. 1911. Kansas State Agricultural College, Bulletin No. 189.
    Milled and Stored Grain Insects, Book No. 5.
Masayasu, Konishi, and Ito Hosiaki. 1973. History of Entomology. Smith,
    Mittler and Smith.
Pepper, J. H., and A. L. Strand. 1935. Superheating as a Control for
    Cereal-Mill Insects, Bulletin No. 297. Montana State College
    Agricultural Experiment Station. Boseman, Montana.
Goodwin, W. H. 1922. Ohio Agricultural Experiment Station, Bulletin 354.
    Heat for Control of Cereal Insects.

AN OVERVIEW OF POST HARVEST INSECT RESEARCH
PERFORMED BY USDA, ARS LABORATORIES

Milton T. Ouye

Editor's Comments

This chapter gives a status report on one of the top,
if not the top, efforts in the world on the control of
stored-product insects, and is the first general reporting.
Past, present, and future research is covered.
Introductions to each section give good brief summaries.

No program mention is made of environmental (household)
species. Problems do arise in industry with cockroaches,
ants, psocids, and the like. Current research within USDA
is diminished but expertise remains. Contact Dr. Richard
Patterson, a research leader, in the "Insects Affecting Man
and Animals Research Laboratory" in Gainesville, FL. The
address is 1600 SW 23rd Drive, P. O. Box 14565 (32604). Dr.
Patterson's phone number is 904-374-5910. Purdue University
has also been very active in research on environmental-type
species. Here the contact is Dr. Gary Bennett, Department
of Entomology, Purdue University, West Lafayette, Indiana
47907, phone number 317-494-4565.

Dr. Ouye did not cover that portion of USDA, ARS's
program dealing with perishable commodities such as fruits
and vegetables. Location and supervision of these
laboratories corresponding to those listed in the chapter
which work on grains, nuts, etc. are shown below.

FRUIT AND VEGETABLE LABS IN ARS, USDA

Subtropical Horticulture
  Research Lab.
USDA, ARS
13601 Old Cutler Road
Miami, FL 33158
Contact: Dr. Donald H. Spalding
Phone: (305) 238-9321

Insect Ecology & Behavior
  Research Lab.
USDA, ARS
3706 W. Nob Hill Blvd.
Yakima, WA 98902
Contact: Dr. Harold
  Moffitt
Phone: (509) 575-5970

U.S. Horticultural
  Research Lab.
USDA, ARS
2120 Camden Road
Orlando, FL 32803
Contact: Dr. Roy McDonald
Phone: (305) 898-6791

Fruit Protection &
  Production Research Lab.
USDA, ARS
509 W. 4th Street
P. O. Box 267
Weslaco, TX 78596
Contact: Dr. William Hart
Phone: (512) 968-3159

Tropical Fruit &
  Vegetable Research Lab.
USDA, ARS
2727 Woodlawn Drive
P. O. Box 2280
Honolulu, HI 96804
Contact: Mr. James E.
  Gilmore, Jr.
Phone: (808) 988-2158

Tropical Fruit &
  Vegetable Research Lab.
USDA, ARS
Stainback Highway
P. O. Box 4459
Hilo, HI 96720
Contact: Dr. H. M. Couey
Phone: (808) 959-9138

Termites can, unfortunately, require special attention. The contact on termites is Dr. Joseph Mauldin, Project Leader, USDA Forest Services Laboratory, Forestry Sciences, P. O. Box 2009, Gulfport, MS 39503, phone number 601-864-3991.

There is no mention in this book on the detection of living insects in food materials, or hidden infestations. One of the more important methods for this need is the use of an infrared detector for evolved carbon dioxide. This technique is faster and more accurate than x-ray. For details on these as well as other procedures, contact Dr. William Bruce at the Savannah laboratory listed in this chapter. I understand from Dr. Bruce that the Eastern and Western Laboratories of the USDA are now active in this area.

Any of the individuals listed above or in the chapter may be consulted for assistance. In the past I have found the members of these organizations to be true professionals, knowledgeable and helpful--true public servants.

A brief word about methyl bromide is in order in view of a recent report on possible carcinogenicity. As of August 1984, no risk assessment is underway in the program portion of EPA. A Registration Standard is scheduled for completion in December 1985. This timing is currently judged as more likely to be delayed than made earlier.

Lastly I want to draw your attention to an upcoming study at all seven laboratories mentioned by Dr. Ouye. It is the study of IPM (Integrated Pest Management). The information gained from this study is bound to be worthwhile since the proper selection and application of the tools to use in each insect control situation is basic to the desired success.

F. J. Baur

AN OVERVIEW OF POST HARVEST INSECT RESEARCH PERFORMED BY
USDA, ARS LABORATORIES

Milton T. Ouye

U.S. Department of Agriculture
Agricultural Research Service
Beltsville, Maryland 20705

INTRODUCTION

Agriculture is the Nation's largest industry, with assets totaling in excess of $1 trillion. Consumers in 1980 spent $260 billion for U.S. farm-produced foods and about $179 billion of this was to move 400 million tons of food from the farm to the consumer's table. The food was assembled, inspected, graded, stored, processed, packaged, wholesaled, and retailed. Most of this food during the marketing process is subject to damage by insects and the loss is estimated to be about 10% (Harein, 1982) or 40 million tons of lost food each year. The marketing system includes agricultural products consigned for export. Farm exports account for about one-fifth of total export earnings and for about one-fifth of U.S. farm income. The fiscal 1983 agricultural estimate was $34.8 billion, and imports were $16.4 billion. This gave the U.S. a net agricultural trade surplus of $18.4 billion. Today, one of every six farmworkers depends on exports for a job. The U.S. exports more than three-fifths of its wheat production, one-half of its soybeans and rice, and more than one-third of its corn and cotton. In fiscal 1982, the U.S. provided nearly 90 percent of the world's soybeans exports; about 55 percent of coarse grain exports; more than 45 percent of wheat exports; and about 22 percent of rice exports. All these products are subject to insect attacks.

The Agricultural Research Service (ARS), USDA, is the leading U.S. Federal agency responsible for research that will lead to new or improved pest control technologies to maintain or improve the quality of agricultural products. As such our programs are balanced between basic and applied research in the areas of chemical control, non-chemical control, and integrated pest management systems. ARS devotes about 45 scientific years annually on stored-product insects research. Because the theme of this book is primarily concerned with quality of long-term, stored, agricultural commodities and emphasizes grains and associated manufactured products, an overview of ARS research that relates to perishable commodities will not be included. Also, research on pests affecting people in homes and public facilities like cockroaches, houseflies, and fleas will not be included.

The ARS laboratories responsible for the stored-product insect research programs and the principal contacts are as follows:

Stored Rice Insects Lab.
USDA, ARS
Texas A&M University
Route 7, Box 999
Beaumont, TX 77706
Contact: Mr. Robert R. Cogburn
Phone: (713) 752-2741

Insect Lab.
USDA, ARS
University of Wisconsin
Dept. of Entomology
Madison, WI 53706
Contact: Dr. Wendell Burkholder
Phone: (608) 262-3795

Horticultural Crops Research Lab.
USDA, ARS
2021 S. Peach Avenue
P.O. Box 8143
Fresno, CA 93747
Contact: Dr. Patrick Vail
Phone: (209) 487-5336

U. S. Grain Marketing Research
  Lab.
USDA, ARS
1515 College Avenue
Manhattan, KS 66502
Contact: Dr. William McGaughey
Phone: (913) 776-2705

Insect Attractants, Behavior and
  Basic Biology Research Lab.
USDA, ARS
1700 SW 23rd Drive
P.O. Box 14565
Gainesville, FL 32604
Contact: Dr. Herbert Oberlander
Phone: (904) 373-6701

Agricultural Storage Insects
  Research Lab.
USDA, ARS
P.O. Box 10125
Richmond, VA 23240
Contact: Mr. Lowell Fletcher
Phone: (804) 771-2551

Stored Product Insects Research
  and Development Lab.
USDA, ARS
3401 Edwin Street
P.O. Box 22909
Savannah, GA 31403
Contact: Dr. Robert Davis
Phone: (912) 233-7981

The ARS program for stored-product insect research which follows is described by various control technologies beginning with chemical controls. Chemical control is subdivided into fumigants, protectants, repellants, and space treatments and residuals. It is followed by non-chemical control which is subdivided into biological control including parasites and predators; insect pathogens and pheromones; host resistance to insect pests; temperature manipulation; controlled or modified atmospheres; radiation; and physical barriers. Finally, the integrated pest management (IPM) systems program is described. ARS devotes a significant portion of its resources to basic research. Such research is described in each appropriate control technology section.

CHEMICAL CONTROL

Chemical insecticides are frequently used to protect agricultural products from insect damage while the commodities are in storage and in marketing channels. Chemical insecticides are the first line of defense to prevent foreign insect pests from entering this country. And conversely, they are used to kill insects in suspect U.S. produced commodities destined for export.

Issues of safety to personnel and the environment are paramount in insect control practices and are particularly significant in post-harvest commodity situations. Manifestly, as the agricultural product moves through processing toward the consumer, product quality becomes more stringent and little, if any, tolerance is allowed to chemical agents that could affect human safety or cause possible side effects. These strict requirements are mandated for candidate chemicals that are tested for their efficacy to control post-harvest insect pests. The following is a list of properties that should be possessed by an ideal chemical pesticide agent:

o Safe for human health and the environment
o Toxic to target pest
o Inexpensive to produce, store, and transport
o Convenient to use
o Harmless to commodities

Chemical insecticides for commodity treatment are categorized into four classes. These are fumigants, protectants, repellents, and space treatments and residuals.

Fumigants

Fumigants are chemicals which at a given temperature and pressure exist in a gaseous state and are toxic to insects (see chapter 11). In addition to the list of ideal chemical characteristics given previously for insecticides, an ideal fumigant should be easily generated, noncorrosive to metals and harmless to fabrics, nonexplosive and nonflammable, insoluble in water, nonpersistent as a residue, and efficient in diffusion and penetration powers. Finally, it should not readily condense to a liquid.

Current Status of Fumigants

Methyl bromide, phosphine, and fumigant formulations that usually contain carbon tetrachloride comprise the major fumigant preparations used in the United States. Methyl bromide is a gaseous fumigant marketed as a liquid under pressure in cylinders or cans. It has been used since the 1930's to disinfest agricultural warehouses and food plants, boxcars, processed commodities, and bulk grain. It is the principal fumigant used for the post-harvest treatment of tree nuts, raisins, and dry edible beans. Methyl bromide is also used as the official quarantine treatment at atmospheric and at reduced pressure for treatment of a wide range of food and nonfood commodities by the U.S. and other countries throughout the world (Thompson, 1966; Monyo, 1969). It conforms to the concentration x time (c x t x product) concept (Hesseltine and Royce, 1960), which makes it especially useful for quarantine fumigation schedules.

Inorganic bromide residues increase in most commodities each time the product is fumigated with methyl bromide. The residues are not known to be toxic, but have potential for reducing commodity quality by production of "off-odor" or "off-flavor." Like all fumigants, sufficient time must be allowed for aeration after fumigation for there is danger from unreacted methyl bromide (Scudamore and Heuser, 1970; Dow Chemical Co., 1957).

Phosphine gas is released from aluminum or magnesium phosphide fumigant preparations. Aluminum phosphide fumigant formulations are made as a powder and placed inside bags of a specific type, and as a solid shaped in form of pellets, tablets, and rounds. Magnesium phosphide is packaged as a solid and in shape of flat plates or rounds. Phosphine gas is released from the formulations whenever they are exposed to moisture such as that normally encountered in atmospheric air. Phosphine fumigation is used throughout the world for disinfestation of a wide variety of stored products. Phosphine is slow acting, but even with this limitation it is internationally the predominant grain fumigant (Lindgren, et al., 1958). Howe (1974) reviewed laboratory and field phosphine fumigation use.

Liquid fumigant formulations marketed in the U.S. usually contain a high percentage of carbon tetrachloride. Other formulation components may be ethylene dibromide, ethylene dichloride, or carbon disulfide. Ethylene dibromide (EDB) has been used to treat harvested horticultural crops to meet quarantine requirements and in many countries is a primary fumigant for control of fruit fly species. It has also been used in combination with carbon tetrachloride for spot treatment of mills and processing plants (see chapter 13). However, the Environmental Protection Agency (EPA) has announced its intention to cancel the use of EDB through regulatory actions initiated in 1977. Ethylene dichloride and carbon bisulfide have multiple uses ranging from treatment of on-farm commodities to treatment of grain loaded in certain classes of ocean-going vessels (Lindgren, et al., 1954; Cotton, 1961; Storey and Davidson, 1973).

Chloropicrin mixed with either methyl bromide or carbon tetrachloride is a fumigant that is used occasionally for treatment of grain. It is a nonflammable liquid that vaporizes to a gas on exposure to air. When used with other fumigants, it also serves as a warning agent and is marketed in pressurized and nonpressurized containers as a space, grain, and soil fumigant. Throughout much of its history, chloropicrin has been referred to as a tear gas. Because of its corrosive action on metals and low vapor pressure, it has been marketed in glass bottles and applied as a liquid fumigant. Chloropicrin is the only fumigant registered for empty grain bins. It is readily sorbed by grain and long aeration periods are required to remove the odor and tear-gas effect following fumigation. Concentrations as low as 1 ppm produce intense irritation of the eyes. These adverse properties make chloropicrin an unlikely alternative to any other fumigant now available for commodity treatment. Physical, chemical, and biological properties of liquid and pressurized fumigants are discussed by Whitney, 1961.

Current ARS Fumigant Program

In ARS, there is a continuing program to evaluate efficacy of candidate fumigant compounds provided by industry. However, during the past 8 years, only four compounds with potential merit for control of stored-product insects have been provided. Of these, two perfluorinated alcohols were more toxic to insects than methyl bromide (Leesch and Sukkestad, 1980). As a consequence, our fumigation research is focused on more effective and safer application methods for treatment of commodities in storage and in transit (Gillenwater, et al., 1980a). Also, special fumigation situations are studied such as treatment of empty grain bins to determine what fumigant is most successful in bins with perforated drying floors (Quinlan and McGaughey, 1983). In other studies, porosity of packaging films and seals to fumigants is determined (Highland, et al., 1979; Childs and Overby, 1983). Some raw or processed agricultural commodities are protected by packaging overwrap. Occasionally these commodities become infested and fumigation is one of the alternatives used to destroy insect life.

Future ARS Fumigant Program

Fumigant research against stored-product insects will be continued. Because new fumigants have not recently been identified, research will be to enhance use of presently registered fumigants. Animal and Plant Health Inspection Service (APHIS), Foreign Agriculture Service (FAS), and Federal Grain Inspection Service (FGIS) emphasize the importance of commodity fumigation, both on farm and in commercial storages as well as in-transit situations. These action agencies frequently depend on fumigation to protect the quality of U.S. produced crops.

Special effort will be put forth to develop new or improved fumigation methods for on-farm grain bins. Vacuum fumigation and aeration of processed and, to some extent, raw agricultural commodities will be explored. Vacuum fumigation kills insects rapidly and is desired by regulatory agencies.

Protectants

Protectants are chemical control agents that are applied to the commodity and provide intermediate or long-term protection against specific insect pests. Many properties applicable to protectants are the same as those required by fumigants except a protectant must have residual action of sufficient duration to protect the commodity during its anticipated storage period.

Current Status of Protectants

Only two chemical insecticides, malathion and synergized pyrethrins, are currently approved in the U.S. for direct application to grain for the purpose of controlling or limiting insect development. These insecticides are generally applied to the grain as a diluted spray as it moves on a conveyor belt into a storage bin.

Malathion is effective against most insect species that attack stored grain but rapidly looses its residual toxicity as grain moisture and temperature increase (Miles, et al., 1971; Kadoum and LaHue, 1979). Also, several species of stored-grain insects have developed resistance to malathion. Resistance has become particularly widespread with the Indianmeal moth, the red flour beetle, and the lesser grain borer (Speirs, et al., 1967; Zettler, et al., 1973; Champ, 1977).

Synergized pyrethrins are limited in their effectiveness, are expensive, are often in short supply, and compared with malathion, offer short periods of commodity protection. In 1958, malathion was the last grain protectant to be registered in the U.S. Since then, many insecticides have been screened as potential stored-grain protectants, and those thought to be promising are chlorpyrifos-methyl, pirimiphos-methyl, and fenitrothion. These compounds have been researched well and are candidates for or are in the process of EPA registration. Chlorpyrifos-methyl and pirimiphos-methyl have long residual action (LaHue and Dicke, 1977; Cogburn, 1981). In Australia, where insect resistance to chemical pesticides is particularly great, carbaryl or bioresmethrin is combined with pirimiphos-methyl or fenitrothion, respectively, to improve protectant efficacy (Bengston, 1980).

Other types of grain protectants include malathion, methoxychlor, and pyrethrins plus piperonyl butoxide applied to bin walls as a spray (Quinlan 1977). Dichlorvos vapor released from strips suspended above grain in a semi-closed bin is occasionally used to control moths (LaHue 1971). Diatomaceous earth, although not used on food and feed grains, is used as a protectant in the seed industry (White, et al., 1966). Generally, grain protectants have been studied for penetration, redistribution, and final location of residues within a grain mass as well as mechanisms and metabolic pathways of insecticide degradation.

Other protectant research effort has been focused on biorational agents such as insect growth regulators. Growth regulators of the juvenile hormone-type control stored-product insects by preventing metamorphosis and embryonic development (Strong and Dickman, 1973; McGregor and Kramer, 1975). This research is demonstrating that juvenile hormone analogues are effective protectants and potential candidates to replace conventional pesticides currently used for control of insects in stored agricultural commodities (Kramer and McGregor, 1978).

Another type of insect growth regulator is chemicals which affect insect physiology by inhibiting cuticle (the outer covering of an insect) formation and embryonic development. This research is establishing that cuticle inhibitors exhibit a broad spectrum of activity against commodity insects and are potential alternatives to protectants that are in use today (Kramer and McGaughey, 1980; Baker and Nelson, 1981).

Curent ARS Protectant Program

The severity of insect resistance to insecticides currently used on stored grain in the U.S., as well as cross-resistance to candidate insecticides, is being monitored. New types of insecticides, synergists, inhibitors, insect growth regulators, and natural products that affect insect response are studied by activity against commodity insects. Avermectin, a natural product insecticide with a novel mode of action, is showing extremely potent, long-lasting, and broad-spectrum activity against these insects. Several types of nitromethylene and pyrethroid analogs and synergists are being tested for activity against pest insects. These new approaches for control of commodity insects are evaluated for effectiveness in laboratory and small bin tests.

Metabolic properties of commodity protectants as influenced by physical, chemical, and biological factors in stored agricultural commodities are being studied. Information on the influence of temperature, moisture, foreign material, protectant application method, etc., is helpful to predict behavior of insecticides in the field and to develop more effective use patterns (Storey, 1972; Anderogg and Madisen, 1983a, 1983b). These studies include investigation on production of insecticide metabolites, discovery and identity of any unextractable residues, and volatiles (Beeman, 1982).

Much of our basic research concerns metabolism and growth of commodity insect pests. Special emphasis is on the formation and degradation of cuticle, schlerotization of cuticle, hormonal systems, enzymatic systems, resistance mechanisms, and cell and tissue transport systems. These research efforts are identifying vulnerable targets for the action of potential commodity protectants.

Future ARS Protectant Program

Despite current control efforts, foreign buyers of our grains have expressed concern about presence of insects at destination. These concerns make the development of new protectants a vital and urgent need. Therefore, most of our protectant research will be continued, especially basic research to develop new types of protectants other than chemical insecticides. It should be emphasized that an effective, persistent protectant will reduce the need for subsequent insecticidal treatments. Also, short-term protectants should be developed for use where only temporary protection of commodity (prior to processing) is desirable.

Synthetic insecticides will continue to play an important role as grain protectants. But, these insecticides must be designed on a biorational basis that takes advantage of unique characteristics and specializations in the insect life cycle. In addition, complex relationships among host commodities, pest insects, and chemical insecticides which determine the net effectiveness of commodity protectants should be analyzed.

Biochemical studies of enzymes and hormones involved in cellular metabolism of egg, larval, pupal, and adult tissues should be undertaken. Key enzymes and regulatory molecules involved in cuticle formation should be investigated with emphasis on tanning, chitin synthesis, and chitin degradation. The role that membranes have in the metabolic processes should be an area targeted for development of new methods of insect control. The specific interference with the cellular export or uptake of key insect proteins would likely provide an effective and specific means of insect control.

In the applied research arena, special attention will be directed towards mixtures of chemical insecticides that control most groups of insects which infest harvested commodities. Insecticides presently used may control one group of insects, but an insect species of another group may be highly tolerant. For example, the lesser grain borer is somewhat insensitive to organophosphate insecticides, but it is easily controlled by pyrethroids and carbaryl insecticides. Protectant formulations containing more than one active ingredient may be preferred and economically sound. However, caution must be exerted to prevent accelerated development of insect resistance to pesticides.

Repellents

Repellents, as the word implies, have characteristics of repelling insect entry or movement across a treated surface.

Current Status of Repellents

Synergized pyrethrins are effective insect repellents in post-harvest situations. They are used as the standard to evaluate candidate compounds. Synergized pyrethrins are applied to multi-walled paper bags and to polyethylene-polypropylene laminate packaging materials to protect commodities from insect invasion (Laudani, et al., 1966; Highland, 1977). Some synthetic pyrethroids, such as permethrin, have excellent insect repellent activity.

Current ARS Repellent Program

An evaluation program for candidate repellents developed by industry and USDA laboratories is a continuing study within ARS. Candidate compound efficiency is rated on a scale of 1 to 5. The pyrethrum standard has a value of 3. Because permethrin is generally rated higher than 3, its potential as a repellent is in test under field conditions as an insect resistant treatment for multi-wall paper bags (McGovern, et al., 1977, 1979; Gillenwater, et al., 1980, 1981b).

Future ARS Repellent Program

The protection of packaged commodities and even bulk-stored grain with repellents is a preferred method for protection of product quality. Commodities free of dead and living insects are not only prone to be undamaged, but also can be processed without the danger of overlooking hidden insect fragments.

Repellents may be useful tools in preventing insect infestation, thus minimizing or even eliminating the need for fumigation and other insecticides. To develop such a system, more information is needed. For example, basic studies on relationship between storage conditions and insect ecology require further study so that repellents can be utilized more effectively. Further research must be conducted on how the producer may apply repellents more uniformly to harvested grain and oilseeds. Life expectancy of repellents under different types of storage conditions must also be studied.

Space treatments include mist sprays, aerosols, and vapors released in the free space of storage facilities to protect raw and processed commodities. These agents control exposed insects and prevent entry of insects into noninfested storages and commodities. Residuals are usually dispensed as coarse sprays and known as "crack and crevice" or "spot" treatments. They are applied to specific areas of storage and processing facilities to protect commodities against invading or hidden insect infestations.

Current Status of Space Treatments and Residuals

Dichlorvos and synergized pyrethrins are the only U.S. registered insecticides for space treatment of agricultural storages (Childs, et al., 1966; Schesser, 1972) (see chapter 10). A number of insecticides are registered as residuals (see chapter 9). Insecticides such as chlorpyrifos, dursban, and diazinon may be used for "crack and crevice" and "spot" treatments. Inside walls of bins may be sprayed with malathion, pyrethrins, or methoxychlor formulations as a residual treatment (Cogburn, 1972; Slominski and Gojmerae, 1972).

Current ARS Space Treatment and Residuals Program

Screening and evaluation of candidate insecticide residuals are continuing with post-harvest insect research laboratories. Insecticide application techniques are studied to improve containment of the spray within the target area and to improve evenness of pesticide residual deposit on bin walls.

Future ARS Space Treatment and Residuals Program

Pyrethrins are not widely used to protect harvested commodities from insects because of expense, and because of occasional availability problems. Dichlorvos is safe when used as directed. However, it is toxic to warm blooded animals. Thus, there is need for new or improved space treatment and residual insecticides. ARS will continue to search for such compounds.

NON-CHEMICAL CONTROL

Research in reproductive biology, pathogens, and insect parasites and predators has indicated these are potential tools for control of stored-product insects (see chapter 17). New insect control technology developed from these research programs is in laboratory and field evaluation stages and in some cases is already incorporated in control programs. For instance, pheromone baited traps (see chapter 7) for detection of hidden infestations of several stored-product insect species are now marketed to enable more efficient and timely application of insect control procedures. The pathogen, Bacillus thuringiensis, has been approved for treatment of grain to control stored-product moths. Other new tools of considerable promise for control of stored-product insects are:

(1) Pheromones as mating inhibitors.
(2) Pheromones to lure insects to pathogen dispensing stations to maximize dispersal of pathogens among the pest population.
(3) Parasites and predators as agents to control stored-product insects.

The innovative application of these biological agents and others still under investigation offers the possibility of revolutionizing the control of stored-product insects. Non-chemical control of insects is categorized in six classes. These are (1) biological, (2) host resistance to insect pests,

(3) temperature manipulation (see chapter 18), (4) controlled or modified atmospheres (see chapter 19), (5) radiation (see chapter 20), and (6) physical barriers (see chapter 23). Biological control is further subdivided into parasites and predators, pathogens, and pheromones.

Biological Control: Parasites and Predators

Stored-product insects are attacked by an array of parasites and predators that exert some degree of natural control. Augumentation or manipulation of these natural enemies offers a potentially new means of controlling storage pests in situations where the presence of insects per se is not objectionable. Examples of such situations include feed and seed grain and raw commodities that are cleaned before processing for human consumption.

Current Status of Parasites and Predators

The potential value of these natural enemies has been confirmed by studies already completed. For example, the parasitic wasp, Bracon hebetor, which attacks late-stage larvae of various moths, has been shown to suppress populations of the almond moth and Indianmeal moth. Its effectiveness is improved when the host population has a high percentage of diapausing (a form of suspended animation in insects) larvae (Hagstrum, 1977, 1978). Therefore, factors that induce diapause also enhance moth control opportunity with parasitic wasps. Parasitic control can also be increased by using insect juvenile hormones to extend life of the last instar (the period or stage between molts in the larval stage) of the host pest in order to increase food supply and ultimately wasp density in the agricultural storage area.

Three predaceous bugs that occur in storage habitats have been studied. One of these, Xylocoris flavipes, was well adapted to agricultural storage conditions and suppressed in numbers many stored-product beetles and moths (Arbogast, 1979).

Current ARS Parasite and Predator Program

A test is in progress to determine effectiveness of a parasitic wasp and a predaceous bug in reducing insect pest populations in small storage-containing packaged commodities. Other, rather long-term, parasite and predator research includes biology and host relationship of two parasitic wasps to control weevils and other stored-product insects; a survey to determine if Trichogramma species occur naturally in peanut warehouses and, if so, a better understanding of the biology and behavior of these parasites in storage situations (Brower, 1983); a study of the parasitic mite, Pyemotes tritici, biology, behavior, methods of mass rearing, and its effectiveness as a biological control agent (Bruce and LeCato, 1979; Bruce, 1983); and a Public Law 480 project in Egypt to locate natural insect enemies in storage habitats and to study their biology and prey relationships.

Basic research is being conducted on parasite biochemistry and development of in vitro rearing for parasites.

Future ARS Parasite and Predator Program

Basic biochemical research on parasites will be continued with emphasis on rearing and self-sustaining populations when hosts are destroyed or removed. Further work will seek to identify additional parasites and predators, and to study their behavioral tendencies as influenced by environmental factors. Later, efforts will be expanded to develop insecticide resistant parasite and predator strains, and to evaluate, on large scale, the effectiveness of the more promising candidates.

211

Biological Control: Insect Pathogens

Many insect pests that infest stored grain, oilseeds, nuts, raisins, and processed products such as cereals are susceptible to microbial pathogens such as certain bacteria, viruses, protozoa, and fungi. These microorganisms are selective in their insect pathogenicity, do not pollute the environment, and are not known to be harmful to man or other mammals. The most important pathogen used today for insect control is a bacterium, Bacillus thuringiensis (BT), (McGaughey, et al., 1980; McGaughey, 1982). BT is effective during the larval feeding stage of most, if not all, lepidopterous (moths and butterflies) insects. BT was registered by EPA in 1979 for control of lepidopterous insects in stored grain. It is the only microbial insecticide registered for use on stored commodities.

Current Status of Insect Pathogens

BT deposited on the surface layer of stored grain effectively controls the Indianmeal moth and the almond moth. It is compatible with chemical insecticides and fumigants which may be used to control coleopterous (beetles and weevils) insect pests. Insect toxicity of BT is due to the protein generated from inclusion bodies, which occur in the larval midgut during feeding and is activated by appropriate conditions of pH and proteolysis. The protoxin and the active (toxic) form of the protein resulting from the insect killing inclusion bodies have been purified and partially characterized according to molecular weight and amino acid composition (Aronson, et al., 1982; Betchtel and Bulla, 1982). These discoveries are helping to determine the mechanism of toxicological action of the crystal and provide basic information that can be used to support development of the bacterium as an ecologically safe insecticide.

Studies, including safety aspects and field trials, of the parasite, Mattesia trogodermae, indicate this protozoan should be further exploited (Pounds, et al., 1981). The protozoan is a parasite of Trogoderma, and holds promise as a potential microbial control agent for the khapra beetle, Trogoderma granarium, (Shapas, et al., 1977).

An insect pathogen effective against the cigarette beetle, Lasioderma serricorne, has been identified. This pathogen is an isolate of Bacillus cereus and was found on tobacco leaves located on the plant stalk near ground level (Fletcher and Long, 1971).

Current ARS Pathogen Program

The biochemical characterization of the toxic protein contained within the BT inclusion body is being pursued. Molecular weight, the amino acid sequence, and activation mechanism are under investigation. Function of the active toxin is studied with the use of insect tissue cultures. The tissue culture method may allow insight to the mode of action of the toxic protein.

Applied BT research is concerned with field studies on the susceptibility of Indianmeal moth and almond moth populations. Also, the pathogen's persistence as it relates to commodity types, storage environment, and different storage systems will be evaluated. These studies will include interactions between biology and behavior of the pest insect species and efficacy of different types of pathogen application methods.

The pathogen, Mattesia trogodermae, is being collected in order to obtain sufficient material for planned studies involving Trogoderma beetles.

Future ARS Pathogen Program

Most of the ARS pathogen research will be a continuation of the studies reported in the above paragraphs on Current ARS Pathogen Program. Granulosis virus as a biological alternative, its use and production, will be given additional attention (Hunter, et al., 1979). Also, we will study efficacy of a protozoan pathogen distributed in food baits fortified with pheromone for control of Trogoderma species, including the khapra beetle (Burkholder, 1981).

Biological Control: Pheromones

Manipulated biological suppression of stored-product insect pests requires basic knowledge of reproductive biology, communication, behavior ecology, and chemistry. Coordinated efforts by scientists in these disciplines have resulted in significant advances in development of pheromones for management and suppression of stored-product insect pests. This research has been discussed by Vick, et al., 1981 and Burkholder, 1981.

Current Status of Pheromones

Research on many of the important stored-product insects has elucidated their reproductive and pheromone biology. Synthetic pheromones have been produced for the almond moth, Angoumois grain moth, Indianmeal moth, Mediterranean flour moth, raisin moth, black carpet beetle, furniture carpet beetle, red and confused flour beetles, cigarette beetle, drugstore beetle, lesser grain borer, and several Trogderma species, including the khapra beetle. Additionally, pheromones released from wing glands of certain male moths are known to cause the female to assume a posture necessary for mating. Because this action is on the egg-producing sex, it holds the possibility of being an effective mating disruptant.

Several studies have shown that pheromone baited traps are effective in detecting hidden insect infestations. Recent studies on pheromone movement in warehouses and determination of insect pheromone thresholds have provided theoretical distance limits over which insects can be attracted to pheromone sources (Cogburn and Vick, 1981). Advances have been made in techniques of formulating and dispensing pheromones from traps. Laboratory studies have shown that pheromones of more than one species can be dispensed from the same trap.

Field tests with the male-produced aggregation pheromone (attracts both sexes) of the lesser grain borer have demonstrated that either probe traps within the grain mass or sticky (adhesive) traps in the headspace of grain bins may be used to detect destructive insects (Khorramshahi and Burkholder, 1981).

Current ARS Pheromone Program

Most of the research reported in the previous section was accomplished cooperatively among ARS and other government scientists and frequently in conjunction with research staffs from industry. Additionally, these groups are attempting to isolate and identify the female cowpea weevil sex pheromone and the male-produced aggregation pheromone of the rice, maize, and granary weevils.

Flight behavior of the male almond moth before and after detecting female-released pheromone was used to assay the male-female communication distance. This type of study will be used to test the hypothesis that permeating the air with pheromone reduces mating by interfering with male-female communication.

Laboratory studies to determine the basic mechanism by which stored-product insects orient to pheromone sources under warehouse conditions and to determine the role of minor pheromone components relative to baited trap efficiency are underway. These studies will assist development of various styles of pheromone baited traps for detection of most insect species destructive to stored agricultural/food commodities.

Also, we are studying the effect of insect age, its diet, and other related factors on production of the maize weevil male-produced aggregation pheromone.

Future ARS Pheromone Program

Most of our future research will be to correlate the number of insects caught in various types of traps with the actual level or density of the infestation. As trap efficiency improves, quality of agricultural commodities can be assured with greater confidence as products move from the producer to the consumer.

Directly related to this goal is such research as: isolation and identification of sawtoothed grain beetle and merchant grain beetle pheromones; detection of oviposition deterring pheromones of stored-product insects; characterization and identification of sweet potato weevil feeding stimulant in sweet potato skins; and identification of the aphrodisiac pheromone of some stored-product moths.

Host Resistance to Insect Pests

For many years plant genetic specialists have sought crop lines that are resistant to damage caused by microorganisms. Plant breeding is the key pest control tool to manage most plant pathogens and some production insect pests. However, little research has been performed to develop plant lines whose grains, oilseeds, or nuts resist attack by stored-product insects.

Current Status of Host Resistance

Most reports of insect resistance in grain involve observations of differential susceptibility among a few commercial varieties. Grain varietal resistance is often shown statistically, but all commercial varieties are susceptible to stored-product insect damage. In order for host-plant resistance to significantly alleviate problems with infestations, exceptionally resistant varieties must be identified and the resistant character(s) must be incorporated into varieties that combine all other characteristics that are required to make them commercially acceptable.

The Angoumois grain moth was used as the test insect against 1000 varieties of rice obtained from the World Collection, of which 200 suggested resistance to moth infestation. In further tests, 24 varieties were found promising, and three of these varieties were crossed with Vista, a very susceptible commercial variety. Resistance was shown in all three of the crossed varieties through four generations, and the resistance was antibiotic, heritable, and dominant (Russell and Cogburn, 1977).

Resistance of barley hulls to insect damage was affected by variety, variety source, year of production, components of the hull, and possibly the level of kernel tightness (Boles and Pomeranz, 1979). Wheat kernels vary in susceptibility to rice weevil damage as related to area of production, year of production, protein levels, amylase inhibitor levels, and fungal invasion. Aging of grain in storage did not affect levels of rice weevil resistance (Yetler, et al., 1979).

Research has shown that cowpeas are infested in the field and suggests host-plant resistance could be used to control cowpea weevil infestations. Tobacco of low alkaloid content and high sugar content is more attractive to the cigarette beetle and the tobacco moth than tobacco with an alkaloid content of more than two percent.

214

Current ARS Host Resistance Program

ARS is continuing to develop breeding lines that will eventually lead to new varieties of rice that are resistant to storage insects (Cogburn, et al., 1980). Since it is known that the Angoumois grain moth and lesser grain borer infest rice before harvest, basic research is in progress to determine resistance of rice varieties to field infestation by storage insect pests.

In addition, research is in progress to determine whether storage of drought-damaged or water-damaged grain is more of a risk than storage of undamaged grains; whether ergosterol levels resulting from fungal invasion make grains more susceptible to insect invasion; and whether aging of grain increases or decreases its susceptibility to insect invasion. Additionally, research is being conducted to monitor grain storability as it is affected by handling techniques, by drying, by turning, by aeration, by types of storage bins, and whether varieties showing different levels of susceptibility should be treated with grain protectants to improve their storability. Tobacco with varying concentrations of alkaloids and sugars will be stored in segregated warehouses within a storage complex and monitored for cigarette beetle and tobacco moth population levels.

Basic studies of insect nutritional requirements have a direct impact of host/plant/insect resistance. Dietary requirements including insect digestive enzymes and naturally occurring enzyme inhibitors have been studied.

Future ARS Host Resistance Program

In view of current resources devoted to host resistance to storage pests, a concerted effort to identify resistant germplasm especially in the major grain and oilseed crops cannot be mounted. However, some effort will be devoted to identify resistant germplasm in some grains. Even moderate resistance could be significantly beneficial, especially when combined with other treatments or cultural practices (Integrated Pest Management or IPM). Full potential of commodity resistance to storage pests cannot be fully understood until plant varieties that contain insect resistance are developed for observation in field or commercial storage situations. Until then, ARS will devote some effort on the basic aspects of host resistance to storage pests.

Temperature Manipulation

The application of high or low temperatures offers a nonchemical means of disinfesting stored commodities. Sublethal temperatures can affect reproduction, growth, development, feeding, and population movement. However, temperatures adverse to survival of storage insect populations must not affect the quality of the commodity.

Oddly, a literature review of temperature effects upon stored-product insects had not been compiled until this book (see chapter 18). Instead it had been implied that stored grain insect pests are susceptible to "low" temperatures because most are of subtropical origin. ARS research effort to investigate thermal effect of stored agricultural commodities upon insect life has been at a modicum. However, this will change as greater research effort will be made to develop computer models that predict insect invasion of commodities as they move from harvest to the consumer or importer.

Current Status of Temperature Manipulation

Medium lethal dosages (LT_{50}) and dosages required to produce 95% mortality (LT_{95}) in a number of stored-product insect eggs including the cowpea weevil, cigarette beetle, sawtoothed grain beetle, and red flour beetle have been established for temperatures ranging from 41°F (5°C) to minus 4°F (-20°C.) (Mullen and Arbogast, 1979). Chilling rates of selected packaged

215

products have also been determined. Temperature manipulation has also been used to control insects in stored tobacco (Fletcher, et al., 1973); and solar energy modeling has been used to show that during the winter, tobacco warehouses can be cooled to 40°F (4.4°C) (Childs, et al., 1983). However, these methods are not presently economically competitive with chemical control. Grain aeration is known to provide insect control in cool climates. Microwaves and infrared radiation have been used experimentally to control insects, in some cases by heating grain, and in other cases by direct action on the insects.

Current ARS Temperature Manipulation Program

Active and passive ventilation are being investigated as means of maintaining low temperature in tobacco warehouses. LT_{50} and LT_{95} values are being determined for each stage of major stored-product insect pests.

Future ARS Temperature Manipulation Program

Research in this area will be continued. Additional information is needed, especially for IPM programs on the use of passive ventilation to maintain low temperatures in warehouses, and aeration and basic information on the effect of cooling of wheat and corn on storage insect pests and on the host. Included in these studies will be the effect of dry air when grain is stored in semi-arid and arid regions of the world.

Controlled or Modified Atmospheres

Killing of insects with modified atmospheres of oxygen, carbon dioxide, and nitrogen has long been recognized; and research programs to develop the use of these atmospheres for storage insect control are in progress in several locations throughout the world. Excellent reviews of this research have been compiled by Bailey and Banks, 1980 and Jay, 1980a.

Current Status of Controlled or Modified Atmospheres

Laboratory studies have shown that atmospheres deficient in oxygen are effective in controlling all life stages of the principal insect species that infest agricultural commodities. Further, continuous storage in these atmospheres for periods up to 1 year did not adversely affect germination or end use properties (milling, malting, brewing, cooking, baking, blanching, and taste) of wheat, rice, barley, malt, almonds, and raisins. These data were used by the Interregional-4 Technical Committee to petition EPA for exemption of a tolerance for controlled-modified atmospheres used to kill storage insect pests infesting raw and processed agricultural commodities. Several carbon dioxide and nitrogen production companies have labeled these gases for control of stored-product insects.

Current ARS Controlled or Modified Atmospheres Program

Current research is to establish better sealing standards for agricultural storages and to improve production and introduction of gases such as carbon dioxide and nitrogen into the storages (Jay, 1980a). Economic assessment of controlled atmospheres, usually an atmosphere of low oxygen content, for protection of stored commodities will be compared with that of chemical insecticides. Further study will be made on the interaction of cold air with carbon dioxide and its decrease in effectiveness (Jay 1980a).

Future ARS Controlled or Modified Atmospheres Program

Further research is needed to develop practical methods of application and determine the economic feasibility of nonproprietary modified atmospheres as a means of disinfestation and commodity quality maintenance. In addition, basic information on population growth and development of surviving insects subjected to these conditions are needed.

Radiation

Extensive data have been developed on the effects of gamma radiation on stored-product insects. Some of this research has been cited by Brower (1980). Watters has reviewed this effort (see chapter 20). Gamma radiation is registered for disinfestation of wheat and wheat products. However, the method is not being used commercially. Ultraviolet light has been found to be an effective miticide, but its development has not been pursued because commodities are rarely infested by mites alone. There are no plans at this time to continue this line of research for the immediate future.

Physical Barriers

Current Status of Physical Barriers

One of the demonstrated non-chemical methods for control of post-harvest pests is the use of physical barriers placed around the commodity. For example, a multi-wall paper bag has been developed to protect military subsistence items and blended cereal foods in the U.S. AID Food for Peace Program. Through package integrity, the multiwall bag effectively excludes invading type insects, and with the addition of synergized pyrethrins on its outer layer, it effectively repels penetrating insects (Highland, et al., 1975). Another physical barrier is spunbonded polyethylene which restricts cigarette beetle penetration. The plastic sheet has high tensile strength, is air permeable, and is used to protect specific types of processed tobacco commodities against insect infiltration (Fletcher and Childs, 1976).

Current ARS Physical Barrier Program

The current research effort is multifaceted with emphasis on selection of packaging materials that offer physical resistance to insect penetration (Highland, 1975; Cline, 1978; Yerington, 1983). Package configurations to emphasize decreased surface to volume ratio are being studied to determine whether reduced quantity of chemical repellents on package surface will also reduce the amount of residue migrating into the commodity. Also in development is the evaluation of different types of containers for the Department of Defense (DOD) to satisfy their needs for an insect-resistant MRE (Meal Ready to Eat) package.

Future ARS Physical Barrier Program

Current research programs will continue. In addition, the following will be developed: insecticides incorporated into plastic foil laminates and other non-paper packaging materials (Fletcher, et al., 1981); a repellent-impregnated package insert; and a physical barrier combined with repellents that will protect packaged subsistence items for two or more years to meet requirements of the DOD packaged food reserves.

INTEGRATED PEST MANAGEMENT (IPM) SYSTEMS

On-farm storage of grain is at a record level and with it the incidence of infestation by storage insects is on the rise (Storey, et al., 1983). Integrated pest management offers an array of procedures to manage post-harvest pests systematically. Through a program of integration either in combination or sequentially, different pest control technologies are made available for the management of pest population(s) which overcome limitations of a single component. Although insecticide combinations have been used against stored-product insects, the selection of treatments or components has usually been on a random basis and biased towards the background and training of the commodity owner. Research is needed to develop an IPM system that has its components technically coordinated on the basis of scientific information. Predictive models will have to be developed. These models will be based on determining how pest populations survive under natural conditions and how they survive when stressed by current pest control factors.

An extensive information inventory on stored-product insect pests has been accumulated. However, there has been no effort to synthesize the existing knowledge into a model which describes the insect populations in a marketing ecosystem. There is no current effort to develop a systems model as a tool for expediting the development and implementation of efficient, practical, and economical IPM programs.

A systems approach to understanding the marketing ecosystem in support of the development and implementation of pest-control programs should have the following objectives:

1. To synthesize available knowledge, identify and fill information gaps, and do the sensitivity analysis and simulations necessary to make the model a research planning tool.

2. To develop descriptive and predictive models to identify the value and the interactions of components of current and projected IPM systems.

3. To build models that predict the likelihood of insect infestation and occurrence and the subsequent need for control.

4. To establish the quantitative relationship between the capture rate of monitoring traps and actual insect population levels.

ARS will initiate and coordinate IPM system programs at all units that conduct research to prevent commodity loss caused by stored-product insects.

SUMMARY

A summary of the ARS programs and the laboratories involved is presented in Table 1.

The objective of ARS postharvest insect research program is to develop the means to reduce losses caused by insects throughout the marketing channel. Within the resources available, our program is balanced between development of chemical, non-chemical, and IPM control technologies.

Further, basic research that will lead to new or improved chemical and non-chemical pest control technologies is being pursued. These technologies include use of fumigants, chemical protectants, repellents, space treatments and residuals, parasites and predators, insect pathogens, attractants, insect resistant commodities, temperature manipulations, controlled or modified atmospheres, radiation, physical barriers, and IPM. IPM utilizes several pest control technologies systematically. Therefore, initially the emphasis will be on developing predictive models that will depict how populations of postharvest insects survive natural environmental stress from preharvest to on-farm storage and throughout the marketing channel. Eventually, a predictive model will be developed that depicts how populations survive when stressed with current and future chemical and/or non-chemical pest control technologies.

TABLE 1 - Summary of ARS research programs by locations

Programs	Beaumont, TX	Fresno, CA	Gainesville, FL	Madison, WI	Manhattan, KS	Richmond, VA	Savannah, GA
Chemicals:							
Fumigants		X			X	X	X
Protectants	X		X		X		X
Repellents				X		X	X
Space Treatments & Residuals					X		X
Non-Chemical:							
Biological	X	X	X	X	X	X	X
Host Resistance	X		X		X	X	X
Temperature Manipulation					X	X	X
Contr/Modif. Atmospheres		X			X	X	X
Radiation							
Physical Barriers						X	X
IPM Systems[1]	X	X	X	X	X	X	X

[1] This program to be initiated.

LITERATURE CITED

Anderegg, B. N., and L. J. Madisen. 1983a. Effective dockage on degradation of ^{14}C malathion in stored wheat. J. Agric. Food Chem. 31:700–704.

Anderegg, B. N., and L. J. Madisen. 1983b. Degradation of ^{14}C-malathion in stored corn and wheat inoculated with Aspergillus glaucus. J. Econ. Entomol. 76:733–736.

Aronson, A. I., D. J. Tyrell., P. C. Fitz-James, and L. A. Bulla, Jr. 1982. Relationship of the synthesis of spore coat protein and parasporal crystal protein in Bacillus thuringiensis. J. Bacteriol. 151:399–410.

Argobast, R. T. 1979. Cannibalism in Xylocoris flavipes (Hemiptera: Anthocoridae), a predator of stored-product insects. Ent. Exp. and Appl. 25: 128–135.

Bailey, S. W., and H. J. Banks. 1974. A view of recent studies of the effects of controlled atmospheres on stored-product pests. Pages 101–118 in: Controlled Atmosphere Storage of Grains, J. Shejbal, (ed.), Elsevier Sci. Pub. Co., Amsterdam.

Baker, J. E., and D. R. Nelson. 1981. Cuticular hydrocarbons of adults of the cowpea weevil, Callosobruchus maculatus. J. Chem. Ecol. 7:175–182.

Bechtel, D. B., and L. A. Bulla, Jr. 1982 Ultrastructural analysis of membrane development during Bacillus thuringiensis sporulation. J. Ultrastruct. Res. 79:121–132.

Beeman, R. W. 1982. Recent advances inmode of action of insecticides. Ann. Rev. Entomol. 27:253–281.

Bengston, M. et al. 1980. Chlorpyrifos-methyl plus bioresmethrin; methacrifos; pirimiphos-methyl plus bioresmethrin; and synergized bioresmethrin as grain protectants for wheat. Pestic. Sci. 11:61–76.

Boles, H. P., and Y. Pomeranz. 1979. Protective effect of barley hulls against the rice weevil. J. Econ. Entomol. 72:87–89.

Brower, J. H. 1980. Irradiation of diapausing and nondiapausing larvae of Plodia interpunctella: Effects on larval and pupal mortality and adult fertility. Ann. Entomol. Soc. Am. 73:420–426.

Brower, J. H. 1983. Utilization of stored-product Lepidoptera eggs as hosts by Trichogramma pretiosum (Riley) (Hymenoptera: Trichogramatidae). J. Kans. Entomol. Soc. 56:50–54.

Bruce, W. A., and G. L. LeCato. 1979. Pyemotes tritici: Potential biological control agent of stored-product insects. Recent Advances in Acarology :213–220.

Bruce, W. A. 1983. Current status and potential for use of mites as biological control agents of stored-product pests. Special Pub. 3304 in: Control of Pests by Mites, M. A. Hoy, G. L. Cunningham, and L. V. Knutson (eds.) U. of California at Berkley.

Burkholder, W. E. 1981. Biomonitoring for stored-product insects. Pages 29–40 in Management of Pests with Semiochemicals, E. R. Mitchell, (ed.) Plenum Press, New York and London.

Champ, R. B. 1977. FAO global survey of pesticide susceptibility of stored-grain pests. FAO Plant Prot. Bull. 25:49–67.

Childs, D. P., G. L. Phillips, and A. F. Press, Jr. 1966. Control of the cigarette beetle in tobacco warehouses with automatic dichlorvos aerosol treatments. J. Econ. Entomol. 59:261–264.

Childs, D. P., and J. E. Overby. 1983. Permeability of coated fiberboard panels, plywood sheets, polyurethane foam and plastic coatings to phosphine. Tob. Sci. 27. 2 pp.

Childs, D. P., L. W. Fletcher, J. T. Beard, and F. A. Iachetts. 1983. Cooling tobacco in warehouses during the winter to kill cigarette beetles, Part I: Relevant physical properties of stored tobacco. Tob. Sci. 27:116–124.

Cline, L. D. 1978. Clinging and climbing ability of larvae of eleven species of stored-product insects on nine flexible packaging materials and glass. J. Econ. Entomol. 71:689-691.

Cogburn, R. R. 1972. Natural surfaces in a gulf port warehouse: Influence on the toxicity of malathion and Gardona to confused flour beetles. J. Econ. Entomol. 65:1706-1709.

Cogburn, R. R., C. N. Bollich, T. H. Johnston, and W. O. Meilrath. 1980. Environmental influences on resistance to Sitotroga cerealella in varieties of rough rice. Environ. Entomol. 9:689-693.

Cogburn, R. R. 1981. Comparison of malathion and three candidate protectants against insect pests of stored rice and advantages of encapsulations. Southwestern Entomol. 6:38-43.

Cogburn, R. R., and K. W. Vick. 1981. Distribution of Angoumois grain moth and Indian meal moth in rice fields and dry storages in Texas as indicated by pheromone baited adhesive traps. Environ. Entomol. 10:1003-1007.

Cotton, R. T. 1961. Controlling insect infestation in milling machinery with spot fumigants. Northwestern Miller 265:30-34.

Dow Chemical Company. 1957. Commodities unsuited for methyl bromide fumigation. Dow Fumifacts no. 4. 2 pp.

Fletcher, L. W., and J. S. Long. 1971. A bacterial disease of cigarette beetle larvae. J. Econ. Entomol. 64:1559.

Fletcher, L. W., W. Knulle, D. P. Childs, R. R. Spadafora, J. S. Long, and D. C. Delamar. 1973. Responses of cigarette beetle larvae to temperature and humidity. USDA, ARS-S-22. 10 pp.

Fletcher, L. W., and D. P. Childs. 1976. Plastic sheets for protecting stored tobacco from the cigarette beetle. Tob. Sci. 20:15-18.

Fletcher. L. W., D. P. Childs, and L. C. Garrett. 1981. Resistance of polyethylene film impregnated with synergized pyrethrins to penetration by the cigarette beetle. Tob. Sci. 25:94-96.

Gillenwater, H. B., L. Jurd., and L. L. McDonald. 1980. Repellency of several phenolic compounds to adult Tribolium confusum. J. Georgia Entomol. Sol. 15:168-175.

Gillenwater, H. B., L. M. Redlinger, J. L. Zettler, R. Davis, L. L. McDonald, and J. M. Zehner. 1981a. Phosphine fumigation of corn intransit on a bulk-dry cargoship. J. Georgia Entomol. Soc. 16:462-475.

Gillenwater, H. B., T. P. McGovern, and L. L. McDonald. 1981b. Repellents for adult Tribolium confusum: alkynl mandelates. J. Georgia Entomol. Soc. 16:106-112.

Hagstrum, D. W., and B. J. Smittle. 1977. Host-finding ability of Bracon hebator and its influences on adult parasite survival and fecundity. Environ. Entomol. 6:437-439.

Hagstrum, D. W., and B. J. Smittle. 1978. Host utilization by Bracon hebator. Environ. Entomol. 7:596-600.

Harein, P. K. 1982. Chemical control alternatives for stored-grain insects. Page 319 in: Storage of Cereal Grains and Their Products, C. M. Christensen, (ed.) Vol. 5, 3rd ed. American Assoc. Cereal Chemists, Inc., St. Paul, Minnesota.

Heseltine, H. K., and A. Royce. 1960. A concentration-time product indicator for fumigations. Pest Tech., 88-92.

Highland, H. A. 1975. Insect resistance of composite cans. USDA, ARS-S-74. 4 pp.

Highland, H. A., M. Secrest, and D. A. Yeadon. 1975. Insect resistant textile bags: New construction and treatment techniques. USDA, ARS Tech. Bull. 1511. 12 pp.

Highland, H. A. 1977. Insect resistant food pouches made from laminates treated with synergized pyrethrins. J. Econ. Entomol. 70:483-485.

Highland, H. A., W. H. Schoenherr, T. F. Winburn, and D. E. Lawson. 1979. Phosphine and methyl bromide fumigation of commodities in woven plastic or paper bags. Cereal Foods World 24:(1), 4 pp.

Howe, R. W. 1974. Problems in the laboratory investigation of the toxicity of phosphine to stored-product insects. J. Stored Prod. Res. 10:167-181.

Hunter, D. K., S. S. Collier, and D. F. Hoffman. 1979. The effect of a granulosis virus on Plodia interpunctella (Hubner) (Lepidoptera: Pyralidae) infestations occurring in stored raisins. J. Stored Prod. Res. 15:65-69.

Jay, E. 1980a. Methods of applying carbon dioxide for insect control in stored grain. USDA, SEA, AAT-S-13. 7 pp.

Jay, E. 1980b. Low temperatures: Effects on control of Sitophilus oryzae (L.) with modified atmospheres. Pages 65-71 in: Controlled Atmosphere Storage of Grains. J. Shejbal, (ed.) Elsevier Sci. Pub. Co., Amsterdam.

Kadoum, A. M., and D. W. LaHue. 1979. Degradation of malathion on wheat and corn of various moisture contents. J. Econ. Entomol. 72:228-229.

Khorramshari, A., and W. E. Burkholder. 1981. Behavior of the lesser grain borer Rhyzopertha dominica (Coleoptera:Bostrichidae) male-produced aggregation pheromone attracts both sexes. J. Chem. Ecol. 7:33-38,

Kramer, K. J., and H. E. McGregor. 1978. Activity of pyridyl and phenyl ether analogues of juvenile hormone against Coleoptera and Lepidoptera in stored grain. J. Econ. Entomol. 71:132-134.

Kramer, K. J., and W. H. McGaughey. 1980. Susceptibility of stored-product insects to chitin inhibitors LY-131215 and LY-127063. J. Kans. Entomol. Soc. 53:627-630.

LaHue, D. W. 1971. Controlling the Indian meal moth in shelled corn with dichlorvos PVC resin strips. USDA, ARS no. 51-52. 9 pp.

LaHue, D. W., and E. B. Dicke. 1977. Candidate protectants for wheat against stored-grain insects. USDA, Mkt. Res. Rep. no. 1080. 16 pp.

Laudani, H., H. A. Highland, and E. G. Jay. 1966. Treated bags keep cornmeal insect free during overseas shipment. Am. Miller and Processor 94:14-19, 33.

Leesch, J. G., and D. R. Sukkestad. 1980. Potential of two perfluorinated alcohols as fumigants. J. Econ. Entomol. 73:829-831.

Lindgren, D. L., L. E. Vincent, and H. E. Krohne. 1954. Relative effectiveness of ten fumigants to adults of eight species of stored-product insects. J. Econ. Entomol. 47:923-926.

Lindgren, D. L., L. E. Vincent, and R. G. Strong. 1958. Studies on hydrogen phosphide as a fumigant. J. Econ. Entomol. 51: 900-903.

McGaughey, W. H., E. B. Dicke, K. F. Finney, L. C. Bolte, and M. D. Shogren. 1980. Spores in dockage and mill fractions of wheat treated with Bacillus thuringiensis. J. Econ. Entomol. 73:775-778.

McGaughey, W. H. 1982. Evaluation of commercial formulations of Bacillus thuringiensis for control of the Indian meal moth and almond moth (Lepidoptera:Pyralidae) in stored inshell peanuts. J. Econ. Entomol. 75:754-757.

McGovern, T. P., H. B. Gillenwater, and L. L. McDonald. 1977. Repellents for adult Tribolium confusum: mandelates. J. Georgia Entomol. Soc. 12:79-84.

McGovern, T. P., Gillenwater, H. B. and L. L. McDonald. 1977. Repellents for adult Tribolium confusum: amides of three heterocyclic amines. J. Georgia Entomol. Soc. 14:166-174.

McGregor, H. E., and K. J. Kramer. 1975. Activity of insect growth regulators hydroprene and methoprene, on wheat and corn against several stored-grain insects. J. Econ. Entomol. 68:668-670.

Miles, J. W., G. O. Guerrant, M. B. Goette, and F. C. Churchill. 1971. Studies on the chemistry, methods of analysis and storage stability of malathion formulations. World Health Org., Geneva. Tech. Rep. Series 475. 72 pp.

Monro, H. A. U. 1969. Manual of fumigation for insect control. Manual 79, FAO, Rome. 381 pp.

Mullen, M. A., and Arbogast, R. T. 1979. Time-temperature-mortality relationships for various stored-product insect eggs and chilling times for selected commodities. J. Econ. Entomol. 72:476-478.

Pounds, J., W. E. Burkholder, and G. M. Boush. 1981. Consideration and proposed experimental design for the safety evaluation of biological pesticides in non-target species. Ecotoxicol. Environ. Safety 5:476-493.

Quinlan, J. K. 1977. Surface and wall sprays of malathion for controlling insect populations in stored shelled corn. J. Econ. Entomol. 70:335-336.

Quinlan, J. K., and W. H. McGaughey. 1983. Fumigation of empty grain drying bins with chloropicrin, phosphine, and liquid fumigant mixtures. J. Econ. Entomol. 76:184-187.

Russell, M. P., and R. R. Cogburn. 1977. World collection rice varieties: resistance to seed penetration by Sitotroga cerealella. J. Stored Prod. Res. 13:103-106.

Schesser, J. H. 1972. Boxcar research with dichlorvos aerosol. Proc. N. Centr. Br., Entomol. Soc. Am. 27:56-57.

Scudamore, K. A., and S. G. Heuser. 1970. Residual free methyl bromide in fumigated commodities. Pestic. Sci. 14:17.

Shapas, T. J., W. E. Burkholder, and G. M. Boush. 1977. Population suppression of Trogoderma glabrum by using pheromone luring for protozoan pathogen dissemination. J. Econ. Entomol. 70:469-474.

Slominski, J. W., and W. L. Gojmerae. 1972. The effect of surfaces on the activity of insecticides. U. of Wisc., Res. Report 2376. 8 pp.

Speirs, R. D., L. M. Redlinger, and H. P. Boles. 1967. Malathion resistance in the red flour beetle. J. Econ. Entomol. 60:1373-1374.

Storey, C. L. 1972. The effects of air movement on the biological effectiveness and persistence of malathion in stored wheat. Proc. N. Centr. Br., Entomol. Soc. Am. 27:57-62.

Storey, C. L., and L. I. Davidson. 1973. Relative toxicity of chloropicrin, phosphine, EDC-CCl$_4$-CS$_2$ to various life stages of the Indian meal moth. USDA, ARS-NC-6. 8 pp.

Storey, C. L., D. B. Sauer, and D. Walker. 1983. Insect populations in wheat, corn, and oats stored on the farm. J. Econ. Entomol. 76:1323-1330.

Strong, R. G., and J. Kiekman. 1973. Comparative effectiveness of fifteen insect-growth regulators against several pests of stored products. J. Econ. Entomol. 66:1167-1173.

Thompson, R. H. 1966. A review of the properties and usage of methyl bromide as a fumigant. J. Stored Prod. Res. 1:353-376.

U.S. Department of Agriculture. 1981. Fact Book of U.S. Agriculture. Rev. Nov. 1981. Misc. Pub. No. 1063. USDA. 133 pp. Agricultural Statistics. USDA. 566 pp. 1982. Agricultural Research Service Program Plan. Misc. Pub. No. 1429. USDA. 73 pp. 1983.

Vick, K. W., J. A. Coffelt, R. W. Mankin, and E. L. Soderstrom. 1981. Recent developments in the use of pheromones to monitor Plodia interpunctella and Ephestia cautella. Pages 19-28 in: Management of Insect Pests with Semiochemicals, E. R. Mitchell, (ed.) Plenum Press, New York and London.

White, G. D., W. L. Berndt, J. H. Schessar, and C. C. Fifield. 1966. Evaluation of four inert dusts for the protection of stored wheat in Kansas from insect attack. USDA, ARS 51-8. 21 pp.

Whitney, W. K. 1961. Fumigation hazards as related to the physical, chemical and biological properties of fumigants. Pest Control 7:16-21.

Yerington, A. P. 1983. Efficiency of various overwraps and seal tightness in excluding insects from dried-fruit cartons. Environ. Entomol. 12:1310-1311.

Yetler, M.S., R. M. Saunders, and H. P. Boles. 1979. Amylase inhibitors from wheat kernels as factors in resistance to post-harvest insects. Cereal Chem. 56:243-244.

Zettler, J. L., L. L. McDonald, L. M. Redlinger, and R. D. Jones. 1973. Plodia interpunctella and Cadra cautella resistance in strains to malathion and synergized pyrethrins. J. Econ. Entomol. 66:1049-1050.

BIOLOGICAL CONTROL OF STORED-PRODUCT
INSECTS: STATUS AND PROSPECTS

Richard T. Arbogast

Editor's Comments

This is probably the technique of the future. Present
applications include Bacillus thuringiensis for moth control
on stored grains, seeds, and nuts, and viruses along with
malathion for corn, wheat, nuts, and raisins. No
registration is required for predators and parasites.
Complete replacement of chemicals is not foreseen but use of
biological measures should steadily grow. Keep your
attention on this area.

F. J. Baur

BIOLOGICAL CONTROL OF STORED-PRODUCT INSECTS:
STATUS AND PROSPECTS

Richard T. Arbogast

Stored-Product Insects Research and Development Laboratory
Agricultural Research Service
Savannah, Georgia   31403

INTRODUCTION

The term "biological control" is used here, as it was first applied by Smith (1919), in referring to the utilization of predators, parasites, and pathogens for control of insect pests, although the term is sometimes used today in a much broader sense to mean any nonchemical method of control that is biologically based. Application of biological control, in the strict sense, involves one or more of the following: (1) introduction and establishment of exotic natural enemies (classical biological control); (2) conservation and augmentation of resident species; and (3) inundative or inoculative releases. Many natural enemies of storage pests have been distributed throughout the world in shipments of insect-infested grain and other commodities, so that use of the first approach in storage situations will necessarily be limited, although it may still play a role. Biological control of stored-product pests is envisioned as comprising primarily four approaches: (1) conservation - altering the environment to minimize adverse effects on natural enemies, such as limiting the use of chemical pesticides; (2) augmentation - manipulation of natural enemies to make them more effective, as for example, development of insecticide-resistant strains or application of supplemental food to sustain natural enemies when prey is scarce; (3) inoculative releases - repeated releases of small numbers of parasites or predators; and (4) application of microbial insecticides. At the time of writing, there are few published studies of conservation and augmentation of natural enemies as means of controlling storage pests, and, consequently, the subject matter of this chapter is limited to the last two approaches. However, when biocontrol methods are fully developed, they may include all four approaches, individually or in various combinations.

Utilization of biological agents to control insect pests in storage situations is not a new concept, but it has only recently received serious attention. There are probably two reasons for this: first, because of the objection that introduction of predators and parasites would increase contamination of a product with insect remains and, second, because of the observation that natural enemies appear in significant numbers only after a product has become heavily infested and serious damage has already occurred. Although the first criticism certainly has some validity, blanket rejection of biological control on the basis of sanitary requirements is unwarranted, because insect remains are of comparatively little concern in some products such as seed grain, animal feed, and raw commodities that will be cleaned during processing. Also, since predaceous and parasitic insects have essentially no ability to penetrate packages, they could be used to advantage in warehouses containing packaged commodities to reduce pest populations and thereby decrease the chances of commodity infestation. With regard to the second objection, it's true that when nature is allowed to take its course, populations of predators and parasites are slow to overtake and suppress pest populations. However, the situation is quite different when natural enemies are deliberately released during the early stages of infestation, as will be illustrated by examples presented in this chapter.

PREDATORS

Predaceous beetles of several families including Carabidae, Staphylinidae, and Histeridae enter the storage habitat and may prey upon stored-product insects. According to Hinton (1945), 14 histerid species have been reported from stored products or storage structures in various parts of the world. Most of these breed only where the local relative humidity is about 90 to 95%, and some have been found in considerable numbers in damp and heated waste grain. These beetles are exclusively predaceous, but their impact upon pest populations has not been assessed. Among the histerid beetles, Saprinus semistriatus (Scriba) and S. semipunctatus (F.) have been found to prey upon larvae of the larder beetle, Dermestes lardarius L., and D. fischii Kugelann in stores of air-dried and smoked fish, and the former species has also been reported from a culture of Tenebrio on bran.

True bugs of the family Anthocoridae are among the predacous insects most commonly encountered in storage situations. The effectiveness of one of these bugs, Xylocoris flavipes (Reuter), in regulating pest populations has been carefully evaluated. The results of this evaluation serve to illustrate the impact that predators can have on pest populations.

The effect of X. flavipes on populations of the red flour beetle, Tribolium castaneum (Herbst), was demonstrated by Press et al. (1975) in an experiment with infested lots of inshell peanuts (farmers' stock). They placed each lot (about 210 liters) in a plywood bin with a capacity of about 1800 liters and added adult beetles at the rate of 4.5/liter of peanuts. One week later they added various numbers of adult predators to all but one bin. Samples taken 15 weeks after the peanuts were infested showed that the predator had suppressed population growth of the beetle and had reduced the number of damaged peanut kernels by 66% (Fig. 1). The beetle population in the control bin, which received no predators, reached a level of 38/liter. This represents a population increase in excess of 800%. The damage level was 37 worm-cut kernels/liter. In comparison, the beetle population in the remaining bins increased by only 20 to 80%, depending upon the initial density of predators, and damage to the peanuts was correspondingly less.

Xylocoris flavipes also had a pronounced impact on populations of the sawtoothed grain beetle, Oryzaephilus surinamensis (L.), in small lots (about 32 liters each) of shelled corn contained in 98-liter fiber drums (Arbogast,

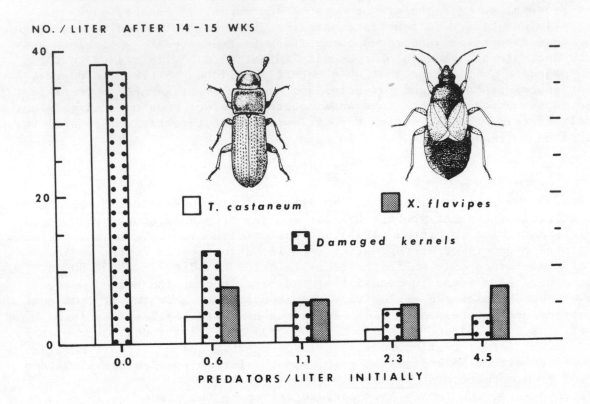

NO./LITER AFTER 14-15 WKS

T. castaneum   X. flavipes

Damaged kernels

PREDATORS/LITER INITIALLY

Figure 1. Effect of Xylocoris flavipes on populations of the red flour beetle
infesting stored farmers' stock peanuts and on the amount of damage
caused by the beetles. (Data from Press et al., 1975).

1976). Suppression in this case ranged from 97 to 99% (Fig. 2). Populations
that were free of predation increased more than 1900-fold in 15 weeks - from
1.2 insects/liter initially to more than 2500/liter. When predators were
added to the drums one week after the beetles at a rate of 0.3/liter,
population increase was only 65-fold, and when predators were added at higher
rates, increase was reduced to about 20-fold.

LeCato et al. (1977) demonstrated that X. flavipes can effectively
suppress residual insect populations in empty storage facilities. They
introduced 15 pairs each of O. surinamensis, T. castaneum, and the almond
moth, Ephestia cautella (Walker), into each of two warehouse rooms that were
empty except for a small quantity of rolled oats scattered on the floor to
simulate food debris and two boards placed on the floor to provide hiding
places. A week later, they added 30 pairs of adult X. flavipes to one of the
rooms. Changes in the almond moth population that was subjected to predation

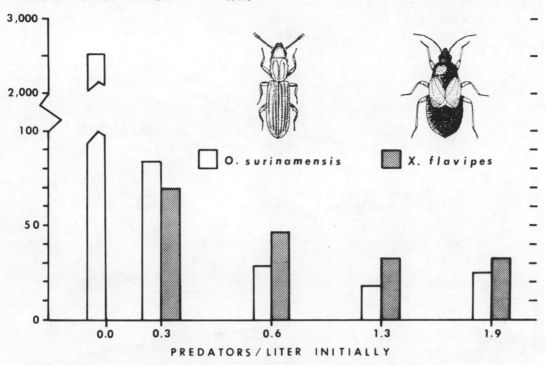

Figure 2. Effect of <u>Xylocoris flavipes</u> on populations of the sawtoothed grain beetle infesting stored corn. (Data from Arbogast, 1976).

are compared in Fig. 3 with changes in the population that was free of predation. The almond moth, and in fact all three pest species, increased steadily in the room that received no predators and reached population levels in excess of 2500 after 100 days. When predators were present, the pest populations increased over the first 25 or 50 days and then declined to levels below those at which they were introduced.

PARASITES

The impact of hymenopteran parasites on populations of stored-product insects has been illustrated by a number of studies. Finlayson (1950a), for example, presented evidence that <u>Cephalonomia waterstoni</u> Gahan, a bethylid wasp that parasitizes species of <u>Cryptolestes</u>, can effectively suppress population growth of its hosts. He compared a population of <u>Cryptolestes</u> that he had studied, and which was under attack by the parasite, with a parasite-free population that had been studied earlier by Oxley and Howe (1944). Both populations occurred in wheat and were concentrated in extensive "hot spots" that were in an advanced stage of development, central temperatures being well over 40°C. The two populations differed markedly in the numbers of beetles at approximately corresponding temperatures, especially at the lower temperatures. At 30°C, for example, Finlayson found a population density of 48 beetles/kg of wheat compared to a density of over 2800 beetles/kg reported by Oxley and Howe. Finlayson concluded that the parasite had reduced the numbers of <u>Cryptolestes</u> considerably and had slowed its spread into the cooler regions of the grain bulk.

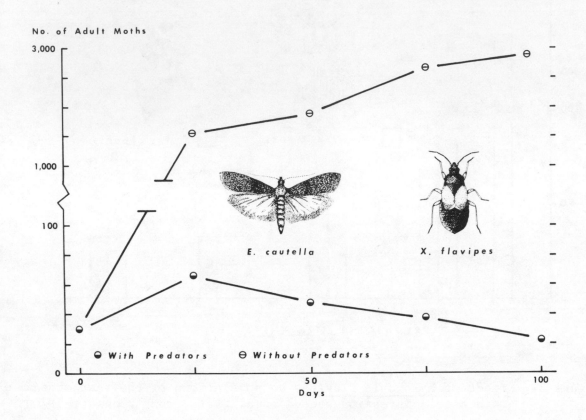

No. of Adult Moths

Figure 3. Trends in an almond moth population exposed to predation by
Xylocoris flavipes and a population free of predation. (Data
from LeCato et al., 1977).

Williams and Floyd (1971) showed that the pteromalid wasps Anisopteromalus
calandrae (Howard) and Choetospila elegans Westwood can reduce population
growth of the maize weevil, Sitophilus zeamais Motschulsky, in stored corn.
They conducted experiments on 250-g replicates of shelled corn, each contained
in a 1.9-liter round fiber carton and infested by the addition of 40 adult
weevils. The cartons were held either in the laboratory or in a farm-type
storage bin exposed to the elements. Five female parasites were added to each
carton 21 days after the weevils were added, and the number of weevils was
determined after 2 months and again after 4 months. Population growth of the
weevil was reduced 25 to 50% by C. elegans and more than 50% by A. calandrae.
Some insight into the effectiveness of the bethylid wasp Holepyris
sylvanidis (Brèthes) in regulating populations of Oryzaephilus can be gained
from information reported by Spitler and Hartsell (1975). In order to evaluate
pirimiphos-methyl (Actellic[R]) as a protectant for inshell almonds, they
placed treated and untreated nuts in drums and stored them in a room where
they were subjected to infestation by several species of insects, including
the merchant grain beetle, O. mercator (Fauvel). When the almonds were first
placed in storage, and at monthly intervals thereafter, 10,000 merchant grain
beetles were released in the room. The beetle population in the untreated
nuts reached a peak of about 1,000 adults/sample after 6 months of storage

230

(Fig. 4). At that time, the control drums were invaded by H. sylvanidis, and
the number of living beetles fell abruptly. After 8 months there were only
200 beetles/sample and the parasite was no longer detectable. The beetle
population then began to increase again, but leveled off following a
resurgence of the parasite. Although the test was terminated after 12 months,
it appears that after overtaking its host population, the parasite was
beginning to hold it at an equilibrium density of about 250 to 300/sample,
even though a large source of immigrant beetles was provided at monthly
intervals.

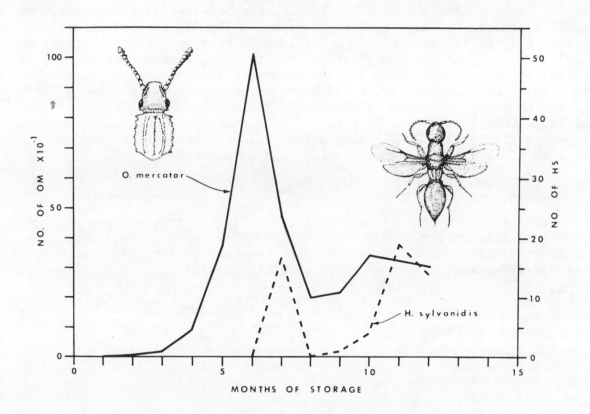

Figure 4. Relationship between population levels of the parasitic wasp
Holepyris sylvanidis and its host Oryzaephilus mercator on
stored inshell almonds. (Data from Spitler and Hartsell 1975).

The braconid wasp Bracon hebetor Say, which parasitizes late-stage larvae
of various moths, can effect significant natural control of moth populations
in warehouses (Hagstrum and Sharp, 1975), suggesting that it may have
considerable potential as a biological control agent. The ichneumonid wasp
Venturia canescens (Gravenhorst) has also shown potential for control of
stored-product moths. The value of both species has been confirmed by
laboratory and small scale warehouse tests. A single introduction of B.
hebetor into laboratory cultures of the Indianmeal moth, Plodia interpunctella
(Hübner), reduced emergence of adult moths by 74% (Press et al., 1974).
Semiweekly releases of B. hebetor into a 42.5-m^3 room containing food debris
infested with E. cautella resulted in 97% suppression of the moth population
and semiweekly releases of V. canescens resulted in 92% suppression (Press et
al., 1982).

Parasitic mites have also demonstrated a marked facility for controlling storage pests, and their potential for this purpose was recently reviewed by Bruce (1983). One of the most promising species is the straw itch mite, Pyemotes tritici (Lagrèze-Fossat and Montané), which was evaluated by Bruce and LeCato (1979). They found that, in the laboratory, this mite inflicted 100% mortality on eggs, larvae, and pupae of O. mercator, and 97% mortality on adults. Although some stages of T. castaneum and of the cigarette beetle, Lasioderma serricorne (Fabricius), were resistant to attack, the mite inflicted 100% mortality on early-instar larvae of both species. Eggs, early-instar larvae, and adults of E. cautella and P. interpunctella were susceptible to attack and suffered mortality ranging from 91 to 100%. Late-instar larvae were apparently immune and pupae suffered only about 50% mortality. Several attributes of P. tritici make it especially attractive as a biocontrol agent: (1) high reproductive potential; (2) short life cycle (4-7 days); (3) females give birth to mature offspring; (4) 98% of offspring are females; (5) females mate at birth and begin host seeking immediately; (6) populations can easily be reared and synchronized; and (7) cosmopolitan distribution. The straw itch mite has one shortcoming as far as biological control is concerned. It does bite humans and produces a rashlike dermititis. However, this drawback is not as serious as it may seem, because the mites usually live no longer than about 10 days without hosts and so will die out soon after a pest infestation has been eliminated.

PATHOGENS

Stored-product insects are subject to infection by numerous pathogenic organisms including protozoa, bacteria, fungi, and viruses. These organisms provide some natural control of insect pests in storage facilities and some show considerable promise as biological control agents.

Protozoa

Protozoa of the Class Sporozoa (Orders Gregarinida, Coccidia, and Microsporidia) are common and widespread among natural populations of stored-product insects. They generally produce chronic debilitative diseases that reduce population growth by increasing mortality, slowing development, and lowering fecundity.

Among the gregarines, most members of the Suborder Eugregarina do little damage to the host tissue and can be considered commensals (commensalism is a relationship in which one organism lives on the surplus food of another but does not injure it), although they may become mildly pathogenic under conditions of dietary or other environmental stress. Harry (1967), for example, showed that when larvae of the yellow mealworm, Tenebrio molitor L., infected with the eugregarine Gregarina polymorpha (Hammerschmidt) (Stein) are reared under optimal conditions, they complete larval development as rapidly and achieve the same pupal weight as noninfected larvae. When infected larvae are grown on a suboptimal diet, however, their ability to complete development and the pupal weight they achieve are reduced. Another example is provided by the work of Dunkel and Boush (1969) who reported that larvae of the black carpet beetle, Attagenus megatoma (F.), infected with the eugregarine Pyxinia frenzeli Laveran & Mesnil, lose weight almost twice as rapidly during starvation as do gregarine-free larvae. They suggested that since feral populations of A. megatoma are frequently infected with P. frenzeli and commonly subjected to partial starvation, P. frenzeli may serve as a continuous check on population growth. Schwalbe and Baker (1976), however, found little difference in weight loss during starvation between A.

megatoma larvae infected with this organism and uninfected larvae. They
suggested that the inconsistency between their results and those of Dunkel and
Boush may have resulted from an improved nutritional state of the larvae in
their tests or from a difference in the degree of infection.

In contrast to eugregarines, gregarines of the Suborder Neogregarina
(Schizogregarina) are virulent pathogens. Finlayson (1950b) reported that
when adult female flat grain beetles, Cryptolestes pusillus (Schoenherr), are
placed in a culture medium containing spores of Mattesia dispora Naville, all
of their offspring die in the larval stage, so that the adult population
declines rather than increases. He also found that this organism infects and
kills the larvae of the rusty grain beetle, C. ferrugineus (Stephens), and a
third, unidentified, species of Crytolestes, but that the larvae of C. turcicus
(Grouvelle) are apparently immune. Species of Mattesia also infect various
species of Trogoderma and are thought to be responsible for the erratic
population trends in these beetles. Schwalbe et al. (1973) showed that the
virulence of one of these pathogens, Mattesia trogodermae Canning, infecting
larvae of T. glabrum (Herbst) is influenced by dosage and temperature. When
they exposed larvae to 100 mg of spore powder per gram of culture medium for
24 h, the median survival time ranged from 20 (35°C) to 29 days (25°C).
Ashford (1970) found that another neogregarine, Lymphotropha tribolii Ashford,
prevents normal larval development and increases larval mortality in T.
castaneum. Furthermore, although the longevity and fecundity of 75% of the
survivors in his tests were normal, the remainder of the survivors were
weakened and sterilized.

The Orders Coccidia and Microsporidia also include virulent pathogens of
stored-product insects. Park (1948), for example, noted that an unidentified
species of the coccidian genus Adelina limits growth of T. castaneum
populations by increasing mortality, especially among the immature stages, and
pointed out that it may also act by reducing fecundity. The microsporidian
Nosema whitei Weiser is a pathogen of Tribolium species, and is also capable
of infecting O. surinamensis but not T. molitor, Palorus ratzeburgi (Wisemann)
(smalleyed flour beetle), Gnatocerus cornutus F. (broadhorned flour beetle),
or P. interpunctella (Milner, 1972b). Milner (1972a) showed that infection
with this organism drastically reduces the rate of development of T. castaneum
and delays molting after the second molt. Infected adults lay few eggs,
although egg viability is unaffected. The pathogenic effect of N. whitei on
T. castaneum is increased by dietary stress. George (1971) showed that
mortality is higher and death occurs sooner among larvae on diets deficient in
protein and cholesterol than among larvae reared on optimal diets. Other
species of Nosema, N. plodiae Kellen and Lindegren and N. heterosporum Kellen
and Lindegren, are pathogens of P. interpunctella (Kellen and Lindegren,
1974). When P. interpunctella was reared on a diet containing spores of these
Nosema, the number of insects that survived to the adult stage decreased as
the spore concentration increased. Nosema heterosporum, with a median lethal
concentration (LC_{50}) of 4.52 X 10^3 spores/g of diet, is more virulent than N.
plodiae, which has an LD_{50} of 8.09 X 10^6 spores/g of diet. All surviving
adults of P. interpunctella reared on a diet containing N. plodiae were
infected. Such adults transmit the organism to other adults during mating and
through the ovaries to the next generation (Kellen and Lindegren, 1971).

Research to evaluate the efficacy of protozoa in controlling stored-product insects has focused primarily on using a combination of pheromone and protozoan spores to attract and inoculate the insects (Burkholder and Boush, 1974). Males attracted and contaminated in this manner return to their natural habitat where they infect others of their kind. Shapas et al. (1977), who evaluated pheromone-baited, spore-transfer sites for suppression of T. glabrum with M. trogodermae, found that a single introduction of spores by this method into dense populations of T. glabrum produced 81% suppression of the F_1 generation and nearly 100% suppression of the F_2 generation.

Bacteria

Numerous species of bacteria occur in association with insects, but relatively few can be considered insect pathogens. Perhaps the best known insect-pathogenic bacterium is Bacillus thuringiensis Berliner, a rod-shaped, spore-forming organism that produces a disease of lepidopterous larvae. Sporulating cells of this organism contain an ovoid endospore and a crystalline parasporal body which are eventually released by autolysis. Insects normally contract the disease by ingesting the spores and crystals. Ingested crystals are transformed to an active toxin that damages the cells of the insect midgut, thus inhibiting feeding and giving entrance to the spores.

Susceptible species may be affected by the spores alone or by the crystals alone, but in most species a mixture of the two produces the greatest mortality. Thus crystals are about 30x more toxic to E. cautella than spores, and the toxicity of a mixture is directly related to the number of crystals in the mixture. On the other hand, crystals are only about 3x more toxic to P. interpunctella than spores, and maximum toxicity is obtained with a 50:50 mixture (McGaughey, 1978a).

Bacillus thuringiensis provides control of E. cautella and P. interpunctella when applied to grain as an aqueous suspension, as a dust, or as a bait. It is effective as a bulk treatment (all grain uniformly treated) or surface layer application (uniformly treated grain layered over untreated grain) and persists under storage conditions without noticeable decrease in insecticidal activity for at least a year (McGaughey, 1976, 1978b; Kinsinger and McGaughey, 1976). Larvae of the Angoumois grain moth, Sitotroga cerealella (Olivier), are also susceptible and although application of B. thuringiensis is less effective in achieving control, application at rates sufficient to control P. interpunctella and E. cautella can be expected to substantially reduce population growth of S. cerealella and may obviate the need for other control measures (McGaughey, 1976; McGaughey and Kinsinger, 1978).

Bacillus thuringiensis is compatible with several grain fumigants (McGaughey 1975a). The toxicity of a formulation consisting of spores and crystals was not reduced by treatment with phosphine, carbon tetrachloride-carbon bisulphide, ethylene dichloride-carbon tetrachloride, or methyl bromide, although methyl bromide killed or otherwise prevented germination of the spores.

Viruses

Insects are affected by a diverse group of viruses, some of which are important pathogens of storage pests. Among these pathogens are granuloses of P. interpunctella (PGV) and E. cautella (CGV) and a nuclear polyhedrosis of E. cautella. Granuloses are rod-shaped nuclear viruses in which each virus particle is enclosed in a protein capsule. Nuclear polyhedroses form cubic, hexagonal, or nearly spherical polyhedra, each containing numerous virus rods, in the nuclei of host cells. When capsules or polyhedra are ingested by a susceptible insect, the protein surrounding the virus particles is digested and the particles enter the host tissue.

Granuloses are usually host species specific, although some show cross infectivity. At a concentration of 8.6×10^4 capsules/g of a bran diet, PGV killed 96% of exposed P. interpunctella larvae in 25 days (Hunter, 1970). CGV killed 74% of E. cautella larvae exposed to a concentration of 1.2×10^5 capsules/g and 100% of those exposed to a concentration of 1.2×10^6 capsules/g (Hunter and Hoffmann, 1970). Ephestia cautella, E. elutella (Hübner) (tobacco moth), and Cadra figulilella (Gregson) (raisin moth) were found to be nonsusceptible to PGV (Hunter, 1970), but P. interpunctella was moderately susceptible to CGV (Hunter and Hoffmann, 1972). Although capsules of CGV from infected P. interpunctella larvae were generally abnormal in form and contained multiple virus particles, the virus remained virulent and developed normally in the natural host.

The efficacy of PGV as a microbial insecticide was investigated by Hunter et al. (1973, 1977, 1979). An aqueous suspension of this virus effectively protected peanuts, almonds, walnuts, and raisins against infestation by P. interpunctella. Inshell almonds were protected for 134 days and feeding damage was so reduced that the number of rejected nuts dropped by as much as 88%. McGaughey (1975b) found that aqueous and dust applications of PGV gave effective control of P. interpunctella on stored corn and wheat when the material was applied as a bulk treatment, a mixed treatment (treated grain mixed with untreated grain), or as a surface layer treatment. PGV was found to be compatible with malathion, and a mixture of the virus and chemical provided better control of P. interpunctella on almonds that either material alone (Hunter et al., 1975).

In contrast to granuloses, many nuclear polyhedroses are cross infective among lepidopterous species. A nuclear polyhedrosis isolated from E. cautella showed an LC_{50} of between 0.25 and 0.50 polyhedra/mm^2 of surface of an agar base diet (Hunter et al., 1973a). Most larvae died of the disease in 8 to 10 days after exposure to the virus at a concentration of 4 polyhedra/mm^2. On a bran diet, the LC_{50} was slightly greater than 3.2×10^4 polyhedra/g, and higher concentrations were required when the inoculum was mixed with the diet than when it was layered on the surface. Plodia interpunctella was also found to be susceptible to this virus, although susceptibility was lower than in the almond moth and the infection developed more slowly. A concentration of 6.4×10^4 polyhedra/g of a bran diet killed 45% of exposed larvae (Hunter et al., 1973b).

CONCLUDING REMARKS

The preceding paragraphs of this chapter constitute a status report on biological control as a means of protecting stored commodities from insect damage. The potential value of this method has been clearly demonstrated, but technologically, it is still rudimentary. Further development of biological control as a storage procedure will require field trials to answer questions concerning: (1) efficacy under actual conditions of commercial storage; (2) compatibility with other storage procedures; (3) cost effectiveness; and (4) undesirable side effects. Routine application of the method will also require improved techniques for mass producing natural enemies suitable for use in the storage environment.

Some progress has already been made along these lines. <u>Bacillus</u>
<u>thuringiensis</u>, for example, is produced commercially and has been used in the
control of lepidopterous pests for some time. It has recently been registered
as a wettable powder and dust formulation for moth control on stored grains,
soybeans, sunflower seeds, crop seeds, and peanuts. The granulosis virus of
the Indianmeal moth (PGV) is still under investigation at USDA's Stored-
Product Insects Research Laboratory, Fresno, California. The next step in
evaluating this pathogen will probably include large scale applications of the
virus under commercial warehouse conditions, but pursuit of safety data for
registration will depend upon future interest by firms engaged in producing
biocontrol agents (W. R. Kellen, personal communication). No registration is
required for use of predators and parasites, but up until now none of these
agents have been evaluated under commercial warehouse conditions. A pilot
test is currently underway at our laboratory to evaluate <u>B. hebetor</u> and <u>X.</u>
<u>flavipes</u> for controlling pests of stored farmers' stock peanuts. This test
will provide the first data on the efficacy and cost effectiveness of
predators and parasites under actual storage conditions.

LITERATURE CITED

Arbogast, R. T. 1976. Suppression of <u>Oryzaephilus surinamensis</u> (L.)
(Coleoptera, Cucujidae) on shelled corn by the predator <u>Xylocoris</u>
<u>flavipes</u> (Reuter) (Hemiptera, Anthocoridae). J. Ga. Entomol. Soc.
11:67-71.

Ashford, R. W. 1970. Some relationships between the red flour beetle,
<u>Tribolium castaneum</u> (Herbst) (Coleoptera, Tenebrionidae) and <u>Lymphotropha</u>
<u>tribolii</u> Ashford (Neogregarinida, Schizocystidae). Acta Protozool.
7:513-529.

Bruce, W. A. 1983. Current status and potential for use of mites as
biological control agents of stored product pests. Pages 74-78 <u>in</u> M. A.
Hoy, G. L. Cunningham, and L. V. Knutson, eds. Control of pests by
mites. <u>Spec. Publ. 3304, Div. Agr. Nat. Resour.</u> Univ. Calif., Berkeley.

Bruce, W. A., and G. L. LeCato. 1979. <u>Pyemotes tritici</u>: Potential biological
control agent of stored-product insects. Pages 213-220 <u>in</u> J. G.
Rodriguez, ed. Recent advances in acarology, Vol. I. Academic Press, New
York.

Burkholder, W. E., and G. M. Boush. 1974. Pheromones in stored product
insect trapping and pathogen dissemination. Bull. OEPP. 4:455-461.

Dunkel, F., and G. M. Boush. 1969. Effect of starvation on the black carpet
beetle, <u>Attagenus megatoma</u>, infected with the eugregarine <u>Pyxinia</u>
<u>frenzeli</u>. J. Invertebr. Pathol. 14:49-52.

Finlayson, L. H. 1950a. The biology of <u>Cephalonomia waterstoni</u> Gahan (Hym.,
Bethylidae), a parasite of <u>Laemophloeus</u> (Col., Cucujidae). Bull.
Entomol. Res. 41:79-97.

Finlayson, L. H. 1950b. Mortality of <u>Laemophloeus</u> (Coleoptera, Cucujidae)
infected with <u>Mattesia dispora</u> Naville (Protozoa, Schizogregarinaria).
Parasitology. 40:261-264.

George, C. R. 1971. The effects of malnutrition on growth and mortality of
the red rust flour beetle, <u>Tribolium castaneum</u> (Coleoptera:
Tenebrionidae) parasitized by <u>Nosema whitei</u> (Microsporidia:
Nosematidae). J.Invertebr. Pathol. 18:383-388.

Hagstrum, D. W., and J. E. Sharp. 1975. Population studies on <u>Cadra cautella</u>
in a citrus pulp warehouse with particular reference to diapause. J.
Econ. Entomol. 68:11-14.

Harry, O. G. 1967. The effect of the eugregarine <u>Gregarina polymorpha</u>
(Hammerschmidt) on the mealworm larva of <u>Tenebrio molitor</u> (L.). J.
Protozool. 14:539-547.

Hinton, H. E. 1945. The Histeridae associated with stored products. Bull. Entomol. Res. 35:309-340.

Hunter, D. K. 1970. Pathogenicity of a granulosis virus of the Indian-meal moth. J. Invertebr Pathol. 16:339-341.

Hunter, D. K., and D. F. Hoffmann. 1970. A granulosis virus of the almond moth, Cadra cautella. J. Invertebr. Pathol. 16:400-407.

Hunter, D. K., and D. F. Hoffmann. 1972. Cross infection of a granulosis virus of Cadra cautella, with observations on its ultra-structure in infected cells of Plodia interpunctella. J. Invertebr. Pathol. 20:4-10.

Hunter, D. K., S. J. Collier, and D. F. Hoffmann. 1973. Effectiveness of a granulosis virus of the Indian meal moth as a protectant for stored inshell nuts: preliminary observations. J. Invertebr. Pathol. 22:481.

Hunter, D. K., S. J. Collier, and D. F. Hoffmann. 1975. Compatibility of malathion and the granulosis virus of the Indian meal moth. J. Invertebr. Pathol. 25:389-390.

Hunter, D. K., S. J. Collier, and D. F. Hoffmann. 1977. Granulosis virus of the Indian meal moth as a protectant for stored inshell almonds. J. Econ. Entomol. 70:493-494.

Hunter, D. K., S. J. Collier, and D. F. Hoffmann. 1979. The effect of a granulosis virus on Plodia interpunctella (Hubner) (Lepidoptera: Pyralidae) infestations occurring in stored raisins. J. Stored Prod. Res. 15:65-69.

Hunter, D. K., D. F. Hoffmann, and S. J. Collier. 1973a. Pathogenicity of a nuclear polyhedrosis virus of the almond moth, Cadra cautella. J. Invertebr. Pathol. 21:282-286.

Hunter, D. K., D. F. Hoffmann, and S. J. Collier. 1973b. Cross-infection of a nuclear polyhedrosis virus of the almond moth to the Indian meal moth. J. Invertebr. Pathol. 22:186-192.

Kellen, W. R., and J. E. Lindegren. 1971. Comparative virulence of Nosema plodiae and Nosema heterosporum in the Indian meal moth, Plodia interpunctella. J. Invertebr. Pathol. 23:242-245.

Kellen, W. R., and J. E. Lindegren. 1974. Modes of transmission of Nosema plodiae Kellen and Lindegren, a pathogen of Plodia interpunctella (Hubner). J. Stored Prod. Res. 7:31-34.

Kinsinger, R. A., and W. H. McGaughey. 1976. Stability of Bacillus thuringiensis and a granulosis virus of Plodia interpunctella on stored wheat. J. Econ. Entomol. 69:149-154.

LeCato, G. L., J. M. Collins, and R. T. Arbogast. 1977. Reduction of residual populations of stored-product insects by Xylocoris flavipes (Hemiptera: Anthocoridae). J. Kans. Entomol. Soc. 50:84-88.

McGaughey, W. A. 1975a. Compatibility of Bacillus thuringiensis and granulosis virus treatments of stored grain with four grain fumigants. J. Invertebr. Pathol. 26:247-250.

McGaughey, W. A. 1975b. A granulosis virus for Indian meal moth control in stored wheat and corn. J. Econ. Entomol. 68:346-348.

McGaughey, W. A. 1976. Bacillus thuringiensis for controlling three species of moths in stored grain. Can. Entomol. 108:105-112.

McGaughey, W. A. 1978a. Response of Plodia interpunctella and Ephestia cautella larvae to spores and parasporal crystals of Bacillus thuringiensis. J. Econ. Entomol. 71:687-688.

McGaughey, W. A. 1978b. Moth control in stored grain: Efficacy of Bacillus thuringiensis on corn and method of evaluation using small bins. J. Econ. Entomol. 71:835-839.

McGaughey, W. A., and R. A. Kinsinger. 1978. Susceptibility of Angoumois grain moths to Bacillus thuringiensis. J. Econ. Entomol. 71:435-436.

Milner, R. J. 1972a. Nosema whitei, a microsporidian pathogen of some species of Tribolium. III. Effect on T. castaneum. J. Invertebr. Pathol. 19:248-255.

Milner, R. J. 1972b. *Nosema whitei*, a microsporidian pathogen of some
    species of *Tribolium*. V. Comparative pathogenicity and host range.
    Entomophaga. 18:383-390.

Oxley, T. A., and R. W. Howe. 1944. Factors influencing the course of an
    insect infestation in bulk wheat. Ann. Appl. Biol. 31:76-80.

Park, T. 1948. Experimental studies of interspecies competition. 1.
    Competition between populations of the flour beetles, *Tribolium confusum*
    Duval and *Tribolium castaneum* Herbst. Ecol. Mongr. 18:265-307.

Press, J. W., L. D. Cline, and B. R. Flaherty. 1982. A comparison of two
    parasitoids, *Bracon hebetor* (Hymenoptera: Braconidae) and *Venturia
    canescens* (Hymenoptera: Ichneumonidae), and a predator *Xylocoris flavipes*
    (Hemiptera: Anthocoridae) in suppressing residual populations of the
    almond moth, *Ephestia cautella* (Lepidoptera: Pyralidae). J. Kans.
    Entomol. Soc. 55:725-728.

Press, J. W., B. R. Flaherty, and R. T. Arbogast. 1974. Interactions among
    Plodia interpunctella, *Bracon hebetor*, and *Xylocoris flavipes*. Environ.
    Entomol. 3:183-184.

Press, J. W., B. R. Flaherty, and R. T. Arbogast. 1975. Control of the red
    flour beetle, *Tribolium castaneum* in a warehouse by a predaceous bug,
    Xylocoris flavipes. J. Ga. Entomol. Soc. 10:76-78.

Shapas, T. J., W. E. Burkholder, and G. M. Boush. 1977. Population
    suppression of *Trogoderma glabrum* by using pheromone luring for protozoan
    pathogen dissemination. J. Econ. Entomol. 70:469-474.

Schwalbe, C. P., and J. E. Baker. 1976. Nutrient reserves in starving black
    carpet beetle larvae infected with the eugregarine *Pyxinia frenzeli*. J.
    Invertebr. Pathol. 28:11-15.

Schwalbe, C. P., G. M. Boush, and W. E. Burkholder. 1973. Factors influencing
    the pathogenicity and development of *Mattesia trogodermae* infecting
    Trogoderma glabrum larvae. J. Invertebr. Pathol. 21:176-182.

Smith, H. S. 1919. On some phases of insect control by the biological
    method. J. Econ. Entomol. 12:288-292.

Spitler, G. H., and P. L. Hartsell. 1975. Pirimiphos-methyl as a protectant
    for stored inshell almonds. J. Econ. Entomol. 68:777-780.

Williams, R. N., and E. H. Floyd. 1971. Effect of two parasites,
    Anisopteromalus calandrae and *Choetospila elegans*, upon populations of
    the maize weevil under laboratory and natural conditions. J. Econ.
    Entomol. 64:1407-1408.

RECENT ADVANCES IN THE USE OF MODIFIED ATMOSPHERES
FOR THE CONTROL OF STORED-PRODUCT INSECTS

Edward Jay

Editor's Comments

Limited field studies have shown that use of modified atmospheres (CO_2, N_2) is cost competitive as an alternative to fumigation. Advantages are that there are no residues, aeration is not required, and eggs are not tolerant. This technique has good potential for in-transit use.

F. J. Baur

RECENT ADVANCES IN THE USE OF MODIFIED ATMOSPHERES FOR THE CONTROL OF STORED-PRODUCT INSECTS

Edward Jay

Stored-Product Insects Research and Development Laboratory
Agricultural Research Service, U.S. Department of Agriculture
Savannah, Georgia 31403

INTRODUCTION

Modified or controlled atmospheres offer an alternative to the use of conventional residue-producing chemical fumigants for controlling insect pests attacking stored grain, oilseeds, processed commodities, and some packaged foods. A modified atmosphere is created by replacing the existing atmosphere in a storage container with one lethal to insects by adding carbon dioxide (CO_2), nitrogen (N_2), or air depleted of oxygen (O_2) through the application of combustion gases from a modified or "inert" atmosphere generator or "burner". These atmospheres also reduce or prevent fungus growth and so they are effective in limiting mycotoxin production such as aflatoxins.

The use of modified atmospheres is an adaptation of the old principle of hermetic storage. This process involves sealing the grain, beans, or oilseeds, generally in underground pits, and allowing the respiration of the stored commodity plus that of any insects present to reduce the O_2 to a level lethal to insects. The decrease in the O_2 level may be accelerated by using wet straw or grass to line the storage container. The low O_2 in the atmosphere of a hermetically sealed storage also protects the commodity from fungal attack and thus maintains the quality at a high level over extended storage periods. Sigaut (1980) states that in pre-industrial times hermetic storage was probably the one means of keeping large quantities of loose grain free from insect attack for significant lengths of time in areas with mild winters. He also reports that the first large scale tests were run in underground pits in Paris from 1819 to 1830. This principle was also used on a large scale in Argentina during and after World War II, when facilities were constructed and used for the underground hermetic storage of over 2.5 million tons (t) of grain. Today, there are active but primitive hermetic storages in operation in India (Girish, 1980), and underground storages are still in use in Yemen, Somolia, Sudan, and Egypt (Kamel, 1980). More modern concrete hermetic storage bins built primarily for famine protection have been constructed in Cyprus and Kenya for corn storage and are continously in operation (DeLima, 1980).

Studies in the 1860s on modifying atmospheres by adding N_2 or "burned air" to grain storages are also reported by Sigaut (1980). However serious interest in using the technique in a practical, routine manner was not pursued until the 1950s and 60s, probably due to the success of conventional fumigants and grain protectants in controlling stored-product pests. During this period, a realization began to develop that the chemicals, if used improperly, left objectional residues, were hazardous to apply, and that there was a potential for the development of insect resistance to them. Research was initiated during this time in Australia and in the U.S. and is ongoing in these and several other countries on the use of modified atmospheres. Research conducted in this country and described in this chapter was used by the U.S. Environmental Protection Agency in granting an exemption from tolerance for CO_2, N_2, and products from an "inert" gas generator when used to control insects in raw (Federal Register 45, pp. 75663-64, Nov. 1980) and processed (Federal Register 46, pp. 32865-66, June 1981) agricultural products.

A symposium on "Controlled Atmosphere Storage of Grains" was held in Rome in 1980. This meeting was attended by almost 100 participants from 26 countries and the proceedings of this symposium have been published (Shejbal ed., 1980). The publication should be consulted by anyone who wants to learn more about this subject than is presented here.

LABORATORY STUDIES

Carbon Dioxide and Nitrogen

Valid laboratory studies can provide guidance in selecting modified atmosphere concentrations, exposure times, and optimum temperature ranges which can be used in conducting field experiments. An example of this can be seen in Table 1 (Jay, unpublished laboratory study). This table shows the effects on mortality (based on adult emergence) of 7 different modified atmospheres on 1-to-5 week (wk)-old immatures of the maize weevil, Sitophilus zeamais Motschulsky, at the same temperature and relative humidity during 1- to 4-day exposures. Increasing the N_2 concentration from 97 to 100% greatly reduces emergence as did increasing the CO_2 concentration from 39 to 62%. Increasing the CO_2 concentration to 99% produced less mortality than that obtained at ca. 60%. Laboratory studies such as these show that CO_2 is more biocidal than N_2 to this species, that there is no need to increase the CO_2 concentration above 60%, and that a longer exposure is needed to obtain complete control of this species.

Table 1.--Mean number of adult maize weevils emerging from wheat infested with 1-to-5 week old immatures and exposed for 1- to 4-days to indicated atmosphere at 80°F and 46% RH.

Atmosphere [1]	Mean Emergence	Atmosphere [1]	Mean Emergence
Air (control)	26.6	39% CO_2	14.9
97% N_2	18.4	50% CO_2	9.4
99% N_2	14.5	99% CO_2	8.9
100% N_2	9.6	62% CO_2	5.8

[1] Balance of the modified atmospheres containing N_2 is O_2; balance of the atmospheres containing CO_2 is air (N_2 plus O_2).

The effect of temperature on the length of time necessary to obtain good control with modified atmospheres is as important as with conventional fumigants. Jay (1971) states that "the temperature of the grain should be above 70°F during the application of CO_2" (to obtain good control). Tables 2 and 3 show the effects of lowered temperatures (from Jay, 1980a). At 35°F (Table 2) the cold air alone is equal to or more effective than the three modified atmospheres in reducing emergence of the rice weevil (Sitophilus oryzae (L.)) under the control emergence at 80°F. Table 3 shows that it took 2-3 wk for the 60% CO_2 atmosphere to give complete kill and that 60°F was not as effective in reducing emergence as was 35°F.

Table 2.--Percent reduction in emergence (PRE) when immature rice weevils were exposed to air or to one of three modified atmospheres at 35°F.[*]

ATMOSPHERE	PRE after exposure of (wk)			
	1	2	3	4
AIR	98.7	100.0	100.0	100.0
60% CO_2	95.6	99.4	99.8	99.9
98% CO_2	99.8	100.0	100.0	100.0
99% N_2	94.5	99.1	99.9	100.0

Table 3.--Percent reduction in emergence (PRE) when immature rice weevils were exposed to air or to one of three modified atmospheres at 60°F.[*]

ATMOSPHERE	PRE after exposure of (wk)			
	1	2	3	4
AIR	67.1	88.5	89.3	89.1
60% CO_2	80.5	99.0	100.0	100.0
98% CO_2	97.4	97.2	99.6	99.9
99% N_2	41.7	86.0	82.7	99.7

[*]Mean adult emergence for all controls at 80°F. was 1 wk - 63.8; 2 wk - 89.6; 3 wk - 172.8; and 4 wk - 239.5.

Other laboratory studies have shown that lowering the relative humidity increases the effectiveness of modified atmospheres. Jay et al. (1971), working with adults of the confused flour beetle, Tribolium confusum J. duVal, the red flour beetle, T. castaneum (Herbst), and the sawtoothed grain beetle, Oryzaephilus surinamensis (L.), found that decreasing the relative humidity in atmospheres containing 99% N_2 (balance O_2) from 68% to 9% gave an increase in mortality of from 3 to 98.5% in a 24-hr exposure of the red flour beetle. The two other insects showed a similar response to reduced relative humidity (Table 4). These three species also exhibited a similar response to mixtures of CO_2 in air at lowered relative humidities.

Table 4.--Mortality of the red (RFB) and confused (CFB) flour beetles exposed 24 hr and mortality of the sawtoothed grain beetle (STGB) exposed 6 hr to 99% N_2 and 1% O_2.

% RH	% Mortality		
	RFB	CFB	STGB
68	3.0	5.2	4.1
54	75.9	39.1	17.0
33	94.8	95.9	27.5
9	98.5	98.1	40.0

Dessication plays a large role in the mortality of stored-product insects when exposed to some modified atmospheres. Jay and Cuff (1981) showed that when larvae, pupa, and adults of the red flour beetle were exposed to varying concentrations of CO_2 or O_2, weight loss was much higher in some of the atmospheres than in others or in air. Table 5 shows that when larvae of this species were exposed to either 99% N_2 or 58% CO_2, weight loss was higher than in those exposed to air, 97% N_2, or 97% CO_2. However, high weight loss cannot always be correlated with high mortality since weight loss was comparatively low but mortality was high in those insects exposed to 97% CO_2.

Other laboratory studies have shown that the susceptibility of different species or strains of the same species varies considerably when insects are exposed to the same concentrations of modified atmospheres (Jay and Pearman, 1971).

Table 5.--Mean larvae weight loss and mean number of deaths for each group of 25 red flour beetle exposed to modified atmospheres.[1]

Atmosphere (%)	Time (hr)		
	24	48	72
Weight loss (10^{-4} g)			
Air	54	66	157
99 N_2	117	190	258
97 N_2	99	118	184
97 CO_2	91	135	185
58 CO_2	149	201	247
Mortality (number out of 25)			
Air	0.2	0.1	1.3
99 N_2	2.8	22.0	25.0
97 N_2	1.7	1.9	3.2
97 CO_2	22.7	22.3	24.7
58 CO_2	3.9	12.0	20.1

[1] From Jay and Cuff, 1981.

Storey (1980a) has conducted mortality studies on several species of insects using combustion gas produced by a laboratory scale generator. This work is best summarized in Table 6.

This table shows the different effect that the same modified atmosphere has on the species and different life stages in each species. Pupae and mature larvae of internally developing species such as the rice weevil complex (Sitophilus spp.), the granary weevil (Sitophilus granarius (L.)), and the lesser grain borer were generally the most tolerant stages while early-instar larvae and adults of externally developing species such as the flour beetles, the Indianmeal moth (Plodia interpunctella (Hübner)), and almond moth (Cadra cautella (Walker)) were the most susceptible. Eggs, which often exhibit a high degree of tolerance to chemical fumigation, were not especially tolerant of the generated atmosphere, but the eggs of some species such as the flour beetles tended to become more tolerant after the first day of their development. For example, 1-day-old eggs of the red flour beetle required only a 25 hr exposure to cause 95% mortality, but 4-day-old eggs required nearly 40 hr of exposure to attain the same level of mortality.

Table 6.—A comparison of time (hr) required for 95 or 100% mortality of various stages of insects exposed at 80°F and 50 $\pm$ 5% RH to an atmosphere produced by a modified atmosphere generator: Composition, 1.0% O_2 and 9.0-9.5% CO_2, the balance principally N_2.[1]

Insect	Eggs[2]	Larvae[3]	Pupae[2]	Adults[4]
Rice weevil	70	79–246	107–241	48
Granary weevil	85	38–137	120–148	55
Lesser grain borer	72	72–192	144–216	36
Angoumois grain moth	48	72–120	120	24
Confused flour beetle	30–40	7–20	24–53	17
Red flour beetle	25–40	8–23	17–47	18
Indianmeal moth	24	8	24	8
Almond moth	48	8	24	8
Cowpea weevil	96	120–192	192	48

[1] Adapted from Storey (1980a).
[2] Range over each day of development.
[3] Range over each week of development.
[4] Minimum exposure time for 100% mortality.

FIELD STUDIES

Although CO_2, N_2, and combustion gases are effective in controlling stored-product insects, one may be more effective than the other two in a particular situation. The use of CO_2 or N_2 involves a small capital outlay for installation of a supply tank and application lines, while a generator requires a large investment. However, once purchased, the generator will continue to supply product with maintenance and fuel being the only costs involved while the use of CO_2 or N_2 involves repeated purchase of the product and the costs involved in leasing the supply tank and other equipment.

Nitrogen can be used effectively in situations where the storage vessel is tightly sealed, but in leaking storages attaining and maintaining the concentration of this gas above the effective level of 99% (Table 1) is difficult and expensive. Carbon dioxide can be used in storage containers where some leaks are encountered because it is effective against stored-product insects at concentrations around 50% (Table 1) and becomes very effective at a concentration of 60% or more. Field research on the use of modified atmospheres at USDA's Stored-Product Insects Research and Development Laboratory, Savannah, Georgia has been concentrated on CO_2 because of this and other factors which will be discussed later.

Carbon Dioxide

It has been known for some time that atmospheres high in CO_2 are lethal to insects. In Australia, a dosage of 0.72 kg (1.6 lb) of CO_2 per ton of grain in an airtight silo was the most effective fumigant when compared to carbon bisulphide and hydrogen cyanide (Anon. 1917). Later in Australia, Frogatt (1921) recommended a rate of 1.4 kg (3.1 lb) of CO_2 per ton of corn in galvanized iron tanks. Little subsequent interest was shown in this technique except in laboratory studies until Jay et al. (1970) attained and maintained effective CO_2 concentrations for 2, 4, and 7 days in an upright concrete silo containing 68,000 bu of in-shell peanuts. Later, Jay and Pearman (1973) controlled a natural infestation of the rice weevil complex and the Angoumois grain moth, Sitotroga cerealella (Olivier) in 28,000 bu of corn in an upright concrete silo. In this test, a CO_2 concentration of about 60% (balance air) was successfully attained and maintained over a 96-hr period. Table 7 shows that over 99.9% reduction of all species of insects was obtained as determined from corn samples taken before and after treatment. There was also a 99% reduction in damaged kernels. During this period Jay (1971) also published suggestions on how to use CO_2 to control insects in grain storage facilities.

Australia began large scale field tests with CO_2 in 1976 (Banks, 1979). In the first test, gaseous CO_2 was released at 3 points into the base of a 7,000 metric t welded metal bin containing wheat and the pressure of the CO_2 eventually pushed the existing atmosphere out of the top of the bin. A second test was conducted in a similar manner except an air pump was used to blend air with the CO_2 so that instead of a 100% concentration going into the bin a

Table 7.--Average number of insects per sample and damage to 28,000-bu of corn. Samples collected before and after a 4-day treatment of the corn with CO_2.

Sample examined	No insects		% damage	
	Before treatment	After treatment	Before treatment	After treatment
Initially	1	1	1.3	0.7
1 month	25	1	2.5	0.2
2 months	204	1	16.3	0.6

70% concentration was released, thus saving CO_2. In the first test, a CO_2 level of 97% was obtained in a 30 hr purge while a level of 72% was obtained in 11 hrs in the second test. After 8.5 days, and without adding CO_2, there was still an average of 42% CO_2 in the bin. Since these tests, Australian researchers have conducted several additional studies on large grain storage facilities including sealing and subsequent treatment of a 16,000 t flat storage with CO_2 (Banks et al., 1979). These researchers have studied sealing extensively, since CO_2 costs are much higher in Australia than in the U.S. (Banks and Annis, 1980), and every effort is made to prevent gas loss. Also, well-sealed bins hold conventional fumigants much better than do leaky ones.

Currently in Australia, all 10 terminal elevators in the State of Victoria have been sealed and modified for the use of CO_2 (Anon., 1980). In Western Australia, the Co-Operative Bulk Handling, Ltd. currently has 25 horizontal grain storages with an average of 25,000 t capacity each sealed for CO_2 use and in operation and many of their vertical steel and concrete storages are being sealed for this purpose. Research or actual storage practices with CO_2 atmospheres are also being conducted in the Australian States of Queensland, New South Wales, and South Australia.

Interest in using CO_2 has also increased in this country but not to the extent it has in Australia. Several large CO_2 producers have applied to EPA for approval for their product and two of these companies have obtained a use label. The Stored-Product Insects Research and Development Laboratory has recently cooperated in tests with two of these companies in applying CO_2 into storage facilities ranging from terminal elevators to inland terminals and on-farm storage situations. These tests and future tests usually involve the cooperation and interest of a grain company. Also, these tests are necessary so that we can further develop methods of sealing and application under U.S. conditions and to study the economics of this method. Future tests can be guided by the publication by Jay (1980b) which details the actual techniques involved in using CO_2 in upright concrete silos, and describes field tests using these application methods. In brief, three methods are discussed for purging the existing atmosphere from a silo with CO_2: (1) applying the CO_2 into the top of a full silo; (2) applying CO_2 into the bottom of a full silo; and (3) adding the CO_2 into the grain stream as the silo is filled.

In one of the more recent studies at a large terminal elevator, an upright concrete silo containing 40,000 bu of wheat was used for two tests. Carbon dioxide was supplied from a 12-t vessel and vaporizers. Naturally infested wheat was used in the first test and a 96-hr treatment produced a 100 percent reduction in emergence (PRE) as determined by insect counts in pretreatment and post-treatment samples about 7 days after the treatment was completed (Table 8). When the samples were examined 15 days after treatment

Table 8.--Treatment of 40,000-bu of wheat having a natural infestation of stored-product insects.

Length of treatment	96 hr
CO_2 concentration	60 + 10%
CO_2 cost	$0.0064/bu
Percent reduction in insect emergence --	
Immediately after treatment	100%
15 days after treatment	68.8%
30 days after treatment	95.4%
60 days after treatment	99.5%

some adults had emerged from the post-treatment (treated) sample and the PRE dropped to 68.8%. These emerging adults were probably pupae during the treatment and survived in this stage due to their low respiration rate. However, 30 and 60 days after treatment the PREs were 95.4 and 99.5%, respectively, indicating a high level of control. Those insects emerging after treatment were not able to reproduce and eventually died.

A probable reason for some survival in this test was discovered in the second study conducted in this silo with caged insects. In a treatment of similar duration (4-days) the PRE was from 75 to 95% in mixed-age immature (larvae and pupae) rice weevils and from 22 to 92% in mixed-age immature lesser grain borers. The cages with the highest PREs were located below the CO_2 injection nozzle in the headspace of the silo while those with low PRE were located above the injection nozzle. Since CO_2 is 1 1/2 times as heavy as air it is assumed that insufficient gas reached the cages in the grain in the higher portion of the silo.

In a test on 32,000 bu of milo (sorghum) in an upright concrete silo at another terminal elevator, a 98.5 PRE of mixed age immature rice weevils was obtained during a 4-day treatment. Concentrations of from 60 to 85% CO_2 were attained and maintained in the silo during this treatment. Tests were also conducted on 32,000 bu of wheat in one silo at this elevator and concentrations of at least 60% CO_2 were attained and maintained throughout a 4-day treatment. Insect bioassay data on a natural population was inconclusive in this test since the pretreatment samples had a very low rate of infestation.

Studies were conducted with naturally infested wheat stored on a farm in South Carolina in 2 Harvestore(R) silos with capacities of 14,000 bu and 6,000 bu, respectively. This wheat was heavily infested with several species and life stages of stored-product insects including the granary weevil, grain beetles, Cryptolestes spp., the red flour beetle, and the black carpet beetle, Attagenus megatoma (F.). Concentrations of 62-80% CO_2 were maintained in these silos for 5 days and the PRE ranged from 95.3 in the top of the 14,000 bu bin to 99.9+ in the bottom of the bin while a 99.9+ PRE was observed in grain samples taken from the 6,000-bu bin. Temperatures of the grain ranged from 82-92°F in these tests.

Carbon dioxide has shown promise for use in situations other than static grain or oilseed storage. Tests were conducted by a large miller and baker in cooperation with the Stored-Product Insects Research and Development Laboratory to determine if CO_2, as dry ice, can replace conventional fumigants for in-transit treatment of hopper cars moving from mills to bakeries (Ronai and Jay, 1982). Dry ice in the form of blocks and pellets (small extruded pieces of dry ice) were placed in the top of the hopper cars after they were filled. Caged red flour beetle larvae and gas sampling lines were distributed through the flour. The cars were shipped approximately 750 miles from the mill to the bakery. Mean mortality among the test insects was 97.9% at the end of the trip and the CO_2 concentrations in the cars at this time, 9 to 10 days after filling, ranged from 25 to 61% (Table 9).

Nitrogen

Interest in the use of N_2 for controlling stored-product insects began in Australia in the early 1970's and continued through 1976. These studies and Australian recommendations for the use of N_2 in grain storage are described by Banks and Annis (1977). Also, considerable research was conducted in Italy on the use of N_2 for long-term grain storage. This research lead to the development of a total marketable system including metal silos, conveyors, and

Table 9.--Pounds of CO_2 (dry ice) applied, percent CO_2 concentration at end of travel, and insect mortality in four 4245 ft^3 hopper cars containing flour.[1]

Car	Lbs CO_2 applied [2]	Range of CO_2 concentrations (%)	Insect mortality (%)
1	200 - 200	28 - 46	99.1
2	200 - 200	25 - 40	99.1
3	200 - 200	32 - 41	95.2
4	400 - 200	33 - 61	98.3

[1] From Ronai and Jay, 1982.
[2] Lbs pellets - lbs block (dry ice).

Based on these and laboratory studies the following temperature and exposure time guidelines are recommended for treating grain or oilseeds with 40 to 60% CO_2[1]:

Temperature of Commodity[2]	Days Exposure No
80°F	4-7
70°F	10-14
60°F	21-28

[1] Add CO_2 if concentration drops below 40%.
[2] Do not use below 60°F.

Nitrogen (Continued)

equipment for applying N_2 and the N_2 itself by Assoreni, a member of the ENI group of companyies for scientific research (Tranchino, et al., 1980). In the U.S. only one recent large scale field study has been published. Barrere (1980) treated 13,000-bu of wheat in an upright concrete silo with N_2. The silo was leaky and large quantities of N_2 were necessary to maintain the concentration above 98.5%. A one week treatment in this test resulted in 100% mortality of adult rice weevils, a life stage of this insect which is very easy to control. Mortality for shorter exposures was not given in this report.

Generated Atmospheres

Although modified or "inert" atmosphere generators ("burners") have been used for many years to produce inert atmospheres for food preservation and storage, little research has been conducted on their use for insect control. The basic principle of burning the O_2 out of the air and then passing it into infested commodities is very simple. However, water is a product of combustion and must be removed from the generator's effluent along with all hydrocarbons before the product is suitable for treatment of grain, oilseeds, or processed agricultural commodities. Removal of these by-products is difficult and may make a generator expensive.

Storey (1973) treated 20,000-bu of wheat with a generated atmosphere in upright concrete silos. The two generators used were capable of producing 15,000 ft^3 per hour of the oxygen-deficient atmosphere, which was composed of 0.1% O_2, 8.5-11.5% CO_2 and the balance N_2. Three tests were conducted and mortality of immature rice weevils ranged from 43 to 70% in a 96 hour exposure. In a 72-hr exposure mortality of adult confused flour beetles was 100% while mortality of immature rice weevil ranged from 39 to 100%.

ECONOMICS OF TREATMENT

Several factors influence the cost of CO_2 or N_2 treatments. The largest of these are the costs of CO_2 or N_2 which are based on yearly usage, transportation costs, and competition among producers. Other costs include equipment rental (tank, vaporizer, etc.) and labor to apply the materials. Loss of gases through leaks or the need to apply more product to maintain the concentration also contributes greatly to the cost. In Australia, where sealing is rigorously practiced before using modified atmospheres, an adequate treatment is considered to be 1 metric t of CO_2 per 1,000 m t of wheat (Banks and Annis, 1980). At a cost of \$75/t for CO_2 this is equivalent to a treatment cost of less than 0.2 cents (c) (U.S.)/bu. However, where sealing costs are high and CO_2 costs are low, as is the case in the U.S., it may not be profitable to do extensive sealing of storage vessels before treatment.

Currently, there is an abundant supply of CO_2 in the United States. Carbon dioxide is a byproduct of the productions of ammonia, ethylene oxide, and alcohol, and is also obtained from natural CO_2 wells. Nitrogen is generally produced by the energy-intensive process of separating it from the other components of air and is generally more expensive per unit than CO_2.

Jay (1980b) estimated CO_2 costs to treat 28,000 bu of corn in an upright concrete silo to be from 1.75 to 3.5 c/bu. However, this was based on a CO_2 cost of from \$104 to \$180/t. Recently, costs have been quoted as low as \$56/t, making the original estimates quite high. At \$56/t, those costs would range from 0.9 to 1.1 c/bu for the CO_2, depending on the method of application.

Costs for CO_2 in the two previously described studies on wheat in the terminal elevators were estimated to be 0.64 (Table 8) to 0.70 c/bu with the price for CO_2 at \$60/t. Treatment costs in the tests with milo and corn were estimated to be 0.89 and 0.69 c/bu, respectively, for a 96-h treatment. Most of these costs are slightly higher than but compare favorably with those quoted by one elevator for the fumigant "80-20" of 0.56 c/bu.

In some situations hidden costs may arise with conventional fumigants which may not be associated with modified atmosphere treatments. For example, in the CO_2 treatment of hopper cars containing flour the cost for the conventional fumigant was \$17/car and for the CO_2 treatment \$24/car. However, prior to unloading a car treated with conventional fumigant 1-1/2 man hours are required for aerating the car. This process is not necessary with CO_2 and reflects a net saving per car of about \$18. Additionally, cars can be unloaded more rapidly since they are not sitting idle at the unloading dock during the aeration period. A large rice processor in Texas has found that portions of the plant undergoing treatment with CO_2 do not have to be shut down as they had to be when they were using the fumigant methyl bromide for insect control.

An analysis of sealing and treatment costs for the use of modified atmospheres in Australia can be found in Connell and Johnston (1981). When reviewing this publication, however, it must be remembered that costs for CO_2,

N_2, and for labor and materials for sealing will be considerably higher in Australia than they are in this country.

QUALITY

Banks (1981) reviewed the effects of modified atmosphere storage on quality. In general, atmospheres high in CO_2 or N_2 with reduced O_2 maintain the quality of grain stored over long periods. This is especially true in the storage of high moisture grain. These atmospheres usually retard or stop the growth of fungus and thereby eliminate the production of mycotoxins, including aflatoxins. However, the atmospheres will not kill the fungus and once grain that would normally grow fungus is returned to a normal atmosphere fungus and toxins can again be produced.

Wilson and Jay (1975) showed that freshly harvested Iowa grown corn having a moisture content (mc) of 29.4% and remoistened corn having a mc of 19.6% could be protected from mold and aflatoxin production (Table 10) and from an increase in free fatty acids when stored in a high N_2 atmosphere or one low in O_2 with 15% CO_2 in the mixture ("burner" atmosphere). However, the free fatty acids of corn held in 60% CO_2 increased when compared with corn stored in the other two atmospheres. Aflatoxin synthesis did not increase to 2 μg/kg in all 3 of these atmospheres during 4-weeks of storage but increased to >1021 μg/kg in the corn held in air for only 2 weeks.

This information provides suggestions for another use for modified atmospheres. When high moisture grains or oilseeds are harvested it could go into silos or bins with modified atmospheres to avoid being stored under conditions conductive to the production of fungus and mycotoxins. In

Table 10.--Aflatoxin content (μg/kg) of high moisture (19.6%) corn innoculated with _Aspergillus_ _flavus_ and exposed to 3 modified atmospheres or to air.[1]/

Weeks Exposed	Atmospheres			
	Burner	N_2	CO_2	Air
1	10	3	2	400
2	2	2	6	>1021
3	2	2	2	>1021
4	2	2	2	-

[1]/ From Wilson and Jay, 1975.

some cases, this could avoid capital outlay for a new drying system to augment existing equipment, particularly in years when high moisture grain is harvested.

Modified atmospheres have no detrimental effects on the overall storability of grain. Germination, milling, and breadmaking characteristics of wheat were unaffected by exposure to a generated atmosphere of less than 1% oxygen for periods of 1/2 to 6 months (Storey 1980b). Similarly, germination, milling, and cooking properties of long-and medium-grain rice exposed as rough, brown, and milled rice were unchanged after 6-months of treatment (Storey 1980b). No adverse effect on the quality of malt produced from barley stored in low O_2 atmospheres was observed and no significant reduction was obtained in the germination of the barley (Storey et al., 1977).

SUMMARY

Laboratory and field research in the U.S. and in other countries has shown that modified atmospheres are a viable alternative to residue producing conventional fumigants for controlling insects attacking stored commodities. Laboratory research has developed modified atmosphere concentration x temperature x time parameters for several species and life stages of stored-product insects when exposed to CO_2, N_2, or product from a modified atmosphere generator. These laboratory tests have also provided some information on the mode of action of these atmospheres and on the effects of relative humidity on exposure time for insect control. Modified atmospheres have been shown effective in reducing fungus growth and subsequent mycotoxin production, thus providing additional protection for stored commodities.

Limited field studies in this country have proven that this control technique is effective in controlling insects and also is cost-competitive with conventional fumigants. However, field studies are still needed to develop data applicable to commodity storage and transport within and from the U.S. This research should include further studies on application and distribution techniques, sealing, and the development of sound economic data so that industry can fully evaluate the method and determine if it fits into their particular marketing programs.

The technique also shows potential for use in in-transit situations such as hopper cars, ocean freight containers, river and ocean-going barges, and ships in the export trade. All of these vehicles have not been thoroughly studied for their potential in insect control systems using modified atmospheres.

LITERATURE CITED

Anonymous. 1917. Grain weevils - a menace to the wheat stacks. J. Dept. Agric. S. Australia. 20:977.

Anonymous. 1980. CO_2 protecting Australian crops. Milling, Feed and Fertilisier. 163:3. 1980.

Banks, H. J. 1979. Recent advances in the use of modified atmospheres for stored product pest control. In. R. Davis and A. Taylor (eds.). Proc. 2nd Int. Working Conf. Stored-Prod. Entomol. Ibadan, Nigeria (1978), pp. 198-217.

Banks, H. J. 1981. Effects of controlled atmosphere storage on grain quality: a review. Food Tech. in Australia. 33: 335-340.

Banks, H. J., and P. C. Annis. 1977. Suggested procedures for controlled atmosphere storage of dry grain. CSIRO Aust. Div. Entomol. Tech. Pap. No. 13, 23 pp.

Banks, H. J., and P. C. Annis. 1980. Conversion of existing grain storage structures for modified atmosphere use. In J. Shejbal (ed.). Proc. Int. Symp. on Controlled Atmosphere Storage of Grains. Elsevier. pp. 461-473.

Banks, H. J., P. C. Annis, and J. R. Wiseman. 1979. Permanent sealing of 16,000 tonne capacity shed for fumigation or modified atmosphere storage of grain. CSIRO Aust. Div. Entomol. Rep. No. 12. 16 pp.

Barrere, J. A. 1980. Use of nitrogen in cereal grain fumigation, storage. Grain Age. 21:9-10, 12, 18.

Connell, P. J., and J. H. Johnston. 1981. Costs of alternative methods of grain insect control. Bureau of Agricultural Economics, Canberra, Australia. Occasional Paper No. 61. 77 pp.

DeLima, C. P. F. 1980. Field experience with hermetic storage of grain in Eastern Africa with emphasis on structures intended for famine reserves. In J. Shejbal (ed.). Proc. Int. Symp. on Controlled Atmosphere Storage of Grains. Elsevier. pp. 39-53.

Frogatt, W. W. 1921. Fumigating maize with carbon dioxide. Agric. Gaz. N.S.W. (Australia). 32:472.

Girish, S. K. 1980. Studies on the preservation of foodstuffs under natural airtight storage. In J. Shejbal (ed.). Proc. Int. Symp. on Controlled Atmosphere Storage of Grains. Elsevier. pp. 15-24.

*Jay, E. G. 1971. Suggested procedures and conditions for using carbon dioxide to control stored-product insects in grain and peanut facilities. ARS 51-46. 6 pp.

*Jay, E. G. 1980a. Low temperatures: Effects on control of Sitophilus oryzae (L.) with modified atmospheres. In J. Shejbal (ed.). Proc. Int. Symp. on Controlled Atmosphere Storage of Grains. Elseiver. pp. 65-72.

*Jay, E. G. 1980b. Methods of applying carbon dioxide for insect control in stored grain. USDA, SEA, Advances in Agricultural Technology, Southern Series, S-13. 7 pp.

*Jay, E. G., and G. C. Pearman. 1971. Susceptibility of two species of Tribolium (Coleoptera: Tenebrionidae) to alterations of atmospheric gas concentrations. J. Stored Prod. Res. 7:181-186.

*Jay, E. G., and G. C. Pearman. 1973. Carbon dioxide for control of an insect infestation in stored corn (maize). J. Stored Prod. Res. 9:25-29.

*Jay, E. G., and W. Cuff. 1981. Weight loss and mortality of three stages of Tribolium castaneum (Herbst) when exposed to four modified atmospheres. J. Stored Prod. Res. 17:117-124.

*Jay, E. G., L. M. Redlinger, and H. Laudani. 1970. The application and distribution of carbon dioxide in a peanut (groundnut) silo for insect control. J. Stored Prod. Res. 6:247-254.

*Jay, E. G., R. T. Arbogast, and G. C. Pearman. 1971. Relative humidity: Its importance in the control of stored-product insects with modified atmospheric gas concentrations. J. Stored Prod. Res. 7:325-329.

Kamel, A. H. 1980. Underground storage in some arab countries. In J. Shejbal (ed.). Proc. Int. Symp. on Controlled Atmosphere Storage of Grains. Elsevier. pp. 25-38.

Ronai, K. S., and E. G. Jay. 1982. Experimental studies on using carbon dioxide to replace conventional fumigants in bulk flour shipments. A.O.M. Tech. Bull., August, pp. 3954-58.

Shejbal, J. (ed.). 1980. Proceedings International Symposium on Controlled Atmosphere Storage of grains. Elsevier. Amsterdam. 608 pp.

Sigaut, F. 1980. Significance of underground storage in traditional systems of grain production. In J. Shejbal (ed.). Proc. Int. Symp. on Controlled Atmosphere Storage of Grains. Elsevier. pp. 3-15.

Storey, C. L. 1973. Exothermic inert-atmosphere generators for control of insects in stored wheat. J. Econ. Entomol. 66:511-14.

Storey, C. L. 1980a. Mortality of various stored product insects in low oxygen atmospheres produced by an exothermic inert atmosphere generator. In J. Shejbal (ed.). Proc. Int. Symp. on Controlled Atmosphere Storage of Grains. Elsevier. pp. 85-92.

Storey, C. L. 1980b. Functional and end-use properties of various commodities stored in a low oxygen atmosphere. In J. Shejbal (ed.). Proc. Int. Symp. on Controlled Atmosphere Storage of Grains. Elsevier. pp. 311-317.

Storey, C. L., Y. Pomeranz, F. S. Lai, and N. N. Standridge. 1977. Effect of storage atmosphere and relative humidity on barley and malt characteristics. Brewer's Dig. 52:40-43.

Tranchino, L., P. Agostinelli, A. Costantini, and J. Shejbal. 1980. The first Italian large scale facilities for the storage of cereal grains. In J. Shejbal (ed.). Proc. Int. Symp. on Controlled Atmosphere Storage of Grains. Elsevier. pp. 445-461.

Wilson, D. M., and E. G. Jay. 1975. Influence of modified atmosphere storage on aflatoxin production in high moisture corn. Appl. Microbio. 29:224-228.

*Copies of these papers may be obtained from the author.

LOW TEMPERATURES TO CONTROL STORED-PRODUCT INSECTS

Michael A. Mullen and Richard T. Arbogast

Editor's Comments

This biological type of control is known to be effective in grain storage. It is a possible alternative to the use of insecticides but certainly should be used as an adjunct.

Remember, insects do acclimatize. It is necessary, therefore, to know what the effective temperature is for any given species (a high of 5°C (41°F) has been suggested).

Additional basic studies are needed to permit optimum use of this approach.

F. J. Baur

LOW TEMPERATURES TO CONTROL STORED-PRODUCT INSECTS

Michael A. Mullen and Richard T. Arbogast

Stored-Product Insects Research
and Development Laboratory
Agricultural Research Service
Savannah, Georgia 31403

INTRODUCTION

Control of stored-product insects using low temperature offers an alternative to chemical control. With the ever increasing restrictions being placed on the use of insecticides because of their potential human and environmental hazards, a need has been created for a simple, safe, and effective means of controlling insect pests in the marketing channels. The technology needed to control pest insects with low temperatures exists, but biological information is needed to implement the technology in an effective and efficient manner. Cooling units of various capacities are common and, with few modifications, could be used in treating grain and packaged products. However, fundamental information is lacking on lethal exposures of each pest to low temperatures; on the effects, if any, on the cooling of each commodity; and on field testing of cold treatments.

GENERAL EFFECTS OF LOW TEMPERATURES

Before we can effectively utilize low temperatures to control insects, it is first necessary to understand the response of insects to cold. Insects are poikilotherms, that is their body temperature follows closely that of the environment. In general, as an insect's body temperature is lowered its activity level decreases until it comes to rest and shows no activity. A further decrease in temperature will result in death. This thermal death point varies with each insect species and is dependent on both temperature and the time spent at that temperature (Ashina, 1966).

Insects fall into three groups according to their resistance to cold. Wigglesworth (1972) identifies these as: (1) insects that are killed by temperatures well above the freezing point; (2) insects that are killed as soon as their tissues freeze; and (3) insects that can withstand subfreezing temperatures by supercooling.

Insects that are of tropical origin, such as many that infest stored-products, are often killed by temperatures well above freezing. The cause of death is not completely understood, but may be attributed to the accumulation of toxic products which otherwise would be eliminated at normal temperatures. Also, in pyralid larvae oxygen consumption continues down to -12°C, but absorption of nutrients from the gut ceases and the larvae die of starvation (Kazhantshikov, 1938). Insects that die as soon as their tissues freeze are probably killed by dehydration or by mechanical injury caused by the formation of ice crystals.

Most insects are able to supercool which accounts for their ability to survive at temperatures below freezing. This is probably the chief factor in cold resistance (Somme, 1968). As the temperature of the insect is lowered, the water in the insect's body becomes supercooled, and ice does not form until the temperature falls to a critical point (-10° to -15°C) when it suddenly jumps up to about -1.5°C through the release of latent heat. At this critical point the temperature proceeds to fall once more and freezing of the tissues begins (Wigglesworth, 1972). The critical point represents the lethal limit of the low temperature and varies among species (Salt, 1961). Full grown larvae of Cadra in the early prepupal stage, and in the pupal stage, have critical points of -5.8, -8.0, and -23.3°C, respectively (Salt, 1936). Newly hatched larvae of Cadra and Sitotroga have a critical point of -27°C and after feeding it is raised to -6°C (Salt, 1936).

The factors which influence supercooling are not well understood. Injury can reduce the capacity of an insect to supercool. In Bruchus sp. the critical point is normally -15 to -20°C. However, if the exoskeleton of the adult is pierced the critical point rises to -8 to -10°C (Salt, 1936). Repeated freezing and thawing eliminates supercooling and freezing occurs as soon as the freezing point is reached (Wigglesworth, 1972).

ACCLIMATION TO LOW TEMPERATURES

Insects infesting stored grain are protected from sudden changes in temperature by the insulation provided by the grain. In cold weather the insects tend to migrate from the outer areas of the grain to the interior where temperature changes are more gradual. Although many insects may die during the cold periods many will survive because of acclimation to the lower temperatures. When non-acclimated adults of the rusty grain beetle, Cryptolestes ferrugineus (Stephens), were exposed to -12°C, there was no survival after 72 hours. However, following 4 weeks of acclimation at 15°C, survival at -12°C increased to 60.8% after 4 weeks (Smith, 1970). The supercooling point was found to be -16.7°C for non-acclimated insects and -20.4°C for acclimated insects. Somme (1968) reported that fewer confused flour beetle adults, Tribolium confusum Jacq. duVal, acclimated at 12°C died when exposed to 0°C than those acclimated at 27°C. Evans (1977) found that the chill-coma temperature (temperature at which activity ceases) of the rice weevil, Sitophilus oryzae (L.) was lowered from 8.4°C for unacclimated adults to 5.3°C for adults previously exposed to 15°C for 8 weeks. Under the same conditions the mean chill coma temperature of the granary weevil, Sitophilus granarius (L.), was lowered from 5.3 to 2.7°C. Ernst and Mutchmor (1969) reported that adults and larvae of the yellow meal worm, (Tenebrio molitor (F.)), confused flour beetle, and the warehouse beetle, (Trogoderma variable Ballion), dispersed more at low temperatures after acclimation at 15 or 23°C than did insects acclimated at higher temperatures. David et al. (1977) found that acclimation of various stages of the rice weevil, the granary weevil and the lesser grain borer, Rhyzopertha dominica (F.), increased survival of all species at low temperatures.

These studies show that insects can acclimate to low temperatures and that this acclimation increases their rate of survival. Before control with low temperatures can be effective we must know precisely what temperatures will kill the insects.

LETHAL EFFECTS OF LOW TEMPERATURES

The effects of low temperatures on stored-product insects have not been widely investigated. Mullen and Arbogast (1979) presented information necessary to control the eggs of stored-product insects in packaged commodities by means of lethal low temperatures (LT). Eggs of the sawtoothed grain beetle (Oryzaephilus surinamensis (L.)), the red flour beetle (Tribolium castaneum (Herbst)), the cowpea weevil (Callosobruchus maculatus (F.)), the cigarette beetle (Lasioderma serricorne (F.)), and the almond moth (Cadra cautella (Walker)) were exposed to temperatures from -20 to +5°C. The LT$_{95}$ for each species was calculated by probit analysis. Exposures of 7-9 hours at -10°C were sufficient to kill the eggs of all but the cowpea weevil and the cigarette beetle, which required 28 and 62 hours, respectively. At -20°C death occurred in a few minutes in all but the cowpea weevil which succumbed in 10 hours. Survival times decreased rapidly as the temperature dropped below freezing.

These results are similar to those reported by Adler (1960) where eggs of the Indianmeal moth, Plodia interpunctella (Hübner), were killed by a 4 hour exposure to -16.7°C. Exposure of Indianmeal moth eggs to 2.4°C for 144 hours reduced egg hatch by 77% (Cline, 1970). Mortality of the Mediterranean flour moth eggs, Anagasta kuehniella (Zeller), was 100% after 43 days at 10°C and after 7 days at -9.5°C (Mathlein, 1961). Larvae and pupae of the Mediterranean flour moth were able to survive with no deleterious after effects at fluctuating temperatures between -4 and -10°C for 12 days, but were killed after 15 days (Mathlein, 1961). Jay (1980) found that 100% of the immature rice weevils exposed to 1.6 or 4.7°C were killed after 2 and 3 weeks exposure, respectively. However, at 10.4 and 15.7°C after 4 weeks 97.9% and 89.1% mortality, respectively, was achieved. Jay (1980) concluded that when grain temperatures were below 10.4°C, no further control techniques against the rice weevil were necessary.

The optimal temperatures which allow stored-product insect populations to increase are between 25 and 39°C and temperatures below this have an adverse influence. However, temperatures must be lowered to below 15°C to prevent feeding and reproduction. Howe (1965) summarized the optimum and minimum temperatures required for population growth (Table 1).

Cox (1974) found that eggs of the raisin moth, Cadra figulilella (Greyson), did not hatch below 12.5°C and that only 10% hatched at 15°C. In addition at 15°C no larval development occurred. Development of the Indianmeal moth and the almond moth was completely arrested at 15°C while development of the tobacco moth, C. elutella (Hübner), and the Mediterranean flour moth was stopped at 10°C (Bell, 1975). A temperature of 5°C has been suggested by Navarro (1974) to control most stored product insects.

METHODS OF CONTROL

Small quantities of foodstuff can be easily disinfested in small home type freezers or in larger commercial units. Mullen and Arbogast (1979) used a home type 0.76m^3 (27 ft.^{3}) freezer and determined the times required for various commodities to reach equilibrium with the freezer at 3 different temperatures. Their results are given in Table 2. Disinfestation of any quantity of a commodity can be accomplished by determining the time necessary for the temperature of the commodity to reach equilibrium with the temperature of the freezer and then adding the time necessary to kill the insects.

Table 1. Estimated optimal and minimal temperatures (°C) for population increase for selected stored-grain insects. (Adapted from Howe, 1965 by Mills, 1978).

Species	Min.	Opt.
Almond moth		
Cadra cautella (Walker)	17	28-32
Angoumois grain moth		
Sitotroga cerealella (Olivier)	16	26-30
Broadhorned flour beetle		
Gnatocerus cornutus (Fab.)	16	24-30
Confused flour beetle		
Tribolium confusum Jacq. duVal	21	30-33
Cowpea weevil		
Callosobruchus maculatus (Fab.)	22	30-35
Flat grain beetle		
Cryptolestes pusillus (Schonherr)	22	28-33
Grain mite		
Acarus siro (L.)	7	21-27
Granary weevil		
Sitophilus granarius (L.)	15	26-30
Indianmeal moth		
Plodia interpunctella (Hubner)	18	28-32
Khapra beetle		
Trogoderma granarium Everts	24	33-37
Lesser grain borer		
Rhyzopertha dominica (Fab.)	23	32-35
Mediterranean flour moth		
Anagasta kuehniella (Zeller)	10	24-27
Merchant grain beetle		
Oryzaephilus mercator (Fauvel)	20	31-34
Red flour beetle		
Tribolium castaneum (Herbst)	22	32-35
Rice weevil		
Sitophilus oryzae (L.)	17	27-31
Rusty grain beetle		
Cryptolestes ferrugineus (Stephens)	23	32-35
Sawtoothed grain beetle		
Oryzaephilus surinamensis (L.)	21	31-34

It would not be economically practical to reduce large masses of stored grain to subfreezing temperatures by refrigeration. However, in colder climates it is possible to exploit natural cold to control insects. In colder climates simple turning of the grain to mix the colder grain found on the outside of the bin with the warmer grain in the center may keep the grain cool enough to prevent development of pests (Watters, 1966).

Aeration is another method to cool grain and to equalize temperatures throughout the grain mass. Aeration may be accomplished by mounting perforated ducts on the floor of the storage bin. At the end of each duct a fan is mounted to either blow cool outside air into the grain or pull cool air down through the mass (Holman, 1960; Calderon, 1974).

Table 2. Chilling times for selected commodities. All commodities were exposed in a 0.76m^3 (27ft.^{3}) freezer filled to capacity. (Mullen and Arbogast, 1979).

Commodity	Freezer setting (°C)	Time to 0°C (h)	Time to equilibrium (h)
Cornflakes (twenty- eight 3.2-lb. (1.45- kg) cases)	-10	7	30
	-15	6	30
	-20	5	35
Flour (seven 100-lb (45.45-kg) bags)	-10	55	160
	-15	29	130
	-20	25	145
Elbow macaroni (fifteen 24-lb (10.91- kg) cases) and			
Blackeyed peas (fifteen 24-lb (10.91-kg) cases)	-10	29	130
	-15	18	95
	-20	19	100

The main advantage of using aeration is that it accelerates the rate of cooling by moving the air through the grain. By lowering and maintaining a low temperature throughout the bulk of the grain some degree of insect control can be achieved. At temperatures below 15°C most insect pests become inactive and may eventually die. Table 3 gives the results following 22 months of aeration during which the grain temperature was lowered from 27-32°C to 10-14°C (Navarro et al., 1969). In addition, aeration can be combined with insecticide treatments (Quinlan, 1972). The initial cost of the equipment is relatively high but the operating costs are low and are comparable to fumigation (Holman, 1966). Costs can be reduced by controlling the fan with a thermostat or intermittently with a timer. Williams (1974) obtained satisfactory control after 18 months in 54 ton silos by using aeration only during the coldest parts of the day.

In many parts of the world aeration has been modified by using refrigerated air (Burrell, 1974). Cool air produced in this manner must be dried to prevent condensation where the cool outside air contacts the warmer grain (Burrell and Laundon, 1967).

Mills (1978) lists several factors that influence the effectiveness of aeration to control insects:

(1) Condition of the grain mass. Sampling and monitoring of the grain should determine as accurately as possible the temperature, moisture content, extent of biological activity, and the organisms causing it, and the variability in density and particle size in the grain mass (broken kernels, dust, chaff, etc.) which influences uniformity of airflow. A hot spot caused by molds or insects may not be cooled adequately.

(2) Air-moving equipment. Design of the duct system in relation to the type of storage structure is important, as well as quality and power of the blower. The objective is an adequate and as uniform airflow as possible, in order to effectively and economically cool all parts of the grain.

(3) Weather conditions. Familiarity with the temperature and humidity characteristics of the area, particularly where adequacy of low temperatures is marginal, will permit the most effective use of the low temperatures that are available.

Table 3. Insect infestation after 22 months of storage, in a bulk cooled to 10 - 14°C by aeration with ambient air (adapted from Navarro et al., 1969).

Sampling area	Free insects in grain sample prior to aeration (mean no./kg grain) and percent survival (shown in parentheses)				
	Lesser grain borer	Rice Weevil	Tribolium sp.	Oryzaephilus sp.	Almond moth
Surface layer	1(0)	14(0)	3(0)	2(0)	14(0)
0.2 - 5-m depth	2(0)	7(0)	1(0)	0(0)	0(0)
5 - 9.5-m depth	2(0)	0(0)	0(0)	0(0)	0(0)
Aeration duct	0(0)	0(0)	2(0)	1(0)	1(0)

(4) <u>Interrelationships of temperature, relative humidity, and moisture content</u>. Understanding of the interrelationships of temperature, relative humidity, and moisture content is important, especially to prevent adding excessive moisture to the grain during aeration.

SUMMARY

The use of low temperatures to control insect losses in the marketing channels is a viable alternative to chemical control. It can be used alone or in conjunction with other control methods. Insects in packaged commodities can be effectively controlled in small home freezer units or in larger commercial size units. Grain losses can be prevented by aeration with refrigerated air or with cooler ambient air. Aeration can be used alone or in conjunction with other control methods. To effectively utilize low temperatures to control insects more basic information on its effects on the pests and on the commodities is needed.

LITERATURE CITED

Adler, V. E. 1960. Effects of low temperatures on the eggs eggs of the Angoumois grain moth, the Indianmeal moth, and the confused flour beetle. J. Econ. Entomol. 53:973-4.

Ashina, E. 1966. Freezing and frost resistance in insects. H. T. Meryman, ed., pp. 451-86 in Cryobiology. Academic Press, London.

Bell, C. H. 1975. Effects of temperature and humidity on development of four pyralid moth pests of stored products. J. Stored Prod. Res. 11:167-75.

Burrell, N. J. 1974. Chilling, in Storage of Cereal Grains and their Products. 422-453. (ed.) C. M. Christensen Amer. Assoc. Cereal Chemists, St. Paul, MN.

Burrell, N. J., and J. H. Laundon. 1967. Grain cooling studies. 1. Observations during a large scale refrigeration test on damp grain. J. Stored Prod. Res. 3:125.

Calderon, M. 1974. Aeration of grain-benefits and limitations. OEPP/EPPO Bull. 6:83-94.

Cline, L. D. 1970. Indianmeal moth egg hatch and subsequent larval survival after short exposures to low temperatures. J. Econ. Entomol. 63:1081-3.

Cox, P. D. 1974. The influence of temperature and humidity on the life-cycles of <u>Ephestia figulilella</u> Gregson and <u>Ephestia calidella</u> (Guenie) (Lepidoptera: Phycitidae). J. Stored Prod. Res. 10:43-55.

David, M. H., R. B. Mills, and G. D. White. 1977. Effects of low temperature acclimation on developmental stages of stored product insects. Environ. Entomol. 6:181-4.

Ernst, S. A., and J. A. Mutchmor. 1969. Dispersal of three species of grain beetles as a function of thermal acclimation, temperature and larval size. J. Stored Prod. Res. 5:407-12.

Evans, D. E. 1977. Some aspects of acclimation to low temperatures in the grain weevils <u>Sitophilus oryzae</u> (L.) and <u>S. granarius</u> (L.). Australian J. Ecol. 2:309-18.

Holman, L. E. 1966. Aeration of grain in commercial storages. USDA Marketing Res. Rep. 178.

Howe, R. W. 1965. A summary of estimates of optional and minimal conditions for population increase of some stored-product insects. J. Stored Prod. Res. 1:177-84.

Jay, E. G. 1980. Low temperatures: Effects on control of <u>Sitophilus oryzae</u> (L.) with modified atmospheres in Controlled atmosphere storage of grains. 65-71. (ed.) J. Shejbal. Elsevier Scientific Publishing Co., Amsterdam.

Kazhantshikov, I. W. 1938. Physiological conditions of cold-hardiness in
    insects. Bull. Entomol. Res. 29:253-62.
Mathlein, R. 1961. Studies on some major storage pests in Sweden, with
    special reference to their cold resistance. National Inst. Plant
    Protect. Contr. Sweden. 12:83, 1-49.
Mills, R. B. 1978. Potential and limitations on the use of low temperatures
    to prevent insect damage in stored grain. 244-59. Proc. 1st Internat.
    Wkg. Conf. on Stored-Prod. Entomol., Savannah, GA. USA. October, 1974.
Mullen, M. A., and R. T. Arbogast. 1979. Time-temperature mortality
    relationships for various stored-product insect eggs and chilling times
    for selected commodities. J. Econ. Entomol. 72:476-78.
Navarro, S. 1974. Aeration of grain as a non-chemical method for the control
    of insects in the grain bulk. 341-53. Proc. 1st Internat. Wkg. Conf. on
    Stored-Prod. Entomol., Savannah, Ga. USA. October, 1974..
Navarro, S., E. Donahaye, and M. Calderon. 1969. Observations on prolonged
    grain storage with forced aeration in Israel. J. Stored Prod. Res.
    5:73-8.
Quinlan, J. K. 1972. Malathion aerosols in conjunction with aeration to corn
    stored in a flat storage structure. Proc. N. Cent. Br. Entomol. Soc.
    Am. 27:63.
Salt, R. W. 1936. Studies on the freezing process in insects. Tech. Bull.
    Minn. Agric. Exp. Sta. 116:1-41.
Salt, R. W. 1961. Principles of insect cold hardiness. Ann. Rev. Entomol.
    6:55-73.
Smith, L. B. 1970. Effects of cold-acclimation on supercooling and survival
    of the rusty grain beetle, Cryptolestes ferrugineus (Stephens)
    (Coleoptera: Cucujidae) at subzero temperatures. Canadian J. Zool.
    48:853-58.
Somme, L. 1968. Acclimation to low temperatures in Tribolium confusum (Col.
    Tenebrionidae). Norsk Entomol. Titsskr. 15:134-36.
Watters, F. L. 1966. The effects of short exposures to subthreshold
    temperatues on subsequent hatching and development of eggs of Tribolium
    confusum DuVal (Coleoptera, Tenebrionidae). J. Stored. Prod. Res.
    2:81-90.
Wigglesworth, V. B. 1972. The Principles of Insect Physiology. Chapman and
    Hall, London. 827 p.
Williams, P. 1974. Grain insect control by aeration of farm silos in
    Australia. Ann. Tech. Agric. Special No. 417-421.

POTENTIAL OF IONIZING RADIATION FOR INSECT CONTROL IN THE CEREAL FOOD INDUSTRY

F. L. Watters

Editor's Comments

Radiation is an alternative to fumigation but high capital costs make it more expensive. Another disadvantage is that it lacks the versatility of fumigation. No residues and effectiveness on eggs indicate radiation may have value along with fumigation and perhaps heat. Keep this approach in mind while awaiting additional study.

Additional study may be forthcoming sooner than had been expected. FDA has proposed use of irradiation as a substitute for ethylene dibromide. FDA would allow increased doses of 15 to 100 and 1,000 to 3,000 krad for fruits and vegetables, and spices, respectively. In an early statement FDA and the National Food Processors Association agreed that in view of the expense widespread acceptance by the food industry may take 5 years or longer.

F. J. Baur

POTENTIAL OF IONIZING RADIATION FOR INSECT
CONTROL IN THE CEREAL FOOD INDUSTRY

F. L. Watters

Agriculture Canada Research Station
Winnipeg, Manitoba R3T 2M9[1]

INTRODUCTION

Ionizing radiation may be used to control stored-product insects by direct treatment of cereals or by genetic control of insect populations. Direct treatment of infested cereals and cereal products provides a residue-free process of controlling insects on or in foods. Genetic control involves the release of irradiated sterile males in sufficient numbers to successfully compete with males in the natural population so that most of the females will lay sterile eggs. The method, first used for the eradication of the screwworm fly from the Caribbean island of Curacao, has served as a model for other attempts to control insect pests of cattle, humans, and stored foods (Smith and von Borstel, 1972). Radiation may therefore be considered as a possible alternative to the use of chemical pesticides for the protection of animals and food from insects.

Two types of ionizing radiation have been evaluated to control stored-product insects: gamma radiation produced by the nuclear disintegration of radioactive isotopes such as cobalt-60; and high speed electrons emitted from a heated cathode and beamed at a thin layer of material moving across an electron scan. Both types of radiation cause the formation of ions in the material irradiated by the ejection or acquisition of electrons from atoms or molecules. Gamma radiation has greater penetrating power than accelerated electrons and can be used to irradiate materials such as grain or packaged products. Accelerated electrons lose energy rapidly when deflected or repelled by charges on orbital electrons of atoms in irradiated material. Consequently electron radiation can only be used for the treatment of material exposed in thin layers, usually up to about 1.7 cm (0.7 in.) deep, depending on the energy output of the accelerator.

Ionizing radiation kills living organisms mainly by injuring the cell nucleus. Thus, radiation is more effective against cells that are forming or dividing, such as reproductive cells or insect eggs during embryonic development, rather than cells that are formed. At high radiation doses, however, both reproductive cells and somatic cells are damaged (Cornwell and Bull, 1960).

Much research has been done in the United States and elsewhere in evaluating ionizing radiation for the control of stored-product insects. Studies on the susceptibility to radiation of various species and stages of development show that it is possible to control insects within the 20 to 50 krad dose range approved in the United States for the irradiation of wheat and wheat flour (Tilton et al, 1971).

Any new pest control agent must be assessed in relation to existing well established practices. The main questions to be answered before its acceptance are: Is it as effective as current methods? Is it as safe or safer from the

[1]Present address: 79 St. Dunstan's Bay, Winnipeg, Manitoba R3T 3H6

standpoint of application and effect on the product? Is it reliable? Can the technique and the required equipment be integrated into existing flow systems in the plant? Is it economically advantageous? Are supplementary measures necessary to ensure effectiveness? Only by providing satisfactory answers to these questions can the potential role of ionizing radiation for insect control in the cereal food industry be properly assessed. Though this review deals with research on the use of ionizing radiation to control stored-product insects in cereals and cereal products, the information is applicable to the control of insects in other foods which would, however, be subject to regulatory clearance before being approved for treatment on a commercial scale.

EFFECTS ON INSECTS

The susceptibility of stored-product insects to various doses of ionizing radiation has been studied extensively. Recent reviews by Cornwell (1966), Banks (1976), Vas (1977), and Tilton (1975, 1979) have listed the advantages and limitations of radiation for insect control in stored foods. A characteristic feature of the radiation process is that irradiated insects take several days or weeks to die at the doses approved for the treatment of wheat and wheat flour.

A comparison of the speed of action of an insecticide, malathion; a flour mill spot fumigant, ethylene dibromide (EDB); and gamma radiation against the confused flour beetle, Tribolium confusum Jacquelin du Val, shows that complete mortality was achieved one day after treatment with malathion or EDB (Tauthong and Watters, 1978; Watters and Smallman, 1953) (Fig. 1). By contrast, 14 days were required to achieve complete mortality of T. confusum after irradiation at 25 krad (Watters and MacQueen, 1967), and 21 days were required for a related insect, the red flour beetle, T. castaneum Herbst.

Delayed mortality of an insect is a function of irradiation dose and the susceptibility of the species and its developmental stages. The resistance of insect stages to radiation varies as development progresses from egg to adult.

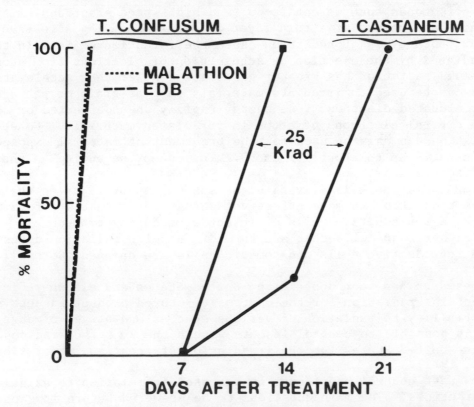

Figure 1. Response time of T. confusum and T. castaneum irradiated with gamma rays compared to response time of T. confusum exposed to malathion or ethylene dibromide (EDB).

Banham and Crook (1966) showed that eggs, larvae, pupae, and adults of T. confusum required doses of 4.4, 5.2, 14.5, and 12.8 krad, respectively, to obtain 99.9 percent kill. By contrast, the corresponding doses for T. castaneum were 10.9, 10.5, 25.8, and 21.5 krad. Irradiated T. castaneum adults died in 3 weeks whereas pupae required 5 weeks to die.

Species of the same genus may show similar sensitivity to radiation. Brower and Tilton (1972) found that eggs and larvae of the stored-grain insects Trogoderma inclusum and T. variable failed to develop to adults when irradiated at 5 krad. Though some pupae formed adults after irradiation at 100 krad, all were sterilized by 30 krad.

Grain mites are more radio-resistant than most stored-product insects. Davis (1972) found that though 5 krad may suppress populations of the grain mite, Acarus siro L., doses of more than 25 krad are required for a complete kill. Irradiation of mites at sublethal doses, however, may stimulate oviposition. Melville (1958) reported that irradiation of the mite Tyroglyphus farinae (Deg.) (a synonym of A. siro), at 5 krad, resulted in a significant increase of eggs laid and hatched when compared with the control. A dose of 40 krad was required to sterilize the population.

Accelerated electrons have been proposed as an alternative to ^{60}Co for irradiation of infested grain at large bulk-handling facilities such as terminal elevators (Cornwell and Bull, 1960). Throughputs of 200 to 400 t/hr are possible with existing accelerators. Comparison of ^{60}Co gamma radiation and accelerated electrons for the disinfestation of corn showed that at 15 or 25 krad the two types of radiation are equally effective in preventing the development and emergence of grain weevil (Sitophilus spp.) adults (Adem et al., 1978). However, a dose of 25 krad was necessary to obtain complete kill of T. castaneum adults. It is possible to regulate the dose rate of accelerated electrons to take advantage of enhanced lethality that might result at doses lower than the 20 krad approved for the treatment of wheat and flour (Table 1). High dose rates caused lower hatches of T. castaneum eggs (Nair and Subramanyan, 1963; Brown and Davis, 1973). However, Adem et al. (1979) found that at 25 krad, dose rates of 3.5 to 30 krad/min produced no significant differences in mortality of the larger grain borer, Prostephanus truncatus (Horn) or T. castaneum. Gonen and Calderon (1973) reported that a dose rate of 11 krad/min from a ^{60}Co source was more effective than doses of 2.85 or 5.0 krad/min in achieving sterility of the almond moth, Ephestia cautella (Walker), at total doses of 20, 30, or 40 krad. They concluded that the greater effectiveness at 11 krad/min was due to the higher number of chromosome breaks per unit time.

The success of the sterile male technique for the control of insect pests of cattle has prompted research on the use of sterile males to control stored-product insects. Bull and Wond (1962, 1963) reported on studies to control the Mediterranean flour moth, Anagasta kuhniella Zeller in an English flour mill. They concluded that, although it was possible to breed sufficiently large populations for the treatment and release of sterile moths, the cost of rearing facilities and staff to manage the project would exceed the cost of fumigation. Further, since the technique is specific against only one species, fumigation would be required to control other flour mill pests such as beetles.

The sterile male technique has been applied to other species of stored-product moths to assess its value as a control agent. Since moth infestations in food storage warehouses are difficult to control with chemical insecticides, the sterile male technique may offer an effective alternative. The almond moth, E. cautella, is a pest of food warehouses in tropical parts of the world. Cogburn et al. (1973) showed that progeny from irradiated adults inherited genetic damage that affected their reproductive ability. In other studies, Calderon and Gonen (1971) found that normal females mated with males irradiated at 45 krad laid 55% of the normal number of eggs. Amuh (1978) suggested that the results of studies with E. cautella were sufficiently promising to undertake field studies in cocoa warehouses in West Africa.

The Indianmeal moth, Plodia interpunctella (Hubner) is an important pest of warehouses containing peanuts, stored flour, and packaged cereal products in

North America. Damage is caused when larvae feed on foods and form webbing. Much of the work of P. interpunctella irradiation has been done at the USDA Research Laboratory, Savannah; Ahmed et al. (1967a) found that irradiation of male P. interpunctella at 50 krad induced complete sterility. Irradiated males were fully competitive with untreated males. Brower (1976) reported that inherited partial sterility of progeny from irradiated males would further increase the effectivenss of ^{60}Co irradiation under warehouse conditions. Moths formed from irradiated pupae were fully competitive with unirradiated males when the ratio of irradiated to unirradiated males was 25:1 (Ahmed et al., 1976b).

Brower (1975) has listed the advantages and limitations of techniques for the genetic control of stored-product insects. Two major disadvantages are that the method is species-specific and, in the case of irradiated beetles, the released insects may continue to feed (Brower and Tilton, 1973). Persistent insecticides that may be used to control other pest species, may kill irradiated sterile insects. Also, release of sterile beetles to control infestations of bulk grain may not be feasible because of the time they require to find females. From studies with flour beetles, Erdman (1974) concluded that the sterile male technique did not seem promising for control of stored-product beetles because large numbers of sterile insects were needed.

Gamma radiation has been combined with other physical or chemical methods in attempts to reduce the effective radiation dose required to achieve control. Pendlebury (1966) found that exposure of the granary weevil, Sitophilus granarius (L.) to 30°C before irradiation sensitized the weevil to radiation damage at 7.7 krad. However, exposure to 30°C during irradiation reduced susceptibility; exposure at 30°C following irradiation at 15.3 krad caused death up to 56 days. Singh (1973) reported similar results for the lesser grain borer, Rhyzopertha dominica (Fabricius).

Gamma radiation combined with infrared radiation was more effective regardless of application sequence for controlling the Angoumois grain moth, Sitotroga cerealella (Olivier) than either treatment used separately (Cogburn et al., 1971). Tilton et al. (1972) found that microwave radiation at 2450 MHz combined with gamma radiation at 10 krad resulted in 96% reduction in emergence of S. cerealella from wheat. Both infrared and microwave radiation raised the temperature of the infested wheat and sensitized the insects to the effects of gamma radiation. It is possible, therefore, that the combination of ionizing and nonionizing radiation would reduce the cost of radiation and increase effectiveness.

Fumigation of T. confusum adults with methyl bromide or hydrogen phosphide after irradiation at 10-50 krad was more effective than either treatment alone if the fumigation treatment was applied one week or more after irradiation (Cogburn and Gillenwater, 1972). Malathion combined with gamma radiation at 10 krad was more effective against T. castaneum than either treatment alone and the insects died sooner (Cogburn and Speirs, 1972). However, irradiation at 5 krad afforded some protection to insects subsequently treated with malathion. Bhatia and Sethi (1980) reported similar results when gamma radiation was combined with DDT, lindane, or malathion.

EFFECTS ON PRODUCTS

Extensive studies in the United States on the wholesomeness of irradiated food have shown that nutritional quality was unimpaired and no toxic or harmful effects were observed following animal feeding tests (Brynjolfsson, 1978). The results of these studies are similar to those carried out in England where animals consumed irradiated food as a substantial part of their diet for two years (Hickman, 1966); protein value of the food was unaffected.

In long-term storage studies of wheat irradiated at 20 krad, Pixton and Hill (1967) reported that wheat quality was unaffected. Other tests on milling and baking qualities of wheat carried out 6 months after irradiation at 6.25 to 150 krad showed that there were no adverse effects (Watters and MacQueen, 1967);

however, germination was reduced at a dose of 25 krad. Deschreider (1966) reported that gamma irradiation generally reduced the proteolytic activity of European wheat flour but baking quality was actually improved. Morad et al. (1978) also found that bread quality was enhanced by irradiation but doses of 50 and 100 krad decreased protein content, wet gluten, and dry gluten and increased gluten absorption, soluble protein, and amino N. Bread baked from wheat irradiated at 125 and 175 krad had a scorched or burned odor when removed from the oven, but no adverse odor or flavor was detected in taste panel studies (Fifield et al., 1967); bread making qualities remained good during storage of wheat for 12 months. Lai et al. (1959) reported earlier that irradiation of wheat at any level did not improve bread properties when an optimum baking formula was used. Roushdy and El-Magoli (1978) found that irradiation of corn flour with 75, 500, or 1500 krad decreased the moisture content and free amino acids of the flour but its total protein, amino acids, farinograph, or extensograph characteristics were unaffected.

Irradiation of barley at doses up to 50 krad produced slight effects on the malting properties (Tipples and Norris, 1963). Germination of barley of 10.6% moisture content was reduced at doses of 8, 12, and 16 krad. However, Watters and MacQueen (1967) found that when barley of 16% moisture content was irradiated at 25 krad, there was no decrease in germination. Ismail et al. (1978) reported that irradiation of milled rice at 50 krad did not affect amino acids or vitamins. Fat acidity, an index of grain storage quality (Sinha and Wallace, 1977), was lower in irradiated brown rice than in controls after 5 months storage under ambient conditions, thus indicating a beneficial effect of irradiation on storability. Aoki et al. (1977) reported a slight organoleptic deterioration of cohesivensss in rice immediately after irradiation at 8-21 krad, but this disappeared after 3 months. Kola nuts irradiated in Nigeria at 3-20 krad and stored in baskets lined with fresh leaves of Mitragna ciliata, according to custom, did not show any detectable change in color, taste, or flavor but germination was adversely affected (Daramola, 1974).

Table 1. Cereals and cereal products cleared by various countries for human consumption following irradiation for insect control.

Country	Product	Radiation source	Dose, Krad	Date of approval
USA	Wheat* and wheat flour* (changed on 4 Mar. '66 from wheat and wheat product)	^{60}Co ^{137}Cs electrons 5Mev	20-50 20-50 20-50	Aug. '63 Oct. '64 Feb. '66
Canada	Wheat* Flour* Whole Wheat Flour*	^{60}Co ^{60}Co ^{60}Co	75 75 75	Feb. '69 Feb. '69 Feb. '69
USSR	Grain*	^{60}Co	30	'59
Bulgaria	Dry food concentrates**	^{60}Co	30	Apr. '72
World Health Organization (FAO/ IAEA/WHO Expert Committee)	Wheat* and ground wheat products* Rice+	^{60}Co ^{137}Cs electrons 10Mev ^{60}Co ^{137}Cs electrons 10Mev	15-100 15-100 15-100 10-100 10-100 10-100	Sept. '76 Sept. '76 Sept. '76 Sept. '76 Sept. '76 Sept. '76
Netherlands	Rice and ground rice products**	^{60}Co	100	Mar. '79

* Unlimited clearance
** Experimental batches
\+ Provisional

Source: Food Irradiation Newsletter, October 1979. Joint FAO/IAEA Division of Atomic Energy in Food and Agriculture, IAEA, Vienna.

LEGISLATION

Extensive testing of the quality of irradiated cereals in various countries has formed the basis for approval of the irradiation process for disinfestation of some cereals and cereal products (Table 1). Nevertheless only a few countries have approved the use of ionizing radiation, thereby placing a restriction on international trade of irradiated cereals and cereal products. The lack of acceptance of the process by major food importing countries, therefore, is a serious impediment to the development and installation of irradiation facilities in the grain and cereal food industries. Vas (1977) has emphasized the need to harmonize legislation among countries to stimulate wider usage of irradiation for the preservation of perishable foods as well as cereals. The Joint FAO/IAEA/WHO Advisory Group on International Acceptance of Irradiated Food has developed model regulations containing suggestions concerning regulatory control of irradiation facilities, irradiation of food, and trade of irradiated food (van Kooij, 1980). Until food importing countries pass legislation approving the acceptability of irradiated food, it is unlikely that irradiation facilities will be installed at export positions in food producing countries.

PILOT PLANTS

Pilot plants have been constructed in several countries to demonstrate the practical value of irradiation with a view to expanding its use if it can be technically and economically justified. Probably the most ambitious undertaking was the construction of a full-scale irradiation plant at Iskenderum, Turkey, for the large-scale disinfestation of wheat to control the Khapra beetle, Trogoderma granarium L. (Libby and Black, 1978). The plant had an initial loading of 163,000 curie (ci) of ^{60}Co to provide a throughput of grain of 30 t/h. Provision was made to increase the source strength to 360,000 ci to accommodate a throughput of 50 t/h (Anon., 1967). Unfortunately, the plant was never used because of adverse publicity of irradiation in Turkey (Libby and Black, 1978).

A bulk-grain and packaged product irradiator was constructed at the USDA Stored-Product Insects Research and Development Laboratory at Savannah, Georgia, to provide a throughput of grain of 4.5 metric t/h (Tilton et al., 1971). The initial source strength of the ^{60}Co source was 26,565 ci. The irradiator was used to determine the susceptibility to gamma irradiation of several stored-product insect species and their developmental stages (Tilton, 1975). The performance of the irradiator was assessed by Cogburn et al. (1972) for the treatment of bulk wheat and by Tilton et al. (1974) for the treatment of flour in metal cans or paper bags. The results of these tests demonstrated the technical feasibility of the process for controlling insect infestations in grain and cereal products.

A pilot plant irradiator has been constructed in Japan for the disinfestation of rice (Aoki et al., 1977). The irradiator was successful in eliminating all stages of the maize weevil, Sitophilus zeamais Molts.

A disadvantage of ^{60}Co irradiators for the disinfestation of grain at terminal elevators is the extensive installation required for throughput rates of 200-400 t/h. The capital costs needed for construction of the plant and the concrete vault required to accommodate a source of high curie strength may prove to be a formidable deterrent.

An alternative source of ionizing radiation is an electron accelerator. A schematic layout for an industrial electron accelerator has been described by Hofmann (1977). Its main advantage is that it can be integrated into terminal elevator flow systems to accommodate treatment rates of 200-400 t/h for the irradiation of infested grain. Brynjolfsson (1971) has listed the technical, logistic, and cost considerations of a practical irradiation facility.

A small pilot plant incorporating an electron accelerator for the treatment of infested maize has been developed and tested at the Institute of Physics,

University of Mexico (Adem et al., 1978). The throughput was rated at 200 kg/h. Comparison of electrons and ^{60}Co radiation for the disinfestation of maize showed that the two types of radiation provided similar control. A larger pilot plant employing a Dynamitron© electron accelerator is under construction. It is expected that the throughput of the new plant will approach the requirements for commercial operation.

ECONOMICS

The research to date attests to the effectiveness of ionizing radiation for the disinfestation of cereal grain and cereal products. Whether or not the process is adopted for the commercial treatment of infested produce will depend on the magnitude of the infestation problem in the products to be irradiated and on the cost of installation and treatment compared with current pest control practices. Although other factors such as consumer acceptance of irradiated products must be taken into account, the decision on whether to install an irradiation plant will hinge largely on its cost effectiveness.

Cornwell (1966) compared the costs of fumigation, gamma irradiation, and electron irradiation for the disinfestation of wheat. Because of the large fixed capital costs of irradiation installations, the estimated costs of irradiation were inversely related to annual throughputs. The cost of ^{60}Co gamma irradiation for an annual throughput of 288,000 t was twice that of irradiating 300,000 t with electrons. Fumigation of 300,000 t of grain with hydrogen phosphide (PH_3) gas applied as tablets or pellets of aluminum phosphide cost the same as irradiation with accelerated electrons. Cornwell (1966) estimated that the installation cost of an electron accelerator was lower than that of a ^{60}Co plant of similar output because less shielding is required. Further, the capital outlay of an electron accelerator plant designed to treat grain at 200 t/h is similar to that of a ^{60}Co plant that handles grain at 70 t/h.

The costs and reliability of electron accelerators have been discussed by Morganstern (1978). Using case studies, it was estimated that treatment of grain for 6000 hr/yr operation was 13 cents/t; 4000 hr/yr would cost 15 cents/t. Flour could be treated for 2.75 cents/50 lb (22.7 kg) bag. It would, however, be necessary to store irradiated grain or flour in insect-free containers or warehouses to prevent reinfestation.

Although cost is a prime consideration in determining whether ionizing radiation is to be used as an alternative to fumigation for disinfestation of cereals, other factors should be also considered. Foremost among these is the frequency with which disinfestation procedures must be applied to production runs. If all cereals or cereal products destined for domestic or export markets are to be treated to control actual infestations or to ensure that possible infestations are eliminated, then irradiation and fumigation may be considered on equal terms, economically. However, if only part of the output run of a plant must be treated, then fumigation may be less expensive than irradiation. Fumigation is more versatile, also, because fumigants may be applied at any point in the product handling or processing sequence depending on need as perceived by quality control procedures. Fumigation dosages have been well researched and established through many years of experience. The dosages recommended will provide immediate control of a wide range of pests, although eggs and pupae of certain species may be tolerant of PH_3 fumigation (Bond, 1980).

By contrast, the substantial differences in radiation susceptibility of stored-product insects (Cornwell, 1966; Tilton, 1975) would necessitate adjustment of the radiation dose needed to control the most resistant species. Eggs of most insect pests appear to be the most susceptible stage to radiation. Other stages may take 2-5 weeks to die depending on the amount of the dose.

Since neither fumigation nor irradiation provides residual protection from reinfestation after treatment, supplementary measures are necessary to ensure continued freedom from infestation. This usually involves storing the product in an insect-free, insect-proof package (Highland, 1975) or a bin, silo, or warehouse that has been disinfested with insecticide.

FUTURE OUTLOOK

Limitations of current pest control practices provide the incentive for critically examining new techniques of pest control. The spectre of insect resistance to insecticides and the on-going scrutiny by regulatory agencies to determine the occurrence of chemical residues in cereals justify the search for other approaches that may counter the shortcomings of chemical pesticides. However, in our quest for alternatives, we must identify both their advantages and disadvantages to determine their possible role in future pest control programs.

Because of high capital costs the installation of radiation facilities would seem to be appropriate for (a) treatment of grain or other foods at the terminal elevator stage; (b) treatment of grain or processed cereal products in cereal processing plants; and (c) treatment of grain or cereal products packed in sacks or packages.

Where large annual throughputs are handled at a terminal facility or a flour mill, the electron beam accelerator would seem to be the most appropriate choice. The decision on whether to install an irradiation facility would be based on (1) the need to treat all or a portion of incoming grain; (2) dissatisfaction with current disinfestation practices; (3) ease of integration of irradiation equipment in current grain handling or milling operations; and (4) ability to arrange for post irradiation protection of grain or cereal products from reinfestation.

Treatment of cereal products in packages or sacks where deep penetration of radiation is desired will require a gamma ray irradiator. Mobile irradiators employing tote boxes are available for this purpose. Fixed plant installations are also available for larger throughputs requiring the use of a pallet or tote box gamma irradiator (Fraser, 1980).

It has been suggested (Bailey, 1979) that radiation disinfestation of grain may be more appropriate for use in food importing countries than in food producing countries. Many importing countries, including less developed countries (LDCs), store large quantities of grain that require disinfestation prior to their transfer to central food depots. However, it must be recognized that irradiation at port facilities is only one step in the disinfestation process. The grain must still be protected from infestation after irradiation, either by storage in insect-proof containers or in warehouses treated with insecticides. Since strategic long-term storage is emphasized in LDCs to provide against famine or crop failure, there is considerable reliance on residual insecticide sprays that will provide extended protection to foods stored in warehouses. Few LDCs have terminal elevators at port areas where electron accelerators could be installed. A ^{60}Co gamma irradiator would be the most appropriate type of installation for disinfestation of imported shipments of grain or flour stored in jute, cotton, or paper bags.

SUMMARY

Ionizing radiation provides a residue-free method for the direct control of insects in cereal foods. Special installations and precautions are needed to ensure the protection of personnel during the irradiation process. Insects do not die immediately after exposure to ionizing radiation. The duration of insect survival following exposure depends on species susceptibility and radiation dose. Irradiated insects take two weeks or longer to die compared to a few hours or days for insects exposed to insecticides or fumigants.

The reliability of ionizing radiation, like that of other pest control methods, depends largely on the applied dose, the efficiency of the delivery system, and the susceptibility of the target species. There is no evidence to date that insects have become resistant to radiation. However, insect resistance to chemical insecticides is a widespread and continuing problem that is causing concern.

Small-scale pilot plant studies and theoretical considerations indicate that the irradiation process can be integrated into existing flow systems in terminal elevators or cereal processing plants. Irradiation of thin layers of free-flowing cereals can be accomplished with electron accelerators designed to irradiate products at flow-rates of 200-400 t/hr. Irradiation of cereals or cereal products in bags or packages can be accomplished better with ^{60}Co gamma rays which have good powers of penetration.

Comparison of the costs of radiation and fumigation indicates that the use of radiation facilitates for 4000 hr/yr for treatment of products considered to be infested would approach the cost of fumigation. However, utilization of facilities for only 500 hr/yr would make fumigation more attractive. In either case, supplementary measures involving chemical or physical methods are required to protect products from reinfestation during subsequent storage.

The decision on whether to install an ionizing radiation device in a terminal elevator or a food processing plant will depend on deficiencies of current control methods, the amounts of chemical insecticides presently used, their frequency of application, analysis of comparative costs, the feasibility of integrating an irradiator into a cereal flow system, and customer or consumer acceptance of irradiated products. The answers to some of these questions can be discovered by installation of industrial pilot plants designed to optimize the benefits of ionizing radiation as a supplement to current pest control methods.

LITERATURE CITED

Adem, E., F.L. Watters, R. Uribe-Rendon, and A. La Piedad. 1978. Comparison of ^{60}Co gamma radiation and accelerated electrons for suppressing emergence of Sitophilus spp. in stored maize. J. Stored Prod. Res. 14: 135-142.

Adem, E., R. M. Uribe, and F.L. Watters. 1979. Dose rate effects on survival of two insect species which commonly infest corn. Radiation Phys. and Chem. 14: 663-670.

Ahmed, M.Y.Y., E. W. Tilton, and J.H. Brower. 1976a. Competitiveness of irradiated adults of the Indian meal moth. J. Econ. Entomol. 69: 349-352.

Ahmed, M.Y.Y., J.H. Brower, and E.W. Tilton. 1976b. Sexual competitiveness of adult Indian meal moths irradiated as mature pupae. J. Econ. Entomol. 69: 719-721.

Amuh, I.K.A. 1978. Control of pest infestation of food by irradiation. in Food preservation by irradiation. Vol. 1. IAEA. Vienna. 197-206 pp.

Anonymous. 1967. Grain irradiation plant. The Engineer 223: 97-98.

Aoki, S., H. Watanabe, T. Sato, T. Hoshi, S. Tanaka, H. Takano, and K. Umeda. 1977. Insect control and organoleptic evaluation of rice irradiated by the pilot scale grain irradiator. Rept. National Food Res. Inst. 32: 351-355.

Bailey, S.W. 1979. The irradiation of grain: an Australian viewpoint. in Australian contributions to the symposium on the protection of grain against insect damage during storage, Moscow, 1978. Div. Entomol. CSIRO, Australia 1979. 136-138 pp.

Banham, E.J., and L.J. Crook. 1966. Susceptibility of the confused flour beetle, Tribolium confusum Duv. and the rust-red flour beetle, Tribolium castaneum (Herbst) to gamma radiation. in P.B. Cornwell (ed.) The Entomology of radiation disinfestation of grain. Pergamon Press, New York. 107-118 pp.

Banks, H.J. 1976. Physical control of insects - Recent developments. J. Aust. Entomol. Soc. 15: 89-100.

Bhatia, P., and G.R. Sethi. 1980. Combined effects of gamma radiations and insecticide treatment (direct spray) on the adults of susceptible strains of Tribolium castaneum (Herbst). Indian J. Entomol. 42: 82-89.

Bond, E.J. 1980. Sorption of tritiated phosphine by various stages of Tribolium castaneum (Herbst). J. Stored Prod. Res. 16: 27-31.

Brower, J.H. 1975. Potential for genetic control of stored-product insect populations. in Proc. First Intl. Working Conf. on Stored-Prod. Entomol. Savannah, Ga. 167-180 pp.

Brower, J.H. 1976. Recovery of fertility by irradiated males of the Indian meal moth. J. Econ. Entomol. 69: 273-276.

Brower, J.H., and E.W. Tilton. 1972. Gamma-radiation effects on Trogoderma inclusum and T. variabile. J. Econ. Entomol. 65: 250-254.

Brower, J.H., and E.W. Tilton. 1973. Weight loss of wheat infested with gamma-radiated Sitophilus oryzae (L.) and Rhyzopertha dominica (F.) J. Stored Prod. Res. 9: 37-41.

Brown, G.A., and R. Davis. 1973. Sensitivity of red flour beetle eggs to gamma radiation as influenced by treatment age and dose rate. J. Georgia. Entomol. Soc. 8: 153-157.

Brynjolfsson, A. 1971. Factors influencing economic evaluation of irradiation processing. in FAO/IAEA-PL-433/2. 13-35 pp.

Brynjolfsson, A. 1978. The high dose and low dose food irradiation programmes in the United States of America. in Food Preservation by irradiation. Vol. 1. IAEA. Vienna. 15-27 pp.

Bull, J.O., and T.J. Wond. 1962. Control of the Mediterranean flour moth Anagasta kuhniella (Zell.) by sterile male release. I. Biological studies related to large scale rearing. U.K. Atomic Energy Authority Rept. AERE-R 3895. 16 pp.

Bull, J.O., and T.J. Wond. 1963. Control of the Mediterranean flour moth, Anagasta kuhniella (Zell.) by sterile male release. II. Susceptibility to gamma radiation. U.K. Atomic Energy Agency Research Group. AERE-R 3967. 44 pp.

Calderon, M., and M. Gonen. 1971. Effects of gamma radiation on Ephestia cautella (Wlk.) (Lepidoptera, Phycitidae) - 1. Effects on adults. J. Stored Prod. Res. 7: 85-90.

Cogburn, R. R., J.H. Brower, and E.W. Tilton. 1971. Combination of gamma and infrared radiation for control of the Angoumois grain moth in wheat. J. Econ. Entomol. 64: 923-925.

Cogburn, R. R., and H.B. Gillenwater. 1972. Interaction of gamma radiation and fumigation on confused flour beetles. J. Econ. Entomol. 65: 245-248.

Cogburn, R.R., E. W. Tilton, and J.H. Brower, 1972. Bulk-grain gamma irradiation for control of insects infesting wheat. J. Econ. Entomol. 65:818-821.

Cogburn, R. R., E.W. Tilton, and J.H. Brower. 1973. Almond moth: gamma radiation effects on the life stages. J. Econ. Entomol. 66: 745-751.

Cogburn, R. R., and R.D. Speirs. 1972. Toxicity of malathion to gamma-irradiated and nonirradiated adult red flour beetles. J. Econ. Entomol. 65:185-188.

Cornwell, R.B. 1966. The entomology of radiation disinfestation of grain. Pergamon Press, New York. 236 pp.

Cornwell, P.B., and J.O. Bull. 1960. Insect control by gamma-irradiation: An appraisal of the potentialities and problems involved. J. Sci. Fd. Agric. 11: 754-768.

Daramola, A.M. 1974. Preliminary studies on the control of Kola weevils by gamma radiation in Nigeria. ISNA Newsletter. 3: 23-26.

Davis, R. 1972. Some effects of relative humidity and gamma radiation on population development in Acarus siro (Acarina: Acaridae). J. Georgia Entomol. Soc. 7: 57-63.

Deschreider, A.R. 1966. Action des rayons gamma sur les elements constitutifs de la farine de Ble. in Proc. Symp. Food Irradiation, Karlsrue, 6-10 June 1966. IAEA, Vienna. 173-185 pp.

Erdman, H.E. 1974. Productivity modifications to flour beetles (Coleoptera: Tenebrionidae) by gamma radiation (^{60}Co) of one or both sexes and by the addition of radiated males or females to population. Researches on Population Ecology 16: 52-58.

Fifield, C.C., C. Golumbic, and J.L. Pearson. 1967. Effects of gamma-irradiation on the biochemical, storage and breadmaking properties of wheat. Cereal Science Today 12: 253-262.

Fraser, F.M. 1980. Gamma radiation processing equipment and associated energy requirements in food irradiation. in Int'l Symp. in Combination Processes in Food Irradiation. IAEA-SM-250/4 8 pp.

Gonen, M., and M. Calderon. 1973. Effects of gamma radiation on Ephestia cautella (Wlk.) (Lepidoptera, Phycitidae) - III Effect of dose-rate on male sterility. J. Stored Prod. Res. 9: 105-107.

Hickman, J.R. 1966. United Kingdom food irradiation programme - wholesomeness aspects. Food Irradiation. in Symposium I.A.E.A. Vienna. 101-117 pp.

Highland, H.A. 1975. The use of chemicals in processing and packaging of stored products to prevent infestation. in Proc. First Intl. Working Conf. on Stored-Prod. Entomol. Savannah, Ga. 254-260 pp.

Hofmann, E.-G. 1977. A multipurpose radiation service center. Rad. Phys. Chem. 9: 613-624.

Ismail, F.A., F.A. El-Wakeil, and S.M. El-Dash. 1978. Tolerance, quality and storability of gamma-irradiated Egyptian rice. in Food preservation by irradiation. Vol. 1. IAEA Vienna. 501-515 pp.

Lai, S.P., K.F. Finney, and M. Milner. 1959. Treatment of wheat with ionizing radiations. IV Oxidative, physical, and biochemical changes. Cereal Chem. 36: 401-411.

Libby, W.F., and E.F. Black. 1978. Food irradiation: an unused weapon against hunger. Bull. Atomic Scientists. 34: 51-55.

Melville, C. 1958. An apparent beneficial effect of gamma-radiation on the flour mite. Nature 181: 1403-1404.

Morad, M.M., S.B. El-Magoli, M. Roushdi, and B. Beshai. 1978. Effect of gamma irradiation on the chemical composition, rheological properties and bread quality of the Egyptian wheat flour. Isotope Radiat. Res. 10:103-110.

Morganstern, K.H. 1978. Economics of electron accelerators in the preservation of food by irradiation. in Vol. 2. Proc. of an Intl. Symposium on Food Preservation by Irradiation Organized by IAEA, FAO and WHO, Wageningen. 267-283 pp.

Nair, K.K., and G. Subramanyam. 1963. Effects of variable dose-rate on radiation damage in the rust-red flour beetle, Tribolium castaneum Herbst. in Radiation and Radioisotopes Applied to Insects of Agricultural Importance. IAEA, Vienna. 425-429 pp.

Pendlebury, J.B. 1966. The influence of temperature upon radiation susceptibility of Sitophilus granarius (L.). in P.B. Cornwell (ed.). The entomology of radiation disinfestation of grain. Pergamon Press: New York. 27-40 pp.

Pixton, S. W., and S.T. Hill. 1967. Long term storage of wheat. - II. J. Sci. Fd. Agric. 18: 94-98.

Roushdy, H.M., and S.B. El-Magoli. 1978. Egypt's policy concerning food irradiation research and technology. in Food preservation by irradiation. Vol. 1. IAEA. Vienna. 29-41 pp.

Singh, H. 1973. Preirradiation temperature-induced radio-sensitivity changes in Rhyzopertha dominica adults. J. Georgia Entomol. Soc. 8: 317-320.

Sinha, R.N., and H.A.H. Wallace. 1977. Storage stability of farm-stored rapeseed and barley. Canad. J. Pl. Sci. 57: 351-365.

Smith, R.H., and R.C. von Borstel. 1972. Genetic control of insect populations. Science 178: 1164-1174.

Tauthong, S., and F.L. Watters. 1978. Persistence of three organophosphorous insecticides on plywood surfaces against five species of stored-product insects. J. Econ. Entomol. 71: 115-121.

Tilton, E.W. 1975. Achievements and limitations of ionizing radiation for stored-product insect control. in Proc. First Intl. Working Conf. on Stored-Prod. Entomol., Savannah, Ga. 354-361 pp.

Tilton, E.W. 1979. Current status of irradiation for use in insect control. in Proc. Second Intl. Working Conf. on Stored-Prod. Entomol. Ibadan, Nigeria. 218-221 pp.

Tilton, E.W., J.H. Brower, G.A. Brown, and R.L. Kirkpatrick. 1972. Combination of gamma and microwave radiation for control of the Angoumois grain moth in wheat. J. Econ. Entomol. 65: 531-533.

Tilton, E.W., J.H. Brower, and R.R. Cogburn. 1971. Critical evaluation of an operational bulk-grain and packaged product irradiator. Intl. J. Rad. Eng. 1: 49-59.

Tilton, E.W., J.H. Brower, and R.R. Cogburn. 1974. Insect control in wheat flour with gamma irradiation. Intl. J. Appl'd Rad. and Isotopes 25: 301-305.

Tipples, K.H., and F.W. Norris. 1963. Some effects of gamma-radiation on barley and its malting properties. J. Sci. Fd. Agr. 13: 646-654.

van Kooij, J.G. 1980. World-wide utilization of Food Irradiation. in International Meat Research Congress, Colorado Springs, Co. Aug. 31-Sept. 5, 1980. IAEA, Vienna. 3-14 pp.

Vas, K. 1977. Food irradiation - technical and legal aspects. Food and Nutrition 3: 2-8.

Watters, F.L., and B.N. Smallman. 1953. Initial and residual effectiveness of spot fumigants in elevator boots. Cereal Chem. 30: 343-348.

Watters, F.L., and K.F. MacQueen. 1967. Effectiveness of gamma irradiation for control of five species of stored-product insects. J. Stored Prod. Res. 3: 223-234.

HEALTH HAZARDS OF INSECTS AND MITES IN FOOD

Joel K. Phillips and Wendell E. Burkholder

Editor's Comments

The fact that stored-product insects can and do cause
illness in humans is well known to entomologists but less so
to the food industry. This book would have been incomplete
if no mention was made of this risk to employees and
consumers. As the authors point out there have been
consumer complaints, court cases, and judgments/settlements.
Since the amount of published information on this topic is
small, a review was indicated. Your attention is directed
to the summary and the facts that some toxic substances
produced are possible carcinogens (quinones) and that some
are not completely inactivated by heat (allergens), as
during processing.

It was noted in the first chapter that FDA has not been
known to cite section 402(a)2, the section of the FD&C Act
that speaks to the illegality of poisonous or deleterious
adulteration (contamination), in instances of insect
adulteration. Although unlikely, such is possible.

F. J. Baur

HEALTH HAZARDS OF INSECTS
AND MITES IN FOOD

Joel K. Phillips and Wendell E. Burkholder

Stored Product and Household Insects Laboratory
Agricultural Research Service, U.S. Department of Agriculture,
Department of Entomology, University of Wisconsin,
Madison, Wisconsin 53706

INTRODUCTION

The presence in food of certain arthropods (invertebrate animals with jointed legs and segmented bodies) or their debris has long been considered an indication of inadequate sanitation. In fact the Federal Food, Drug, and Cosmetic Act (FD&C Act) assumes a cause-and-effect relationship between "filth" (including certain forms of insect and mite contamination) and disease (Gorham, 1975; Gorham, 1979). Modern control measures against both field and storage pests have greatly reduced this problem. Despite the best efforts, however, some contamination persists. This has prompted the food industry and government to accept the premise that for practical purposes completely pure food is unattainable.

The American consumer enjoys a diet of relative purity and high nutritional value. This is due in part to the technologically advanced state of our food production industry. In addition, it would seem unlikely that producers would risk their corporate reputations and profits by marketing dangerously contaminated food in this age of heightened consumer awareness and activism. Nonetheless, regulatory agencies continue to issue enforcement actions against producers of adulterated food (see the Editor's introduction summarizing FDA Weekly Enforcement Reports). In addition, actions are possible involving occupational hazards to food production workers whose health may be compromised in badly contaminated work areas. Enforcement actions can be costly to the producer in terms of production losses, fines, or penalties, and quite possibly, legal suits. In the following discussion, the medical hazards associated with arthropod contaminated food will be outlined. The purpose of this review is to heighten awareness to these potential problems.

INSECTS AS DIRECT FOOD HAZARDS

Non-allergic Disorders

Cases have been reported of people, involved in the handling and preparation of foods, suffering insect-related mechanical irritations of the skin, eyes, and upper respiratory tract. This is hardly surprising when we consider that many insects have an abundance of hair-like setae, spines, scales, and other abrasive projections including chitinous plates, legs, and wings. Some spines and hairs are equipped with poison glands, thus compounding the problem (Harwood and James, 1979). These anatomical structures may also elicit allergic reactions, a topic that will be covered separately in another section.

Beetles of the family Dermestidae (dermestids) are found in many storage environments. Where larval skins and hairs are reduced to powder, an active mechanical irritation of the skin can occur. Nasal mucous membranes can also be affected, and the body fluids and excreta of some beetles can have a severe vesiculating (blistering) effect when rubbed on the skin or eyelids (Sheldon and Johnston, 1941). Dock workers, unloading a cargo heavily infested with Dermestes frischii, developed itching of the neck and shoulders, a dry and constant cough, a state of nausea, and lesions to the sides of the eyes. The irritant proved to be the powdered remains of dermestid larvae. These remains contained enormous quantities of detached larval hastisetae, barbed hair-like structures noted for their urticating (stinging) properties. Responsibility for the workers' lesions was determined by authorities to lay with the owner of the ship (Loir and Legangneux, 1922).

Other reported cases of apparent non-allergic dermatosis involve straw itch mites (Pyemotes tritici). These mites normally attack the larvae of insects, including food pests like the bean weevil (Acanthoscelides obtectus), pea weevil (Bruchus pisorum), and Angoumois grain moth (Sitotroga cerealella). During times of rapid mite population growth, or when natural hosts are scarce, these mites are likely to attack other organisms including man. Since these mites are quite susceptible to dehydration, grain and dock workers are often attacked. By inserting their mouthparts into the skin, the mites cause irritations that are sometimes severe and often extensive (Fine and Scott, 1965; Leclercq, 1969).

Allergens

Any discussion of the role of insects as allergens in food is complicated by the fact that food itself can be the cause of allergies. Generally, food allergens are associated with those foods that are high in protein and consistently ingested, often from infancy (Matsumura, 1982). For some people, ingestion of foods as common as milk, legumes, and cereals can be troublesome. For others, inhalation or simple contact with certain foods can elicit an allergic response. Wilbur and Ward (1976) discuss a case of "baker's asthma", an occupational allergy to wheat, that resulted in sinusitis, rhinitis, contact dermatitis, and asthma for an unfortunate baker. Jarvinen et al. (1979) established that 19 of 21 asthmatic workers in one Finnish bakery had no symptoms of the disease until they started work in the bakery. Definite indications of the role of flour dust in the causation of asthma were obtained in 14 of these cases. However, we can not ignore insects and related arthropods (i.e., mites) as potential allergens in food. Since allergens are offending proteins, we are reminded that approximately 60% of the dry weight of insects is protein. Human antibody production, in response to the extracts of common stored product pest insects, has been observed (Bernton and Brown, 1964).

Injectant Allergens

The most popular image of arthropod allergy is that associated with the bites or stings of venomous species. Indeed, over one hundred people per year in the U.S. are reported to die as the result of reactions to arthropod stings (Anonymous, 1979). The vast majority of victims suffer varying degrees of non-fatal discomfort generally associated with histamine induced swelling. In all the cases, complex venoms are injected into the flesh, mostly by means of abdominal stingers or cephalic fangs.

Injectant-type allergies can also be caused by stored-product pests. Straw itch mites can, in the course of biting man, introduce allergens that result in a form of dermatitis (Gorham, 1975). Acarus siro, the common grain mite or cheese mite causes "baker's itch" and also "vanillism", a rash afflicting vanilla pod handlers, although it is uncertain whether these reactions are caused by bites or simple allergy. The same applies to the mold mite (Tyrophagus putrescentiae) which causes "grocer's itch", "copra itch" among copra handlers, and a dermatitis of dock workers who handle cheese (Harwood and James, 1979; James and Harwood, 1969).

The attached hastisetae of dermestid beetle larvae are allergenic. On contact, these hairs can penetrate the skin resulting in urticarial or edematous (swelling) dermatitis. Conjunctivitis of the eye has also been reported (Leclercq, 1969). Mechanical irritation plays a role in the irritation caused by hastisetae in addition to the effect of allergens. Regardless of their mode of action, workers and consumers should not be exposed to large numbers of hastisetae in view of their proven irritant qualities.

Contactant Allergens

Sometimes mere contact with arthropods or their body parts or waste can induce allergic reactions. In Bulgaria, workers shelling and cleaning walnuts developed eczema, dermatitis, and pruritus (intense itching of the skin with no eruption) associated with exposure to the larvae and excreta of the Indianmeal moth (Plodia interpunctella) (James and Harwood, 1969). Transient dermatitis can also be caused by an allergic reaction resulting from direct contact with food-associated mites, especially body fluids released upon crushing. These mites can infest many food products including cheese, bran, dried fruits, jams, and sugars (Gorham, 1975).

Inhalant Allergens

The most commonly reported food-related insect allergies seem to be those in which the allergen is inhaled. When insects or related arthropods infest food storage and processing facilities, their carcasses, cast skins, and excreta are often dried and pulverized creating thousands of respirable, airborne, microscopic particles. This situation can, with repeated exposure, eventually result in allergic disorders among those who work in such an environment. A situation paralleling a food facility heavily infested with insects is an insectary, where insects are reared for research. In his survey of occupational (insect rearing) allergies to arthropods, Wirtz (1980) found 67% of his respondents attributing their symptoms to direct or airborne contact with lepidopteran (moth and butterfly) scales, with respiratory allergic reactions a major problem. Two laboratories had 53% and 75% of their personnel develop allergies to moth scales despite the routine use of exhaust hoods and protective masks and coats. Wittich (1940) reported respiratory allergies occurring where the Indianmeal moth infests shelled seed corn releasing large numbers of wing epithelial particles. Fuchs (1979) stated that symptoms of inhalant lepidopteran allergy include bronchial asthma and rhinitis. One can easily envision problems of this kind occurring wherever workers are repeatedly exposed to great numbers of the scales or body parts of stored-product moths.

Members of the insect order Orthoptera, including the cockroaches and locusts, can also be the cause of inhalant allergic reactions. Locusts are field pests that are often harvested along with the crop on which they happen to be feeding. In the course of harvesting, processing, and storing food, the bodies of locusts and other field pests are often dried and crushed, again resulting in an abundance of respirable dust. Leclercq (1969) notes a case where inhalant allergies occurred among workers rearing locusts. These people suffered rhinitis, itching skin, bronchitis, and ultimately asthma, in general sequence. Wirtz (1980) reported that all workers involved in a study of migratory locusts (Locusta migratoria) developed allergies to the insect. Fuchs (1979) implicated locusts and cockroaches as sources of inhalant allergens leading to bronchial asthma and allergic conjunctivitis. Kang (1976) found evidence that components of cockroach carcasses and excreta in house dust were a probable cause of bronchial asthma. Fuchs (1979) detected allergens common to the various life stages of the offending insect and to various species within the same order. The latter phenomenon was also noted by Perlman (1961) and suggests that locusts, grasshoppers, cockroaches, and crickets, as well as other groups of taxonomically related insects, produce common allergens. Ominously, the three cases of anaphylactic shock reported in the Wirtz (1980) survey involved reactions to orthopterans.

Obligate beetle and weevil pests of stored grains and milled products (i.e., flour) have also caused inhalant allergies. Workers involved in rearing granary weevils (Sitophilus granarius) (Frankland and Lunn, 1965) and dermestid beetles (Trogoderma spp.) (Okumura, 1967) suffered allergic rhinitis and bronchial asthma. Leclercq (1969) noted that inhalation of urticating dermestid hastisetae has caused dyspnea (labored breathing), presumably the result of an allergic reaction. The present authors report the case of a researcher who developed skin itching, hives, rhinitis, and dyspnea as a result of repeated exposure to maize weevil (Sitophilus zeamais) cultures. A co-worker suffered chronic rhinitis and asthma-like symptoms as a result of culturing cowpea weevils (Callosobruchus maculatus). These cases are not surprising since stored-product beetles and weevils produce large quantities of potentially allergenic, respirable dust as a result of feeding and the disintegration of their body parts.

Flies, especially those of the genus Musca, can be a serious problem where foods are stored or processed. Aside from their roles as vectors of known human pathogens, there is evidence of flies causing inhalant allergies. Frazier (1969) listed house flies among known sources of inhalant allergens. Perlman (1961) and Feinberg et al. (1956) noted positive reactions among human subjects to house fly extracts during skin testing studies. In addition, house fly induced nasal allergy has been reported by Jamieson (1938). House flies can reproduce prodigiously in suitable environments, and in the process, generate considerable filth including excreta, cast skins, and carcasses. When dried and blown about, this matter poses an obvious threat as a source of inhalant allergens.

Ants will often flourish where foods are stored or processed, especially in spillage areas. There is some evidence that ants are also allergenic. Hellreich (1962) reported positive skin reactions to extracts of ant pupae among subjects with a history of seasonal and perennial allergic rhinitis and bronchial asthma. In a study of asthma and rhinitis from insect allergens, Feinberg et al. (1956) recorded 51 of 79 patients (65%) giving positive skin reactions to extracts of ant eggs. Interestingly, patients who gave positive reactions to one insect usually reacted to a number of insects, and to some more often than to others.

In addition to insect allergens, dust containing mites, especially household dust, can cause rhinitis, bronchitis, and bronchial asthma by inhalation (Fuchs, 1979). Dermatophagoides spp. mites have been implicated as the cause of house dust allergy, and their excreta, exoskeletons, and scales seem to have a common allergen (Harwood and James, 1979). Wittich (1940) found that extracts of flour mites (Tyroglyphus farinae) produced strong allergic reactions in some grain mill dust-sensitive cases. In addition, Wittich implicated psocids or book lice (Troctes divinatorius) as the cause of dust-related respiratory allergy among grain bin cleaners. Two mite species generally associated with food, Acarus siro (common grain mite, or cheese mite) and Glycyphagus domesticus, an opportunistic mite accustomed to dry material of plant or animal origin, may also be the source of inhalant allergens (Wraith, 1969). Mites have also been recovered from the sputum of patients suffering lung disorders (Baker et al., 1956).

Ingestant Allergens

Insect or arthropod allergies of the ingestant type seem to be reported less often, yet generate the most concern. This is not surprising since ingestant allergies would seem to occur most often among food consumers. Again, some contamination of food is to be expected despite the efficiency of our food industry. We can also assume that the small amounts of arthropod debris consumed with our foods are, more often than not, medically inconsequential. There is evidence, however, that certain food-contaminating insects and arthropods are potential sources of ingestant allergens.

Bernton and Brown (1967) provide the best case for ingestant allergens of insect origin. Utilizing dialyzed extracts of seven common grain and grain-product pests, they skin-tested both allergic and normal (non-allergic) individuals. Test insects included the rice weevil (Sitophilus oryzae), fruit fly (Drosophila melanogaster), Indianmeal moth (Plodia interpunctella), sawtoothed grain beetle (Oryzaephilus surinamensis), red flour beetle larvae and adults (Tribolium castaneum), confused flour beetle (Tribolium confusum), and lesser grain borer (Rhyzopertha dominica). Although irritational reactions could have accounted for some of the positive results of these tests, positive skin reactions among 230 allergic patients numbered 68, or 29.6%. Among 194 normal subjects, 50 positive reactions occurred, or 25.8%. Where positive reactions occurred, the degree of sensitivity to test extracts was practically the same for both groups of subjects. Reactions to extracts of Indianmeal moths (24.0%) and red flour beetle larvae (23.1%) were by far the most common among 333 positive reactions. Then, in decreasing order, came red flour beetle adults (14.1%), rice weevils (9.6%), fruit flies (9.0%), confused flour beetles (8.1%), sawtoothed grain beetles (6.3%), and lesser grain borers (5.7%). The authors also heat treated (100°C for 1 hour) five of the extracts and noted that these treatments also yielded positive skin reactions. These reactions, however, were less vigorous than those of the unheated extracts. Nonetheless, cooking temperatures failed to completely deactivate these insect allergens, presenting another case for limiting, as much as possible, insect debris in food.

In an earlier study, Bernton and Brown (1964) tested the extracts of American and Oriental cockroaches (Periplaneta americana and Blatta orientalis) on 367 people. Positive skin responses were noted in 28.1% of the allergic and 7.5% of the normal subjects. In addition, 22 of 23 blood sera samples of positive reactors indicated the presence of a skin sensitizing antibody. The cockroach allergen also resisted deactivation when heated to 100°C for one hour. Since human contact with cockroaches is common, regardless of economic status, these insects must comprise an important part of the whole subject of ingestant allergens (Frazier, 1969). We can also safely assume, on the basis of the evidence presented thus far, that traces of many other insects in food elicit ingestant allergic reactions.

Whole Insects and Mites In Food

It is safe to say that most arthropods in food are killed by processing or home cooking before the food is consumed. If not, then most of the remaining offenders are killed by mastication or digestion and pass safely out of the body. Nonetheless, medical problems do occur, some of a very serious nature.

A common malady caused by the presence of whole arthropods in foods is intestinal upset or dyspepsia. Kenney (1945) experimentally induced nausea, abdominal cramps, and diarrhea in human volunteers by having them ingest live house fly (Musca domestica) larvae. The role that psychological or allergic factors may have played in this trauma is unclear. Clear-cut cases of dyspepsia have been proven, however, where insects or their body parts cause mechanical irritations or even obstructions. Migratory locusts are often consumed as a dietary supplement by Congolese peoples (Leclercq, 1969). Apparently, the locusts are consumed in quantity so that the legs, amply endowed with spines, may become plugs leading to complete occlusion of the intestine. Unless treated surgically the condition is fatal. Although it would be rare to find such hazardous insect matter in our foods, inspections are warranted to maintain high standards of food purity.

In some instances, living arthropods infest the organs or tissues of man or animals. Where flies are the offending organism, the term "myiasis" is used. According to Harwood and James (1979) about fifty species of fly larvae have been implicated, some questionably, in cases of accidental enteric myiasis in man. Most belong to the families Muscidae (includes the house flies) and Sarcophagidae (flesh flies). This phenomenon has been noted for other food-infesting fly families including the Calliphoridae (blow flies), Drosophilidae (pomace flies, common fruit fly), and Piophilidae (skipper flies) (Leclercq, 1969). Infestation occurs mainly by ingestion of young fly larvae or eggs with food. In the highly acidic and mostly anaerobic environment of the digestive tract, most insects are carried through passively without further development (Harwood and James, 1979). Nonetheless, genuine enteric myiasis is, on occasion, likely to occur. In severe infestations, the subject may at times suffer depression and malaise, with vomiting, nausea, vertigo, and violent abdominal pain. Bloody diarrhea may occur as a result of damage to the intestinal mucosa by rasping larval mouth hooks. Living or dead larvae may be evident in either vomit or stools. Such trauma also causes loss of appetite, gastric pains, bloating, dizziness, and violent headaches. In children, violent spasms and epileptic seizures may occur. Symptoms seem to disappear with the passing of the larvae (Harwood and James, 1979; James, 1947). Internal injury would seem to account for many of the described symptoms, although the effects of injury are likely synergized by allergic factors attributable to the presence of insects and their by-products.

In rare cases, beetles will invade the tissues or organs of the body. The term "canthariasis" is used to describe larval infestations. "Scarabiasis" is used to designate an invasion by adult beetles (Harwood and James, 1979). As with accidental enteric myiasis, most beetle "pseudoparasitisms" occur as a result of ingesting food contaminated with stored-product beetles. Leclercq

(1969) stated that this is especially the case with mealworms, and more particularly, the larvae of the cosmopolitan yellow mealworm, Tenebrio molitor. Okumura (1967) described the case of an infant suffering ulcerative colitis as the result of having eaten baby cereal infested with dermestid (Trogoderma spp.) larvae. Similarly, Ebeling (1975) reported the case of another baby suffering abdominal pain after eating cereal contaminated with fragments of Trogoderma ornatum. In both cases hastisetae may have caused injury to the intestinal mucosa. However, allergic reactions to hastisetae and other insect matter can not be ignored as contributing factors.

The extremely rare and accidental ingestion of moth or butterfly larvae is termed "scoleciasis" (Harwood and James, 1979). The phenomenon is rarely reported in medical or veterinary literature, although Gorham (1975) stated that since the Indianmeal moth frequently infests candy bars, the public accidentally ingests more of its larvae, dead or alive, than any other arthropod. He cited as evidence the findings of Bernton and Brown (1967) where the highest rate (24.0%) of positive skin sensitivity reactions was to P. interpunctella.

Mites can occur in enormous numbers on foods especially where grains or grain products are stored. Therefore, cases of accidental enteric acariasis are certainly possible. In view of the vulnerability of some mites to desiccation, it would be difficult to imagine mites surviving the acidic and mostly anaerobic environment of the human digestive tract. Yet mites have proven to be severe contactant and inhalant allergens. With this in mind, we must regard mites as potential ingestant allergens, especially if they are consumed in quantity.

INSECTS AND MITES AS INDIRECT FOOD HAZARDS

Arthropod Toxins

Tenebrionid beetles produce highly reactive alkylated benzoquinones as a defense against predation (Tschinkel, 1975). Grain and flour infesting tenebrionids known to produce quinones include the red and confused flour beetles (Tribolium spp.), mealworms (Tenebrio spp.), lesser mealworm (Alphitobius diaperinus), longheaded flour beetle (Latheticus oryzae), broadhorned and slenderhorned flour beetles (Gnathocerus cornutus and G. maxillosus), and smalleyed and depressed flour beetles (Palorus ratzeburgi and P. subdepressus). Impressive amounts of quinones, as much as 384µg per beetle, have been found (Ladisch et al., 1967). These chemicals impart a noxious taste and odor to the flour or grain in which the beetles are thriving rendering the product unpalatable.

Human toxicity studies have shown that quinones can cause dermatitis with skin discoloration, erythema (reddening), and the formation of papules and vesicles. Exposure to quinone vapors can result in eye disorders such as conjunctivitis and corneal ulceration. Discoloration of the conjunctiva and cornea are also reported (Merck Index, 1968). In addition, quinones, including those produced by tenebrionids, are suspected of being carcinogens and have been implicated as wide-spectrum toxicants (Ladisch, 1963; Ladisch et al., 1967; Hueper, 1965). The American and Oriental cockroaches, common food pests, also secrete a benzoquinone derivative chemically related to known carcinogens (Ladisch, 1963). Quinones are known to react rapidly with certain food components such as proteins (Ladisch, 1966) possibly reducing quinone toxicity. Nonetheless, quinone producing insects are a potential health hazard. They should therefore be strictly monitored and controlled at all levels of food production and distribution.

The potential for the transmission of pathogens by food infesting arthropods has long been recognized. Of primary concern are those species whose habits bring them into close contact with known sources of pathogens, i.e., feces, sewers, drains, refuse, and floors, and then food and food-handling surfaces. Flies, cockroaches, and ants fall into this category. Of lesser concern are species whose habits confine them in or near the food they happen to be infesting, thus reducing their vector potential. Flour beetles, stored-product moths, and mites comprise this group (Gorham, 1975). In the following section, we will treat classes of human pathogens separately, relative to known arthropod vector-pathogen interactions.

Bacterial Pathogens

Most processed foods are routinely pasteurized or sterilized to eliminate the risk of bacterial spoilage or pathogen transfer. Problems arise when bacteria-carrying arthropods contact exposed foods or penetrate the product's packaging.

Among potential vectors perhaps none have been the object of more suspicion and research than flies. This is not surprising since flies frequently breed and feed in filth, including feces. As opportunistic feeders, flies will also visit human food. Harwood and James (1979) cited a study where the number of bacteria on a single fly averaged 1.25 million and was as high as 6 million. Flies also harbor bacteria in their digestive tracts and frequently expel them in vomitus (digestive secretions expelled at the onset of feeding) or feces. The house flies (Musca domestica and other members of the family Muscidae) have been implicated in the transmission of three genera of bacterial pathogens that cause enteric infections: (1) Shigella spp., important cosmopolitan agents of dysentery and diarrhea (shigellosis); (2) Salmonella spp., including the agents of typhoid fever, as well as those causing acute gastroenteritis with diarrhea (salmonellosis); and (3) pathogenic serotypes of Escherichia coli often associated with infant diarrhea and "travelers' disease" in adults (Harwood and James, 1979). Blowflies (family Calliphoridae) have habits similar to the house fly, and thus, the potential for disease transmission (Chandler and Read, 1961). In addition, flies have been found to harbor the bacterial agents of cholera and anthrax (Harwood and James, 1979; Pelczar and Reid, 1965). Every effort should be made to prevent these insects from contacting food during processing, storage, and home use.

There exists much circumstantial evidence to implicate cockroaches in the spread of human bacterial pathogens. Like the flies, their habits bring them into close contact with sources of potentially harmful bacteria. Cockroaches frequent drains, sewers, refuse dumps, and decaying matter, then scatter over foods or food preparation surfaces contaminating both with an abundance of undesirable bacteria. These insects will thrive on human food if poor sanitation makes it readily available. A high priority should be placed on cockroach control since they have been shown to harbor 48 strains of pathogenic bacteria, mostly Enterobacteriaceae including the agents of shigellosis and salmonellosis (Harwood and James, 1979). Infective organisms such as Salmonella spp. have been found alive in cockroach feces ten to twenty days after experimental feedings in one study, and after 199 days in another (Frazier, 1969). Experimentally, cockroaches have also been shown to harbor the bacteria causing cholera, anthrax, undulant fever, cerebrospinal fever, pneumonia, diptheria, tetanus, and tuberculosis (Frazier, 1969; Harwood and James, 1979).

No one can deny the ants' fondness for human food. Since ants are quite mobile, often traveling the same routes as cockroaches in homes and buildings, they deserve our attention as potential vectors of bacterial pathogens. Harwood and James (1979) cited studies where ants were shown to harbor the agents of dysentery and other harmful bacteria. Several other food-infesting insects may likewise be incidental vectors of human bacterial pathogens, and therefore, all such pests should be controlled.

Viral Pathogens

The role of insects as potential carriers of viral pathogens is similar to their role as vectors of bacteria. Again, insects whose habits bring them into frequent contact with human or animal filth are of most concern. Flies have been found to harbor the viral agents of hepatitis and poliomyelitis (Harwood and James, 1979; Chandler and Read, 1961), although it is uncertain if they play more than an incidental role in the spread of these diseases. Cockroaches have a similar capacity for disease transmission, having been shown to carry the viruses causing poliomyelitis, Coxsackie, mouse encephalitis, and yellow fever (Frazier, 1969; Harwood and James, 1979). Any human environment badly infested with flies, roaches, or other arthropod pests, is a threat from a human health standpoint. That such pests may be harboring human viral pathogens is all the more reason to maintain high standards of sanitation.

Protozoan Pathogens

Several protozoa cause human intestinal diseases, the most important being Entamoeba histolytica, the main cause of amoebic dysentery. In 1933, an epidemic originating in Chicago at the time of the World's Fair spread across the country resulting in 1400 cases and four deaths (Pelczar and Read, 1965). Although the disease is transmitted in contaminated food and water and by direct contact, there is evidence to suggest that flies and cockroaches harbor the pathogen. Both will contact human feces and become contaminated with infective protozoan cysts. In addition, Harwood and James (1979) stated that cockroaches have been shown experimentally to carry Trichomonas hominis, Giardia intestinalis, and Balantidium coli, all suspected or proven protozoan agents of dysentery and diarrhea.

Toxoplasma gondii, the cause of toxoplasmosis in man, affects a number of wild and domestic animals. The domestic cat often acquires the protozoan by feeding on infected birds and rodents, and then passes the infective oocysts in its feces. Humans are thought to contract the disease by handling the contaminated feces of domestic animals. Retransmission among humans can then occur transplacentally. However, cockroaches, flies, and beetles may acquire the parasite by feeding on the feces of infected domestic animals and then transmit the pathogen to man by contaminating human food (Wallace, 1973).

Helminth Parasites

Grain infesting beetles are often intermediate hosts for certain tapeworms (cestodes) that normally parasitize rodents and in many cases man. Where rats and mice are present in grain storage, grain beetles such as Tenebrio spp. and Tribolium spp. are also found. Hymenolepis diminuta, the primary cosmopolitan worm parasite of domestic rats, and Vampirilepis (Hymenolepis) nana, the dwarf tapeworm, utilize these beetles for the development of infectious cysticercoids (larvae). The beetles acquire the parasites by ingesting eggs in contaminated rodent feces. Human infection can occur by direct ingestion of cestode eggs, or when the intermediate host beetles are ingested with contaminated dried cereals or other such products (Chandler and Read, 1961). Additional treatment of beetles as developmental hosts of helminths is provided by Harwood and James (1979).

Because of their habit of feeding or breeding in human or animal feces, cockroaches and flies are suspects in the transmission of helminth parasites. Although evidence to support this relationship is largely circumstantial, it deserves our attention. The American cockroach (Periplaneta americana) will serve as an intermediate host for the acanthocephalan Moniliformis dubius, the spiny-headed worm parasite of rats and occasionally man (Chandler and Read, 1961). Flies, especially the house and blow flies, have been found to carry the eggs or infectious stages of a number nematodes (roundworms) and cestodes. These include Ancylostoma duodenale, Ascaris lumbricoides, A. equorum, Dipylidium caninum, Diphyllobothrium latum, Echinococcus granulosus, Enterobius vermicularis, Hymenolepis diminuta, Necator americanus, Taenia solium, T. hydatigena, T. pisiformis, Toxascaris leonina, Trichuris trichura, and Vampirilepis nana (Harwood and James, 1979). Obviously, the potential for this type of transmission is highest where filth and human infection occur concurrently, conditions we would not expect to find in the American food industry.

Mycotoxins

Fungal spores are ubiquitous, often comprising a significant portion of the airborne particles in the home and work place. Thus, it is not surprising that the bodies of insects, especially those infesting stored grain and grain products, are often contaminated with fungal spores. We have noticed many spores clinging to the setae and waxy secretions of grain weevils. Many of these spores were 2 to 10 μ in diameter. This would be well within the range of diameters for spores of Aspergillus and Penicillium spp. Various fungal species within these genera are known to produce mycotoxins, highly toxic compounds of dietary, medical, and veterinary importance. Aspergillus flavus and A. parasiticus produce aflatoxins (Stoloff, 1976), among the most potent naturally occurring hepatotoxins and carcinogens known to man. Certain Penicillium strains also produce mycotoxins known to occur in foods or feeds (Stoloff, 1976). In addition, certain Fusarium fungi produce toxins that cause contact skin lesions and, if ingested, internal hemorrhaging (Smalley and Strong, 1974). Toxicoses can occur when these and other toxigenic fungi contaminate various animal feeds or human foods and produce mycotoxins as natural metabolic by-products.

The role of insects in disseminating spores resulting in mycotoxicosis is uncertain. Given favorable air currents, fungal spores are capable of contaminating exposed foods without the help of an insect vector. Experimental evidence, however, has indicated that some insects are incidental spore vectors. Kantack and Staples (1969) described the ability of dermestid beetle larvae to easily transport spores of stored grain fungi on their bodies throughout a grain mass. They also noted that T. glabrum larvae and young adults passed viable spores of Aspergillus fungi in their feces, indicating another potential route of pathogen dispersal. McMillian et al. (1980) infested ears of corn with maize weevils (Sitophilus zeamais), that had been exposed to spores of A. flavus, and observed almost twice as much fungal infection as ears infested with unexposed weevils. Increased A. flavus infestation on the corn was attributed to maize weevils damaging the corn kernels and thereby providing a portal of entry for spores carried on the insects' bodies. However, A. flavus must have adequate moisture to grow and produce aflatoxins. Sauer and Burroughs (1980) found that moisture levels below 85% relative humidity and 17% corn moisture content resulted in limited A. flavus growth and no aflatoxin production. We therefore assume that if insects are involved in the transfer of fungal spores to foods resulting in the production of mycotoxins, certain minimal conditions of moisture must be met. Thus, dry storage conditions would reduce the threat of certain products being infested by fungi originating from airborne or insect-borne spores. However, exposed foods that provide a liquid or moist substrate for fungal growth, would seem vulnerable to either mode of spore introduction, and subsequently, the risk of mycotoxin production.

CONCLUSIONS

There exists a strong case for limiting, and eliminating whenever possible, the presence of pest insects and related arthropods in food. As we have noted, several different types of arthropods can be associated with food-related health hazards. Some like dermestid beetles and their larvae may cause allergic or mechanical irritations of the skin, eyes, respiratory system, or digestive tract. Food-related contactant, inhalant, and ingestant allergies can also be linked to the Indianmeal moth and its larvae. The defense secretions of red flour beetles and their relatives chemically irritate body tissues, possibly in combination with allergic reactions. In some cases, whole insects or mites are ingested with food. The ingestion of fly larvae, for example, may lead to gastrointestinal disorders or broader systemic problems. And we have seen convincing evidence that various arthropod food pests, such as flies and cockroaches, are potential vectors of human parasites and microbial pathogens. In some cases, the best evidence linking arthropods in food to human health hazards is experimental. Additional case histories linking arthropods to food-related illnesses and injuries are also needed. However, many cases have undoubtedly gone unreported since many of the medical symptoms we can now attribute to the presence of food-infesting arthropods resemble those of more common maladies.

The evidence linking arthropod food pests to human health disorders is likely to increase with an expanding awareness of these problems. Therefore, continued vigilance is justified in monitoring and controlling these pests, as well as a strong commitment to sound sanitation practices at all levels of food production.

ACKNOWLEDGEMENT

Research was supported by the College of Agricultural and Life Sciences, University of Wisconsin, Madison, and by a cooperative agreement between the University of Wisconsin and the ARS, USDA. Review of the manuscript by Dr. Janet Klein, Dept. of Entomology, and Prof. Steve Taylor, Dept. of Food Microbiology and Toxicology, University of Wisconsin - Madison, is greatly appreciated.

REFERENCES

Anonymous. 1979. FDA approves insect venom products. FDA Drug Bull. 9(3):15-16.

Baker, E. W., T. M. Evans, D. J. Gould, W. B. Hull, and H. L. Keegan. 1956. A Manual of Parasitic Mites of Medical or Economic Importance. Nat. Pest Control Assoc., New York, 170 pp.

Bernton, H. S., and H. Brown. 1964. Insect allergy - preliminary studies of the cockroach. J. Allergy 35:506-513.

Bernton, H. S., and H. Brown. 1967. Insects as potential sources of ingestant allergens. Ann. Allergy 25:381-387.

Chandler, A. C., and C. P. Read. 1961. Introduction to Parasitology. John Wiley & Sons, Inc., New York, 822 pp.

Ebeling, W. 1975. Urban Entomology. University of California Division of Life Sciences, 695 pp.

Feinberg, A. R., S. M. Feinberg, and C. Benaim-Pinto. 1956. Asthma and rhinitis from allergens. J. Allergy 27(5):437-444.

Fine, R. M., and H. G. Scott. 1965. Straw itch mite dermatitis caused by Pyemotes ventricosis: Comparative aspects. South. Med. J. 58:416-420.

Frankland, A. W., and J. A. Lunn. 1965. Asthma caused by the grain weevil. Brit. J. Indust. Med. 22:157-159.

Frazier, C. A. 1969. Insect Allergy. Warren H. Green, Inc., St. Louis, Mo. 493 pp.

Fuchs, E. 1979. Insects as inhalant allergens. Allergol. Et Immunopathol. 7:227-230.

Gorham, J. R. 1975. Filth in foods: Implications for health. J. Milk Food Technol. 38(7):409-418.

Gorham, J. R. 1979. The significance for human health of insects in food. Ann. Rev. Entomol. 24:209-224.

Harwood, R. F., and M. T. James. 1979. Entomology in Human and Animal Health. Macmillan Publishing Co., Inc., New York, 548 pp.

Hellreich, E. 1962. Evaluation of skin tests with insect extracts in various allergic diseases. Ann. Allergy 20:805-808.

Hueper, W. C. 1965. Experimental studies on 8-hydroxyquinoline in rats and mice. Arch. Pathol. 79:245-250.

James, M. T. 1947. The flies that cause myiasis in man. U.S. Dept. of Agric. Publ. 631. 176 pp.

James, M. T., and R. W. Harwood. 1969. Herm's Medical Entomology. Macmillan Publishing Co., Inc., New York, 484 pp.

Jamieson, H. C. 1938. The house fly as a cause of nasal allergy. J. Allergy 9:273-274.

Jarvinen, K. A. J., V. Pirila, F. Bjorksten, H. Keskinen, M. Lehtinen, and S. Stubb. 1979. Unsuitability of bakery work for a person with atopy: A study of 234 bakery workers. Ann. Allergy 42:192-195.

Kang, B. 1976. Study on cockroach antigen as a probable causative agent in bronchial asthma. J. Allergy Clin. Immunol. 58(3):357-365.

Kantack, B. H., and R. Staples. 1969. The biology and ecology of Trogoderma glabrum (Herbst) in stored grain. Nebr. Agric. Exp. Stn. Res. Bull. 232. 24 pp.

Kenney, M. 1945. Experimental intestinal myiasis in man. Proc. Soc. Exp. Biol. Med. 60:235-237.

Ladisch, R. K. 1963. Quinone toxins and allied synthetics in carcinogenesis. Proc. Penn. Acad. Sci. 37:144-149.

Ladisch, R. K. 1966. Tetritoxin: binding with amino acids and proteins. Proc. Penn. Acad. Sci. 39:48-56.

Ladisch, R. K., S. K. Ladisch, and P. M. Howe. 1967. Quinoid secretions in grain and flour beetles. Nature 215:939-940.

Leclercq, M. 1969. Entomological Parasitology. Pergamon Press, New York, 158 pp.

Loir, A., and Legangneux, H. 1922. Accidents de travail occasionnes par des Coleopteres. Bull. Acad. Med. Paris 88:68-72.

Matsumura, T. 1982. Food allergy in children and adults. p. 339-349, in E. F. P. Jelliffe and D. B. Jelliffe (eds.) Adverse Affects of Foods. Plenum Press, New York, 614 pp.

McMillian, W. W., N. W. Widstrom, D. M. Wilson, and R. A. Hill. 1980. Transmission by maize weevils of Aspergillus flavus and its survival on selected corn hybrids. J. Econ. Entomol. 73(6):793-794.

Merck Index: An Encyclopedia of Chemicals and Drugs. 1968. Quinone and Hydroquinone (p. 547, 907). Merck and Co., Inc., Rahway, NJ., 1713 pp.

Okumura, G. T. 1967. A report of canthariasis and allergy caused by Trogoderma (Coleoptera: Dermestidae). Calif. Vector Views 14(3):19-22.

Pelczar, M. J., and R. D. Reid. 1965. Microbiology. McGraw-Hill Book Company, New York, 662 pp.

Perlman, F. 1961. Insect allergens: their interrelationship and differences. J. Allergy 32(2):93-101.

Sauer, D. B., and R. Burroughs. 1980. Fungal growth, aflatoxin production, and moisture equilibration in mixtures of wet and dry corn. Phytopathology 70(6):516-521.

Sheldon, J. M., and J. H. Johnston. 1941. Hypersensitivity to beetles (Coleoptera). J. Allergy 12(5):493-494.

Smalley, E. B., and F. M. Strong. 1974. Toxic Trichothecenes. p. 199-228, in I. F. H. Purchase (ed.) Mycotoxins. Elsevier Scientific Publishing Co., New York. 443 pp.

Stoloff, L. 1976. Occurrence of mycotoxins in foods and feeds. p. 23-50, in J. V. Rodricks (ed.) Mycotoxins and Other Fungal Related Food Problems. Advances in Chemistry Series 149. American Chemical Society, Washington, D. C., 409 pp.

Tschinkel, W. R. 1975. A comparative study of the chemical defensive system of tenebrionid beetles: chemistry of the secretions. J. Insect. Physiol. 21:753-783.

Wallace, G. D. 1973. Intermediate and transport hosts in the natural history of Toxoplasma gondii. Amer. J. Trop. Med. Hyg. 22:456-464.

Wilbur, R. D., and G. W. Ward. 1976. Immunologic studies in a case of baker's asthma. J. Allergy Clin. Immunol. 58(3):366-372.

Wirtz, R. A. 1980. Occupational allergies to arthropods - documentation and prevention. Bull. Entomol. Soc. Amer. 26(3):356-360.

Wittich, F. W. 1940. The nature of various mill dust allergens. Journal - Lancet 60:418-421.

Wraith, D. 1969. Mites and house dust allergy. Health 6(4):14-16.

INSECTICIDES AND OCCUPATIONAL HEALTH IN THE FOOD INDUSTRY

Gary W. Olmstead

Editor's Comments

It is true that episodes of poisoning or illness of employees or applicators from the use of insecticides in plant sanitation programs are infrequent. However, they do occur. This chapter outlines the measures for prevention of such episodes and for proper management should they occur. Follow the author's recommendations. In so doing, regulatory compliance requirements will be met.

F. J. Baur

INSECTICIDES AND OCCUPATIONAL HEALTH
IN THE FOOD INDUSTRY

Gary W. Olmstead

Health and Safety Department
General Mills, Inc.
P.O. Box 1113
Minneapolis, Minnesota 55440

BACKGROUND

Pesticides are usually defined as chemical substances used to kill pests such as weeds, microbes, insects, rats and mice, nematodes, or other destructive forms of life. Insecticides are one category of pesticides intended to prevent infestation or destroy any insects that may be present in a certain environment. These chemicals are used to meet sanitation requirements involved in Good Manufacturing Practices and to help comply with government regulations. Thus, insecticides are poisons intended to kill insects but which may also represent a potential health hazard to pesticide applicators and other persons who may come into contact with these chemicals. Consequently, there have been many questions raised about the health risks of exposure to insecticides by employees, unions, government agencies, and the general public.

TYPES OF INSECTICIDES USED IN THE FOOD INDUSTRY

Insecticides used in the food industry may be grouped into four classes - general fumigants, spot fumigants, grain fumigants, and sprays, mists, and aerosols. General fumigation involves an application to a large volume such as an entire building. Two common general fumigants are methyl bromide and phosphine. Spot fumigation is an application to limited spaces where insects are likely to harbor such as behind walls or inside equipment. The only currently registered spot fumigant is a combination of carbon tetrachloride and ethylene dichloride. Grain fumigants are chemicals applied directly to raw agricultural commodities to kill insects. Grain fumigants are composed of various combinations of carbon tetrachloride, ethylene dichloride, and carbon disulfide. Finally, sprays, mists, and aerosols are insecticides dispersed by foggers, crack and crevice sprayers, misters, aerosol devices, or vapor dispensers for control of flying insects and exposed crawling insects. Malathion, pyrethrins, dichlorvos, Dursban, Baygon, Ficam, and other organophosphates and carbamates are examples of typical chemicals used in sprays, mists, and aerosols.

USAGE IN THE FOOD INDUSTRY

There are several operations in a food facility where insecticide exposure could occur. Incoming raw materials may have been treated. Insecticides may also be applied to processing equipment, stored ingredients, and the food plant premise. Finally, insecticides may also be used on outgoing shipments of food products.

All insecticides are toxic or they wouldn't be used to kill insects. The real question is how might any particular insecticide affect people. To understand an insecticide's danger, it is important to know the difference between toxicity and hazard. The toxicity of an insecticide describes the nature, degree, and extent of undesirable health effects potentially caused by exposure to that chemical. Toxicity is a basic biological property of that material and reflects its inherent capacity to produce injury. Hazard, on the other hand, relates the likelihood of this toxicity being manifest. Thus, hazard is the probability of injury resulting from actual use of the substance in the quantity and manner proposed. The severity of any particular insecticide's hazards depends on:

> Exposure concentration.
> Duration of exposure.
> Previous exposure.
> Frequency of exposure.
> Individual susceptibility.
> Personal protective equipment used.
> Work practices of the applicator.

Hazards increase with higher exposure concentration, longer duration of exposure, recent previous exposure, many frequent exposures, individual health status, lack of (or improper use of) personal protective equipment, and poor work practices.

Fortunately, most (if not all) insecticide applications in the food industry involve exposure to relatively low concentrations of chemicals for short periods of time. In addition, most insecticides are not used that frequently so there is little problem with previous exposures resulting in a cumulative effect. Of course, all people are somewhat unique and have slightly different responses to various environmental stresses. This is one reason why the use of personal protective equipment and medical monitoring programs is so important.

ROUTES OF ENTRY AND PROTECTION TECHNIQUES

Before an agent can exert its toxic effect, it must enter the body. There are three main routes of entry for any toxic agent - ingestion, skin absorption, and inhalation.

Although ingestion is not usually a major problem in the use of insecticides in the food industry, accidental ingestion in other settings is the most common safety problem with these chemicals. One of the more frequent scenarios occurs when an insecticide is placed in a beverage bottle and someone accidentally swallows it before realizing the contents are toxic. Thus, insecticides should never be measured, transported, or stored in beverage containers. All insecticide containers should be labeled and appropriately marked as to hazard. It should also be noted that there can be incidental ingestion by allowing contaminated gloves or hands to touch pencils, cigarettes, or food items. Thus, gloves should be removed and hands washed after insecticide application and pencils, cigarettes, or food items should never be carried into, or stored in, a treated area. In addition, insecticides should not be stored in a lunchroom nor should persons eat or drink near insecticides. All pesticides should be kept under lock and key with restricted availability.

Many insecticide formulations contain chemicals which are either primary irritants or can be absorbed through the skin. Primary irritants affect the skin or eye in the immediate vicinity of the site of exposure. The skin or eye may become irritated and painful. Some insecticide chemicals can be absorbed through the skin especially the back of the neck and the genital area. Gloves, goggles, faceshields, aprons, and suits can be used to prevent skin exposure. However, not all rubber or plastic material is impervious to all chemicals. Manufacturers should be consulted to determine which type of rubber or plastic material will provide the greatest protection from various chemicals. Gloves, for instance, can be manufactured using rubber, neoprene, nitrile butadiene rubber (NBR), polyvinyl chloride, polyvinyl alcohol, etc. Glove manufacturers recommend the following actions:

a. Choose the type of glove material that provides the least breakthrough for the chemical and physical conditions involved.
b. Select unsupported gloves for extra dexterity and sense of touch. If cut, snag, puncture, or abrasion resistance are important, pick a fabric-lined style.
c. Select the palm finish to provide the grip needed for the job-smooth, sprayed, dipped, or embossed. Sprayed and dipped finishes grip best when wet.
d. Select thin gauge gloves for jobs demanding sensitive touch and high flexibility. If greater protection or durability is wanted, choose a heavy duty style.
e. Choose the glove size or sizes that will assure optimum wear, dexterity, working ease, comfort, and employee satisfaction.

It should be noted that if the inside of the glove or suit becomes contaminated with the insecticide, the hazard can increase due to the more prolonged and intense exposure. If punctures occur or the interior of the glove or suit becomes contaminated; they should be removed promptly, the area affected should be thoroughly washed, and a new pair of gloves or suit should be obtained.

Inhalation is the most significant route of potential hazardous exposure in food plants. The label on each insecticide container should describe what kind of respiratory protection is recommended. There are several different types of respiratory protection ranging from cartridge respirators to gas masks to supplied air respirators to self-contained breathing apparatus. These various kinds of respiratory protection along with their advantages and disadvantages are summarized in Table 1. It should be noted that the air purifying respirators do not supply oxygen. Thus, there must be sufficient oxygen (>19.5%) for their use. In addition, there are several different purifying cartridge or canister media which must be chosen based upon the particular chemical exposure.

The useful life of a canister/cartridge is a function of:

The exposure time and concentration of a particular chemical(s).
The rate of respiration of the person wearing the mask.
The size and content of the cartridge/canister.
The temperature and humidity.

TABLE 1
TYPES OF RESPIRATORY PROTECTION

1. Air-Purifying Devices
   a) mechanical filter respirators
   b) chemical cartridge respirators
   c) combination of a & b
   d) gas masks

2. Supplied Air Devices
   a) air - line respirators
       constant flow
       demand
       pressure demand
   b) hose masks
       with blower
       without blower

3. Self-Contained Breathing Apparatus
   a) oxygen cylinder rebreathing type
   b) chemical oxygen rebreathing (self-generating) type
   c) demand type
   d) pressure demand

Respirator Type	Advantages	Disadvantages
Air-Purifying Respirators (using appropriate cartridge/canister)	1) Protection against low concentrations of gases and vapors 2) Good protection against particulates 3) Lightweight 4) Compact construction 5) Simple maintenance 6) Low initial cost	1) Discomfort 2) Resistance to breathing 3) Fatigue 4) Interference with vision 5) Interference with communication 6) Limited protection to high concentrations of gases or vapors 7) Negative air pressure in respirator
Supplied-air respirators	1) Positive air pressure in most respirators 2) Minimal breathing resistance 3) Light weight 4) Moderate initial cost 5) Relatively low operating cost 6) Can be used for long continuous period	1) Loss of source of respirable air eliminates protection 2) Restricted movement due to airline or hose 3) Limited vision and communication 4) Not recommended for situations that may be immediately dangerous to life or health
Self-contained breathing apparatus	1) Worker carries his own supply of respirable air so he can move about with great freedom	1) Equipment is heavy and bulky 2) Limited service life 3) Expensive 4) Difficult to maintain 5) Fatigue 6) Limited vision and communication

The useful life of a canister/cartridge can be determined by breakthrough time or by calculation. If the chemical has good warning properties (i.e. can be smelled or tasted at concentrations less than its threshold limit value (TLV), the respirator can be worn until the applicator detects the chemical coming through. If there are no good warning properties, you may be able to calculate breakthrough time if you know the airborne concentration and have assistance from the respirator manufacturer.

The shelf life expiration date is shown on each cartridge/canister and it should not be used after expiration of this date.

The following are basic requirements of an effective respiratory protection program:

Written standard operating procedures governing the selection and use of respirators should be established.

Respirators should be used only for the purpose intended and no modification of the equipment should be made. Parts from different respirator manufacturers should not be used interchangeably. For example, cartridges and canisters that are designed for a particular respirator shouldn't be mixed or matched with another manufacturer's respirator.

Respirators approved by National Institute of Occupational Safety and Health (NIOSH)/Mine Safety and Health Administration (MSHA) should be used when available.

Selection of the proper type of respiratory protective equipment should be based on the following procedures.
    a. Identify the substance or substances against which protection is necessary.
    b. Know the hazards and significant properties of each substance.
    c. Determine the conditions of exposure and level of air contaminant present.
    d. Ascertain whether any individual's capabilities and limitations are essential to the safe use of the device.
    e. Fit the respirator carefully and instruct the worker in its use. A positive and negative pressure fit test as well as a challenge exposure to irritant smoke, isoamyl acetate, or saccharin mist should be used to determine proper fit.

In areas where the wearer, with failure of the respirator, could be overcome by a toxic or oxygen-deficient atmosphere, at least one additional person should be present. Planning should be such that one individual will be unaffected by any likely incident and have the proper rescue equipment to be able to assist the other in case of emergency.

Workers must be trained and educated in the proper use of respiratory equipment. Training should include the following:
    a. Instruction in the nature of the hazard, whether acute, chronic, or both, and an honest appraisal of what may happen if the respirator is not used.
    b. A discussion of the respirator's capabilities and limitations.
    c. Instruction and training in actual wearing of the respirator and close and frequent supervision to insure that it continues to be properly used.
    d. Instruction as to where and when respiratory protection is required.

A program for maintenance and care of the respirators should be adjusted to the type of plant, working conditions, and hazard involved. It should include the following basic services:

   a. Periodic inspection for defects (including a leak check).
   b. Cleaning and disinfecting after each use.
   c. Repair.
   d. Storage. Respirators should be stored to protect against dust, sunlight, heat, extreme cold, excessive moisture, or damaging chemicals.

Respiratory equipment should never be abused because it can mean saving someone's life. <u>In an emergency rescue, self-contained breathing apparatus should be used since exposure concentrations are unknown.</u>

Medical evaluation must be performed by a qualified physician on a periodic basis to assure the applicator is physically able to wear the respirator for the task assigned.

Appropriate surveillance of work area condition and degree of employee exposure should be maintained. The continued effectiveness of the program should be regularly evaluated. (i.e. air sampling may be needed to determine employee exposure and to verify that the type of respirator being worn gives adequate protection.)

Thus, only experienced and capable persons should be in charge of the selection, training, use, and maintenance of respiratory protective equipment.

TOXICOLOGICAL CONCEPTS

There are several toxicological concepts which are basic to understanding the occupational health concerns with insecticides - local versus systemic effects, acute versus chronic effects, measures of acute toxicity, and interaction with alcohol.

   Local effects are those which occur near the exposure site of the chemical while systemic effects are those which affect other organs of the body remote from the site of contact.

   Acute effects are those which occur relatively quickly after exposure to a chemical while chronic effects are those which occur only after frequent exposures over a long period of time.

   Acute toxicity of a chemical is often defined as the LD_{50} or LC_{50} for that particular agent (see Table 2). The dosage of a chemical necessary to kill 50 percent of a certain number of treated animals by ingestion or dermal absorption is referred to as the lethal dose for 50 percent of that population (LD_{50}). If the route of exposure is inhalation, this value is referred to as the lethal concentration for 50 percent of that population (LC_{50}). The lower the LD_{50} or LC_{50}, the more toxic the chemical is. The higher the number, the safer the insecticide, since this means that it takes more material to kill a test animal.

300

TABLE 2
CATEGORIES OF ACUTE TOXICITY

Categories	Signal Word Required on The Label	LD50 Oral mg/kg	LD50 Dermal mg/kg	LC50 Inhalation mg/l	Probable Oral Lethal Dose for 150 Pounds man
I Highly Toxic	DANGER- skull and crossbones POISON	0-50	0-200	0-2,000	A few drops to a teaspoonful
II Moderately Toxic	WARNING	over 50 to 500	over 200 to 2,000	over 2,000 to 20,000	Over one teaspoonful to one ounce.
III Slightly Toxic	CAUTION	over 500 to 5,000	over 2,000 to 20,000	......	Over one ounce to one pint or one pound.
IV Relatively Non-toxic	None	over 5,000	over 20,000	......	Over one pint or one pound.

There are no standard measures like LD50 for chronic toxicity studies. Often the length of the experiment in days, months, or years and the amount of each dose is stated. Thus, LD50 relates to acute toxicity but is no measure of chronic toxicity. Some insecticides (i.e. halogenated hydrocarbons) may accumulate in body tissues (i.e. fat) as long as exposure continues and could result in a chronic health hazard. The major routes of absorption and toxic effects of the various types of insecticides are given in Table 3.

The ingestion of alcohol shortly before or after the application of certain insecticides (i.e. halogenated hydrocarbons) may affect the body's metabolism of that insecticide resulting in a greater health effect than would otherwise have occurred.

TRAINING

Insecticides should only be applied by persons trained in their safe use. Applicators must understand the label which describes the hazards of the chemical(s). The proper protective equipment must also be provided and worn. All unauthorized people should be kept away from application areas and warning signs posted. It is also recommended that the local fire/police departments be notified and guards posted when significant treatment is to be done. Workers should be alert for symptoms of poisoning and receive immediate medical treatment when indicated. A chart outlining symptoms and treatment is given in Table 3. A stand-by person should be provided where necessary. By carefully pre-planning the insecticide application and being prepared for any possible emergency, the probability of preventing accidents, or at least minimizing their impact, increases greatly.

TABLE 3

EMERGENCY MEDICAL TREATMENT FOR INSECTICIDE POISONING

Chemical Basis	Methyl Bromide	Phosphine	Halogenated Hydrocarbons	Organophosphorus Compounds	Pyrethrins	Hydrocarbons
ROUTE(S) OF ABSORPTION	Inhalation, Skin	Inhalation	Inhalation, Skin, Ingestion	Inhalation, Skin, Ingestion	Inhalation, Skin, Ingestion	Inhalation, Ingestion
TARGET ORGAN(S)	Liver, Kidney, CNS, Lungs	Lungs	CNS, Kidney, Liver	Anticholinesterase (irreversible)	CNS, Skin	CNS, Lung, Kidney, Liver
SYMPTOMS	Symptoms include dizziness, headache, nausea, vomiting, abdominal pain, weakness, slurred speech, staggering gait, mental confusion, tremors, convulsions, rapid respiration, pulmonary edema, cyanosis, collapse, and death. Late manifestations may include bronchopneumonia, pulmonary edema, and respiratory failure. Methyl bromide may produce cutaneous blisters and kill via dermal exposure. Chronic poisoning may be characterized by blurred vision, papilledema, numbness of the extremities, confusion, hallucinations, fainting attacks, and bronchospasm.	Symptoms include nausea, vomiting, diarrhea, great thirst, headache, vertigo, tinnitus, pressure in chest, back pains, dyspnea, feeling of coldness, stupor, or attacks of fainting. Patient may develop hemolytic icterus and cough with sputum of a green fluorescent color. Chronic poisoning may be characterized by anemia, bronchitis, G.I. disturbances, dental necrosis, and disturbances of vision, speech and motor functions.	Symptoms include abdominal pain, nausea, vomiting, dizziness, confusion, restlessness, apprehension, cyanosis, convulsions, fall of blood pressure, coma, and respiratory failure. If patient recovers consciousness, he may have coma, liver damage, or kidney damage in 1 day to 2 weeks. Chronic poisoning may also be characterized by the symptoms above but less severe. Other symptoms of chronic poisoning include fatigue, anorexia, weakness, vertigo, and tremors. Dermatitis follows repeated skin exposure.	Symptoms include: **Mild:** Anorexia, headache, dizziness, weakness, anxiety, tremors of the tongue and eyelids, miosis, and impairment of visual activity. **Moderate:** Nausea, salivation, lacrimation, abdominal cramps, vomiting, sweating, slow pulse, muscular tremors. **Severe:** Diarrhea, pinpoint and nonreactive pupils, respiratory difficulty, pulmonary edema, cyanosis, loss of sphincter control, convulsions, coma, and heart block. In chronic poisoning, the cholinesterase inhibition can persist 2-6 weeks. Thus, an exposure which would not produce symptoms in a person not previously exposed, might produce severe symptoms in a person previously exposed to smaller amounts.	The toxicity of pyrethrins, or synergists commonly found in pyrethrin formulations, are low to man. Poisoning is usually due to the petroleum distillates contained in these insecticides: Symptoms may include skin inflammation, nausea, coughing, vomiting, bloody sputum, numbness of tongue and lips, diarrhea, headache, tinnitus, incoordination, and stupor. Death due to respiratory paralysis. Heavy ingestion may cause coma or convulsion.	Symptoms may include dizziness, weakness, euphoria, headache, nausea, vomiting, tightness in the chest, and staggering. Higher exposures may cause visual blurring, tremor, shallow, rapid respiration, paralysis, unconsciousness, and convulsions. Other symptoms are cough and pulmonary irritation progressing to pulmonary edema, and bronchobloody sputum, and bronchopneumonia. Chronic poisoning may be characterized by headache, drowsiness, nervousness, and pallor.
ONSET	If exposure concentration is high, symptoms occur after 4-6 hours. At low concentrations, symptoms may not appear for 12-24 hours.	If exposure concentration is high, symptoms occur immediately. In smaller exposure, symptoms may appear up to 24 hours later.	If exposure concentration is high, symptoms usually appear within 20 minutes but may be delayed up to 4 hours.	Acute reactions occur in 30-60 minutes and are at a maximum in 2-8 hours.		Onset depends on exposure concentration, but usually occurs in less than 2 hours.
TREATMENT	1. Remove from exposure and observe for 48 hours. 2. Remove contaminated clothing and wash skin with water. Vaseline dressing should be applied to blistered areas. 3. Before symptoms appear, give BAL at 3-4 mg/kg (or 0.3-0.4 ml/10kg) every 4 hours for the first two days and every 12 hours for ten days. 4. Restrain hyperactive patients and use barbiturates for convulsions. 5. May require specific therapy for acidosis, pulmonary edema, bronchospasm (use 0.3 ml of 1:1000 epinephrine solution subcutaneously), respiratory paralysis, or renal failure.	1. No specific antidote. 2. Remove from exposure. 3. If exposure was mild, have patient rest 1-2 days and keep warm and quiet. 4. I.V. glucose for nausea and vomiting. 5. Saline or Ringers solution for hyperglycemia. 6. Inhalation of O² or O²/CO². 7. Consider cardiac and circulatory stimulants. 8. In case of pulmonary edema: a) 500-1000 mg Prednisolone b) Heart glycosides (I.V.) c) Consider venesection d) Intubation with positive pressure breathing e) Check electrolytes f) Dialysis if necessary g) In extreme cases, blood transfusion.	1. Remove from exposure. 2. Remove contaminated clothing. 3. If swallowed, remove by gastric lavage with water or saline solution. 4. Maintain blood pressure by giving 5% glucose I.V. 5. Do not give stimulants. Epinephrine or ephedrine may induce ventricular fibrillation. 6. Respiratory stimulants such as caffeine or nikethamide should be tried if CNS depression is severe. 7. Before severe kidney damage is apparent, treatment with an osmotic diuretic may be useful. If no diuresis ensues, fluids and electrolytes should be administered only cautiously. 8. Give high carbohydrate diet to attempt to restore optimal liver function.	1. Remove from exposure. 2. For extreme symptoms of poisoning inject massive doses of atropine I.V. (2 to 4 mg. or 1/30 to 1/15 grain) every 5-10 minutes until signs of atropinization occur. A total of 25 to 50 mg. or more may be necessary during the first day. Watch for reduction in salivation. Interruption of atropine therapy may be rapidly followed by fatal pulmonary edema or respiratory failure. Do not give atropine to a cyanotic patient. Give artificial respiration first then administer atropine. Oral atropine is never used and atropine prophylaxis is not recommended. 3. 2-PAM (Protopam chloride), 1 gram I.V. slowly over a period of 5 minutes. Give a second dose of 500 mg in 30 minutes if muscle	1. Remove from exposure. 2. Remove contaminated clothing and wash skin with soap and water. 3. Activated charcoal followed by gastric lavage with tap water. 4. Symptomatic and supportive treatment with oxygen, artificial respiration, barbiturates and parenteral fluids are beneficial. 5. See treatment for hydro-carbons.	1. Remove from exposure. 2. Remove contaminated clothing and wash skin with soap and water. 3. Use no emetics. 4. Use gastric lavage if more than 4 mg/kg (1/2 pt (150 pounds) has been ingested. Use extreme care to prevent aspiration(cuffed endotracheal tube) 5. Give 250 ml (8 oz.) of liquid petrolatum orally to dissolve kerosene and slow absorption. 6. Oxygen and corticosteroids may be needed in severe case. Antibiotics are sometimes indicated. 7. Do not give epinephrine or related drugs. They may induce ventricular fibrillation. Monitor ECG to detect ventricular abnormalities foreshadowing possible cardiac arrest.

M
E
N
T
	9. After acute nervous symptoms subside and before visceral lesions appear, hypothermia should be tried to reduce hepatic and renal damage.
	10. Treat acute renal failure.
	  a) Oliguirisis lasts 7-10 days followed by diuresis which may last up to 3 weeks.
	11. Hepatic coma:
	  a) Control blood ammonia.
	  b) Limit protein intake to 20-30 gm/day.
	  c) Prevent absorption of ammonia from stool by daily milk of magnesia or sodium sulfate.
	  d) 8 gm of Neomycin daily to reduce ammonia formation in bowel.
	12. Give KCL for alkalosis.
	13. Hemodialysis may be necessary to control blood electrolytes.
		No residual toxic effect would be expected from normal use of pyrethrin.		weakness persists. Use only with maximum atropine administration.
				4. Avoid morphine, theophyllin, aminophyllin, barbiturates, or phenothiazines.
				5. Keep airway open. Aspirate, use oxygen, insert endotracheal tube. Artificial respiration and tracheostomy in severe cases.
				6. For ingestion, lavage stomach with 5% sodium bicarbonate.
				7. For skin contact, wash with soap and water. Wear rubber gloves while washing contact areas.
				8. Keep patient under constant observation for 24-36 hours.	

PROGNOSIS	Survival for 48-72 hours usually brings complete recovery.	Survival for 4 days is ordinarily followed by recovery.	Survival for 48 hours usually brings complete recovery. In anuria, return of kidney function may begin 2-3 weeks after poisoning. Complete return of liver and kidney function requires 2-12 months.	The first 4-6 hours are the most critical in acute poisoning. Improvement of symptoms after treatment is instituted means that the patient will survive if adequate treatment is continued.	After the first 24 hours, the extent of pulmonary involvement indicates severity. Infiltration of more than 30% of the lungs requires 2-4 weeks for resolution. Death may occur up to 3 days after poisoning. Rapid progression of symptoms and lack of response to removal of hydrocarbon, indicate poor outcome.
LABORATORY TESTS	Blood samples should be collected for bromide ion concentration: 1 mg% = normal 5 mg% = definite exposure but no immediate danger 15 mg% = symptoms of poisoning will appear	None	Kidney and liver function tests.	Cholinesterase test. Send 10cc heparinized blood to lab for plasma and red cell cholinesterase. Draw blood for cholinesterase test preferably before 2-PAM is given. Levels 30-50% of normal indicate exposure, although symptoms may not appear until the level falls to 20% or less.	Red and white blood cell count.

Selected References:
1. Emergency Medical Treatment for Acute Pesticide Poisoning. United States Navy. 1974.
2. Handbook of Poisoning. 8th Edition. R. H. Dreisbach. Lange Medical Publications. 1974.
3. Clinical Toxicology of Commercial Products. 3rd Edition. M. N. Gleason, R. E. Gosselin, H. C. Hodge, and R. P. Smith. Williams and Wilkins. 1969.

DETERMINING INSECTICIDE EXPOSURE CONCENTRATIONS

There are several methods to determine airborne concentrations of the various insecticides. Detector tubes, adsorbent media (i.e. charcoal), or continuous reading instruments (i.e. photoionization or infrared devices) can be used for air monitoring.

Detector tubes are an easy, inexpensive method to determine approximate concentrations of insecticides. They are sealed tubes which contain chemicals that react with specific insecticide chemical ingredients. The tubes are broken open and air is drawn through the tube by a hand-operated pump. The tubes are specific to the chemical being measured and produce a color reaction proportional to the airborne concentration of that chemical. Major manufacturers of detector tubes are Drager, Bendix, and MSA. It should be noted that detector tubes are usually only accurate to ±25%. They are useful for determining instantaneous exposure but not daily time weighted average (TWA) exposure.

Another method involves charcoal (or other media) which can be placed in a tube through which air is drawn or put in diffusion monitors for sampling airborne chemicals. This is an accurate method for determining exposure concentrations but requires laboratory analysis.

There are also continuous reading instruments which operate by monitoring the absorption of infrared or ultraviolet light to determine insecticide concentrations. These instruments can be very useful if properly calibrated and maintained. Their major drawback is cost.

An occupational health professional should be consulted if questions arise regarding proper measurement of insecticide exposure or the interpretation of that data in terms of health hazard.

MEDICAL EXAMINATIONS

Plant locations using insecticides should use medical consultants who are made aware of information concerning potential employee exposure and medical guidelines for treatment of hazardous exposure. Employees frequently exposed to insecticides should be given a pre-placement medical examination and, thereafter, offered periodic examinations. These examinations should include:

      Medical history.
      Physical examination.
      Urinalysis.
      Pulmonary function test.
      RBC, CBC with differential.
      Blood chemistry profile (i.e. SMA-12).
      RBC cholinesterase (if working with organophosphates).
      Chest X-ray based on age or if clinically indicated.

FIRST AID TREATMENT

In the event of overexposure to insecticides, first aid treatment should be limited to the following:

> Ingestion
>   Induce vomiting if the label specifically so directs, do
>   not give victim any food or fluids.
> Skin contact
>   Remove contaminated clothing and wash exposed skin with
>   soap and water.
> Eye contact
>   Flush eyes with tap water for minimum of 15 minutes.
> Inhalation
>   Remove victim from exposure, give artificial respiration if
>   required.

EMERGENCY MEDICAL TREATMENT FOR INSECTICIDE POISONING

Despite all precautions taken, an overexposure or accident may occur which could result in a person becoming ill. Table 3 presents the route(s) of absorption, target organ(s), symptoms, onset, recommended treatment, prognosis, and laboratory tests for emergency medical treatment of insecticide poisoning. Unfortunately, most of the symptoms of insecticide poisoning can be confused with those of other illnesses. Thus, the chemical(s) may not be the cause of the illness but it is prudent to seek medical assistance irrespective. Be sure that the label from the insecticide container is brought to the physician so that appropriate medical care can be administered. Further information on treatment can also be obtained by calling the insecticide manufacturer, the local Poison Control Center, the State Health Department, or the Center for Disease Control in Atlanta.

GOVERNMENT REGULATIONS

There are several government agencies which have regulations influencing the use of insecticides. The two most important are EPA and OSHA:

> Environmental Protection Agency (EPA). Under the Federal
>   Insecticide, Fungicide, and Rodenticide Act (FIFRA), EPA
>   is charged with responsibility for providing for
>   registration, labeling, and classification of pesticides
>   (see Chapter 26). EPA sets competency standards for certifying
>   pesticide applicators who want to work with restricted use
>   pesticides. In addition, EPA and the Food and Drug
>   Administration (FDA) develop residue tolerances for various
>   pesticides.

> Occupational Safety and Health Administration (OSHA). OSHA
>   establishes and enforces employee health and safety
>   standards which are designed to prevent or minimize health
>   and safety hazards. EPA has jurisdiction over the
>   application of pesticides while OSHA regulates occupational
>   exposure in all other instances. The most common insecticides
>   with their respective OSHA permissible exposure limit

(PEL), the American Conference of Governmental Industrial Hygienists (ACGIH) threshold limit value (TLV), and National Institute of Occupational Safety and Health (NIOSH) recommendations are listed in Table 4. ACGIH is a non-governmental organization that reviews and develops health guidelines for chemical exposure while NIOSH is a government agency located in the Department of Health and Human Services which recommends health standards to OSHA. Neither ACGIH's threshold limit values or NIOSH recommendations have the force of law. All these values, however, refer to airborne concentrations of substances and represent conditions under which it is believed that nearly all workers may be repeatedly exposed day after day without adverse effect. Threshold limits are based on the best available information from industrial experience, from experimental human and animal studies, and, where possible, from a combination of the three. These standards should be used as guides in the control of health hazards and should not be used as fine lines between safe and dangerous concentrations. In spite of the fact that serious injury is not believed likely as a result of exposure to the threshold limit concentrations, the best practice is to maintain concentrations of all contaminants as low as is practical.

TABLE 4
HEALTH STANDARDS FOR VARIOUS
INSECTICIDE CHEMICALS

	OSHA PEL	ACGIH TLV	NIOSH RECOMMENDATION
carbamates	none	none	none
carbon disulfide	20 ppm	10 ppm	1 ppm
carbon tetrachloride	10 ppm	5 ppm	2 ppm
dichlorvos	none	0.1 ppm	none
ethylene dibromide	20 ppm	none	0.13 ppm
ethylene dichloride	50 ppm	10 ppm	none
malathion	15 mg/m3	10 mg/m3	15 mg/m3
methyl bromide	20 ppm	5 ppm	none
phosphine	0.3 ppm	0.3 ppm	none
pyrethrins	5 mg/m3	5 mg/m3	none

SUMMARY

All insecticides are toxic, however, they need not be hazardous. The hazard is determined by how they are used. Insecticides are a very important part of all food operations and should represent no significant occupational health risk when used by trained applicators following recommended safe handling procedures. On the other hand, insecticides can be used improperly creating certain health hazards. Thus, the food industry should be aware of and implement a personal protective equipment program, training program, air sampling program, medical examination program, first aid program, and a medical emergency treatment program in order to prevent (or minimize) insecticide-related occupational health hazards and ensure continuing compliance with all relevant government regulations.

REFERENCES

Edmont-Wilson: Job Fitted Gloves and Protective Clothing. Coshocton, Ohio, 1982.

Hayes, W.J.: Pesticides Studied in Man. Williams and Wilkins, Baltimore, Maryland, 1982.

Hayes, W.J.: Toxicology of Pesticides. Williams and Wilkins, Baltimore, Maryland, 1975.

Gleason, M.N., R.E. Gosselin, H.C. Hodge, and R.P. Smith: Clinical Toxicology of Commerical Products. Williams and Wilkins, Baltimore, Maryland, 1969.

Dreisbach, R.H.: Handbook of Poisoning. Lange Medical Publications, Los Altos, California, 1974.

U.S. Navy: Emergency Medical Treatment for Acute Poisoning, 1974.

INSECT INFESTATION OF PACKAGES

Henry A. Highland

Editor's Comments

Continuing product integrity in the distribution chain relies upon packaging which is resistant to insect entry and penetration. Even though the products are no longer under control of a company, that company can and will be held responsible if insects are found in the product, including possible entry in the home. Flexible type packaging is particularly vulnerable. This chapter gives some insights into which insects present the greater risks. It was the editor's experience that raw materials, particularly packaging materials, can be the greatest source of possible insect problems while the products are still under the manufacturer's control and oversight.

Research continues within USDA and the packaging industry to make films more penetration resistant by means of film physical properties and the judicious use of insecticides.

F. J. Baur

INSECT INFESTATION OF PACKAGES

Henry A. Highland

Stored-Product Insects Research and Development Laboratory
Agricultural Research Service
U.S. Department of Agriculture
Savannah, Georgia 31403

INTRODUCTION

Any comprehensive study of insect control in the food industry must consider the packaging used in modern food storage and distribution systems. Most non-perishable foods are shipped to the consumer in packages and with the exception of canned foods most are subject to attack by stored-product insects. The selection and handling of packages must include consideration not only of protection and sales appeal, but also factors involving storage, distribution, merchandising, public health, and legal requirements as related to the total food storage and distribution system. These factors should be considered in selecting packaging for all infestible items, including pet food and seeds, which are stored and distributed with food.

New and modified food handling procedures, increasingly stringent sanitation standards, and increasing international trade impose a need for systems that will protect food, feed, and seed from infestation from the time it is packed until the package is opened by the consumer. Knowledgeable selection of packaging materials, packages, and packaging equipment can help produce packages that resist infestation. Such packages can be economically feasible since there would be reduced losses of the product that has already incurred all costs of growing, harvesting, transporting, processing, and storing. Direct losses are not only reduced – company image is preserved, because the consuming public usually holds the producer responsible no matter how or where the packaged product becomes infested.

INSECT PESTS OF PACKAGED FOODS

Most stored-product insects are cosmopolitan in that any given species is usually found world wide in areas with similar climatic conditions. For example, Highland (1978a) found up to 24 species in warehouses in 1 to 8 countries, and all of the species are also found in the United States. Warm humid environments promote insect growth; consequently, insects are likely to be more troublesome in packages stored in warmer warehouses than in cool warehouses. Obviously, there would be few problems of insects in packages stored in refrigerated warehouses. For descriptions of the effects of cold on stored product insects, see Chapter 18.

The extent and incidence of package infestations increase greatly as summer progresses in the United States, and generally as one moves south. In some southeastern areas, for instance, it was once customary to remove corn meal from retail shelves at 2-3 week intervals during the summer to avoid development of noticeable populations of insects within the packages. Longer, warmer, more humid summers provide longer favorable growth periods during which insects can multiply and spread from infested to uninfested packages stored under ambient conditions. Problems with psocids increase noticeably as summer progresses. These insects can be found in or on empty paper packages (usually corrugated cases) used as secondary containers for empty primary containers such as jars or cans. The psocids contaminate the empty jars or cans before they are filled and sealed. Thorough decontamination, by wet cleaning or sufficient heating, may be required.

Although there is little evidence to show that packaging influences the attractancy of foods to insects it is an established fact that some foods are more likely to become infested than others. Infestation probably is a consequence of searching, exploratory activities of the insects, and the subsequent discovery of food that is suitable for nourishment and reproduction. Products that become infested can then be the foci for infestation of other products stored nearby. Dry pet food is seldom packed in insect-resistant packages. Because it is a suitable food medium for several species of insects, populations are likely to build up during storage, especially during the summer. The pet food then serves as a reservoir of insects when stored in grocery warehouses containing other infestible foods. Obviously there is direct relationship between length of storage, severity of infestation, and resulting cross contamination of other products if no insect control measures are utilized.

Food warehouses are, of course, not the sole site of infestations in marketing channels. Packaged foods can become infested during shipment in trucks, railcars, ships, or during storage at wholesale and retail levels.

The problem these insects present is exacerbated by the ability of some species (penetrators) such as Rhyzopertha dominica (F.) (lesser grain borer), Tenebroides mauritanicus (L.) (cadelle), Lasioderma serricorne (F.) (cigarette beetle), Trogoderma variabile Ballion (warehouse beetle), Dinoderus minutus (F.) (bamboo powder post beetle), and Corcyra cephalonica (Stainton) (rice moth) to bore through one or more of most flexible packaging materials in use today. Other species, including Ephestia cautella (Walker) (almond moth), Trogoderma glabrum (Herbst), and Dermestes maculatus (DeGeer) (hide beetle) also penetrate, but less frequently.

Other common species (invaders) usually do not enter packages unless there is an existing opening. These include, but are not limited to, Tribolium castaneum (Herbst) (red flour beetle), Tribolium confusum Jacquelin duVal (confused flour beetle), Oryzaephilus surinamensis (L.) (sawtoothed grain beetle), Cathartus quadricollis (Guerin-Meneville) (squarenecked grain beetle), and Cryptolestes pusillus (Schonherr) (flat grain beetle). These species, however, can easily find the small openings in many packages as currently designed and manufactured. Sawtoothed grain beetle adults, for instance, can enter an opening less than 1 mm in diameter. Red flour beetle adults can pass through openings that are 1.35 mm in diameter (Cline and Highland, 1981). Newly hatched larvae can, of course, enter much smaller openings. Yerington (1978) showed a direct correlation between package seal quality and the extent and swiftness of infestation. Thus package seals are critical in the design and use of packages that are expected to protect foods from insect infestations.

These categories of penetrators and invaders, though useful, are somewhat artificial, because some of the invaders will penetrate given the proper circumstances and environment. Cline (1978) found that larvae of red flour beetle, flat grain beetle, square-necked grain beetle, and merchant grain beetle would not penetrate any of 8 packaging materials, but when starved, larvae of red flour beetle penetrated cellophane and polyethylene. Adults of these species penetrated some of the least resistant films, and recently it was found that extremely crowded red flour beetles will penetrate 10-mil polyethylene film (Highland, unpublished). In general, the more important penetrators, based on the frequency with which they are found in packaged foods and their ability to penetrate packages, are lesser grain borers, cigarette beetles, almond moths, rice moths, drugstore beetles (L.), Plodia interpunctella (Hubner) (Indianmeal moth), and warehouse beetle. Cadelle larvae are strong penetrators but they are rarely found in processed food handling or storage facilities. Although lesser grain borers are infrequently found in these storage facilities, their ability to penetrate most flexible packaging materials makes them an important pest. Many dermestid (Trogoderma and Dermestes species) larvae are also good penetrators.

It may be necessary at times to determine whether the insect formed ingress or egress holes. The direction of such penetrations can be determined with reasonable certainty by the characteristics of the bored hole (Brickey et al., 1973). The hole on the exit side of the penetrated packaging material has a clean-cut perimeter; the diameter is usually smaller than the entrance side; and there is no surface fraying, scratches, or depression around the hole. The perimeter of the hole on the entrance side on the other hand, is tapered, may be terraced, usually is scratched or roughened, and on plastics the perimeter is usually upturned. Although the direction of penetration may be determined by this method, it may be difficult or impossible to use such a determination to ascribe the origin of the infestation. This difficulty arises from the ability of insects to enter even minute openings that are detectable only with great difficulty. Perforated thumb notches on paperboard cartons and partially sealed carton overwraps are examples of openings that are difficult to detect but large enough to allow oviposition or the entry and exit of small larvae or adults.

Most species of stored-product insects attack packaged foods. Exceptions are the grain weevils, Sitophilus granarius (L.) (granary weevil), S. oryzae (L.) (rice weevil), and S. zeamais Motschulsky (maize weevil), which are restricted mostly to rice or bulgur, and moths that attack woolens (Tineola bisselliella Hummel (webbing clothes moth) and Tinea pellionella (L.) (casemaking clothes moth). Some species are found in packaged foods more frequently than others. These include red flour beetles and confused flour beetles, which attack most grains and farinaceous foods, and cigarette beetles, drugstore beetles, Indianmeal moths, almond moths, and sawtoothed grain beetles which attack a wide variety of packaged foods. In recent years some dermestids, particularly warehouse beetles have increased in importance as pests of packaged foods.

PACKAGING MATERIALS

Paper and cellophane are probably the least resistant to penetrating insects of all flexible packaging materials. Depending on conditions and species some insects can penetrate kraft paper in less than one day, and multi-ply construction adds little to the resistance.

Almost all flexible polymer films or combinations of polymer films can be penetrated by one or more species of insects. However, there are differences between packaging films in their susceptibility to penetration by insects. Highland and Jay (1965) found that polycarbonate film is very resistant to penetration by insects, but this film is not adaptable to food packaging. In laboratory tests polyester, cellulose diacetate, nylon, urethane, unplasticized polyvinyl chloride, and some polypropylene films resisted penetration by lesser grain borers (Highland and Wilson, 1981). Most other films, including cellophane, polyethylene, plasticized polyvinyl chloride, ionomer, saran, cellulose propionate, and ethylene vinyl acetate were penetrated within 18 hours.

Rao et al. (1972) used small pouches in a beaker and found that insects, both penetrators and some invaders, could penetrate one or more of 18 films and laminates including polyethylene, cellophane/polyethylene, cellophane, polyethylene/paper, saran/cellophane, polyethylene/jute, paper/foil/polyethylene, and foil/vinyl; however, no penetrations were found in polyester/polyethylene, polyethylene/canvas, foil/polyethylene, or paper/foil/polyethylene. This latter foil was twice as thick as the foil in the penetrated paper/foil/polyethylene. Gerhardt and Lindgren (1954, 1955) found that insects easily penetrated single thicknesses of polyethylene, cellophane/saran, cellophane laminates, and saran/pliofilm. However, laminates of saran plus polyester, with the polyester side exposed to the insects, were not penetrated.

Yerington (1975) showed that when tested as pouches against mostly penetrators, polypropylene/polyethylene combination films were generally more resistant than were other combination films. Also, coextruded polypropylene-ethylene vinyl acetate and coextruded polypropylene-polyethylene films showed more resistance to penetration than saran-coated polypropylene and polyethylene-coated cellophane films. Oriented polypropylene films were more resistant than non-oriented polypropylene films. Highland et al. (1968) showed that biaxially oriented polypropylene film provided excellent insect-resistant properties when used as an overwrap on paperboard cartons, but was ineffective as a pouch.

Table 1.--Resistance of various materials to insect penetration.

	Excellent	Good	Fair	Poor
Polycarbonate	x			
Polyester	x			
Polyurethane	x			
Cellulose diacetate		x		
Nylon		x		
Polyethylene (10-mil)		x		
Polypropylene (biaxially oriented)		x		
Polyvinyl chloride (unplasticized)		x		
Acrylic			x	
Ethylene tetrafluoroethylene			x	
Fluorinated ethylene propylene			x	
Polyethylene (5-mil)			x	
Cellophane				x
Cellulose propionate				x
Corrugated paperboard				x
Ethylene vinyl acetate				x
Ethylene vinyl acetate/polythylene (coextrusion)				x
Ionomer				x
Kraft paper				x
Paper/foil/polyethylene laminate (pouch material)				x
Polyethylene (1, 2, 3, 4-mil)				x
Polyvinyl chloride (plasticized)				x
Saran				x
Spunbonded polymers				x

Tests have clearly demonstrated that aluminium foil can also be penetrated, although foil packages are generally more resistant to penetration than are film or paper packages. Indeed, Batth (1970) reported tht dried soup mix pouches of 4-mil foil were penetrated within 4 days by cadelles, and within 2 weeks by Trogoderma inclusum (LeConte), but not by sawtoothed grain beetles. Paper/foil/polyethylene pouches of cocoa powder and dried soup mix were easily penetrated in simulated warehouse storage tests (Highland et al., 1977).

Thicker films are more resistant to penetration than are thinner films made of the same polymer resin. Although polyethylene film is not notably resistant to penetration, unpublished packaging tests conducted at the Savannah Laboratory have shown that some resistance is exhibited by films of 5-mil or greater thickness. Thus, films vary in susceptibility to penetration depending on thickness, on the basic resin from which the film is made, on the combination of materials, on the package structure, and on the species and stage of insects involved.

The relative resistance to insect penetration of common packaging materials is given in Table 1. Where no thicknesses are given these estimations are based on thicknesses commonly used in food packaging. Absolute values are difficult to determine because resistance to penetration is influenced by factors such as package configuration and the presence or absence of folds, tucks, and other harborage sites. Therefore, after candidate packaging materials have been selected they must be evaluated in situ for insect resistance.

PACKAGE STRUCTURE

Shipping Containers

Although packaging materials can be carefully chosen and chemically treated to make them resistant to insect penetration, no treatment is effective if the package is not insect tight. Highland et al. (1964) showed that paper bags treated with methoxychlor and having tight, permanent seals remained insect-free during 24 months of exposure to heavy insect populations, but similarly treated bags with sewn, untaped closures were infested within 3 months.

Multiwall paper shipping bags are made insect-resistant by using a folded-over, stepped-end closure which is sealed to the opposite face of the bag with a heat-activated adhesive (Laudani et al., 1966). Insect resistance was enhanced with a synergized pyrethrins coating on the outer ply. Another type of insect-resistant multiwall paper shipping bag, also treated with synergized pyrethrins, has synergized pyrethrins-treated paper tape heat-sealed over the stitching on both ends of the flat, non-gusseted bag. These bags have been used to carry millions of pounds of processed cereals to overseas destinations in the U.S. foreign aid programs. Fabric shipping bags made of cotton, jute, or woven plastic provide little protection from infestation.

The corrugated paper case is a major container for shipping foods produced in the United States and in other developed countries. It is difficult, however, to make these cases resistant to infestation. Well-sealed tape can be placed over all six flap junctures, or the case can be overwrapped with a heat-shrunk, polyethylene film. However, neither method protects from penetrating insects for long storage periods because both the film and the paper can be penetrated by insects. Yerington (1979) showed that cases of bulk-packed raisins can be made more resistant to insects by inverting them so that the folded closure of the polyethylene liners are on the bottom, held tightly closed by the weight of the raisins.

Heat-shrunk polyethylene film overwraps, along with a polyethylene deck sheet between the load and pallet, can be applied to pallet loads of bags or cases to provide protection from invading insects. During the heat-shrunk process, the deck sheet is partially bonded to the film overwrap. Use of a black deck sheet and an infrared heat source can provide an insect-tight seal with the overwrap.

Consumer-Sized Packages

Although it is more economical to make the outer shipping container insect-resistant, it is often more desirable to insect-proof the consumer-sized package. This provides protection from infestation throughout the distribution channel, including retail outlets, until the package is opened. The paperboard carton is a typical, widely used consumer package in developed countries and in developing countries where food is imported. The cartons can be made insect-resistant by the use of tight overwraps such as those provided by polypropylene film (Highland et al., 1968). Also, laboratory tests and reports from food-processing companies show that fully adhered paper overwraps are effective, and Collins (1963) showed that fully adhered laminated paper/foil overwraps are quite resistant to insect invasion.

In the absence of an insect-tight overwrap, the flaps of cartons must be sealed so they are insect-proof, a difficult procedure in modern high-speed operations. Beads of adhesive along the folds and edges of end flaps, or extended folded-down tabs at the sides of the end flaps (the 'Van Buren ear') provide some insect resistance. Yerington (1971) described a polyethylene-coated carton having heat-sealed flaps and polyethylene-overwrapped cartons that protected raisins from infestation. Composite cans, made of various combinations of kraft paper, aluminum foil, polymer films, and glassine paper and having metal ends provide excellent protection from boring and invading insects (Highland, 1975).

Insecticide Treatments

Insect-repellent treatments are only applied to packaging materials for food and other infestible items in very restricted areas of usage and with carefully controlled procedures. Thus, these procedures have a favorable cost-benefit ratio in terms of their impact on the environment when compared with procedures used to control such insects as mosquitoes or field-crop forest insects. Nevertheless, the spatial and temporal proximity of chemicals to products destined for human consumption makes this use of pesticides potentially hazardous. These factors must be weighed against the benefits accrued by protecting the commodity from infestation when the commodity has already incurred the full cost of growing, harvesting, shipping, storing, and processing. We must also add the costs of packing, shipping, storing, and distributing the processed, packaged commodities. It becomes obvious that to achieve the desired protection without objectionable residues, package treatments must be used judiciously, at minimal dosages, and with all available expertise.

Obviously, insecticides or repellents used on packages must not migrate from the treated surface and through the package walls to produce residues in the commodity that would exceed safe, legal tolerances. This migration can be reduced or prevented by the use of barrier plies positioned between the treated ply and commodity. Highland et al. (1966), found that four plies of kraft paper prevented excessive residues of piperonyl butoxide in cornmeal, flour, rice, and beans stored for long periods in synergized pyrethrins-treated bags. Also, saran-coated paper and greaseproof paper were effective barriers to migration of this chemical (Highland et al., 1968a,

1970). The copolymer ethylene vinyl acetate in a coating containing synergized pyrethrins reduced the movement of piperonyl butoxide into the packaged food (Highland et al., 1968b).

In the United States and England much of the early work on insect-resistant packaging was done with the so-called safe insecticides and synergists - methoxychlor, pyrethrins, allethrin, piperonyl butoxide, sulfoxide, and similar materials. More recently, other compounds have been used successfully in laboratory tests, including synthetic pyrethroids (Highland and Merritt, 1973), an experimental antifeeding compound (Loschiavo, 1970), an inert silica gel (Watters, 1966), and carbaryl (Highland, 1967). However, in the United States only pyrethrins synergized with piperonyl butoxide may be used as an insect-resistant treatment. Currently this treatment is registered for use only on multiwall paper bags or on cotton bags (Anon., 1975), or in the adhesive of a cellophane/polyethylene laminate for dried fruit packages (Anon., 1974). Other insecticides such as malathion, DDT, dieldrin, trichlorfon, dichlorvos, lindane, and aldrin have been evaluated as treatments for food and feed packages (Joshi and Kaul, 1965; Lal et al., 1961; Langbridge, 1970). However, many attempts had only limited success because of the failure to utilize insect-tight constructions. Also, it was shown earlier by Butterfield et al. (1949) that chemicals migrated from the treated package into the contents and produced unacceptable residues.

Highland et al. (1977) found that pouches made of laminated glassine-foil-polymer and treated with synergized pyrethrins in the laminating adhesive protected cocoa powder and dried soup mix from infestation by mixed populations of common stored-product insects. The glassine prevented the occurrence of detectable residues in the cocoa powder or dried soup mix. Heselev (1978) described an insect-resistant film in which pyrethrins and piperonyl butoxide were incorporated into the polyethylene polymer used to manufacture the film.

Foil-wrapped meat cubes have been protected from infestation by _Ptinus ocellus_ Brown (Australian spider beetle), _Dermestes lardarius_ L. (larder beetle), drugstore beetles, and confused flour beetles and by packaging them in synergized pyrethrins-treated cartons (Dennis, 1962). Likewise, Mallis et al. (1961) recommended a mixture of 1,5a,6,9,9a,9b-hexahydro-4a(4H)-dibenzofuran-carboxaldehyde (MGK Repellent 11[R]) and N-(2-ethylhexyl)-5-norbornene-2,3-dicarboximide (MGK 264[R]) sprayed on corrugated shipping cases to prevent _Blattella germanica_ (L.), German cockroaches, from infesting beer cases. These shipping cases are not sealed tightly enough to prevent the entrance of cockroaches, so short-term prevention of infestation is entirely dependent upon the repellent activity of the chemicals. These treatments are not approved for use in the United States. As mentioned, the only approved insect-resistant treatments are pyrethrins combined with piperonyl butoxide on cotton or multiwall paper bags containing at least 50 pounds of product or between film laminations for dried fruit packages (Anon., 1974). All insect-resistant treatments on food packages must be approved in accordance with the Federal Insecticide, Fungicide, and Rodenticide Act and the Federal Food, Drug and Cosmetic Act as implemented by the Environmental Protection Agency.

SUMMARY

Agriculture-related industries process and package infestible foods that are shipped and stored under a wide variety of physical and biological environments, both foreign and domestic. Protecting these foods from insect attack is necessarily a concern of the packager since he is almost always held responsible for the quality of the food. This is a truism even though the

317

packager may lose control of his products after they leave the packing site. Insect-resistant packaging can help maintain the high quality of such foods.

Selection of packaging materials and packages (including seals, closures, configurations) should be based on a knowledge of the behavior of stored product insects as well as the distribution system through which the packages move. If the food packages have a very short shelf-life (a few weeks) and would potentially encounter only invading insects, a tightly sealed package may suffice to protect the food from invading insects up to the point of purchase. However, where longer storage and transit periods are anticipated or where penetrating insects may be encountered, it would be necessary to also protect from the penetrating insects, which can bore through most common packaging materials. If insect-repellent chemical treatments are used they must not produce unsafe, illegal residues due to migration of the chemical into the food from treated package surfaces. Such contamination can be avoided by the use of proven, appropriate barriers.

In the current environment of consumerism the packager/producer is more and more held responsible for the quality of packaged foods. Packages and packaging materials vary in resistance to insect attack. Therefore, candidate materials and packages should be evaluated for insect resistance as part of the package development phase when new products are being developed, or when existing packaging changes are anticipated.

LITERATURE CITED

Anonymous. 1974. Piperonyl butoxide and pyrethrins. Federal Register. 39:38224-5.

Anonymous. 1975. Tolerances for pesticides in food and animal feed administered by the Environmental Protection Agency. Federal Register. 61:14156-64.

Batth, S. S. 1970. Insect penetration of aluminum foil packages. J. Econ. Entomol. 63(2):653-5.

Brickey, P. M., J. S. Gecan, and A. Rothschild. 1973. Method for determining direction of insect boring through food packaging materials. J. Assoc. Anal. Chem. 56(3):640-2.

Butterfield, D. E., E. A. Parkin, and M. M. Gale. 1949. The transfer of DDT to foodstuffs from impregnated sacking. J. Soc. Chem. Ind. 68:310-13.

Cline, L. D. 1978. Penetration of seven common flexible packaging materials by larvae and adults of eleven species of stored-product insects. J. Econ. Entomol. 71(5):726-9.

Cline, L. D., and H. A. Highland. 1976. Clinging and climbing ability of adults of several stored-product beetles on flexible packaging materials. J. Econ. Entomol. 69(6):709-10.

Cline, L. D., and H. A. Highland. 1981. Minimum size of holes allowing passage of adults of stored-product coleoptera. J. Ga. Entomol. Soc. 16(4):525-31.

Collins, H. E. 1963. How food packaging affects insect invasion. Pest Control. 31(10):26-9.

Dennis, P. O. 1962. Insect infestation. The protection of foodstuffs by synergized Pybuthrin. Food Proc. Packag. 31(267):131-3.

Gerhardt, P. D., and D. L. Lindgren. 1954. Penetration of various packaging films by common stored-product insects. J. Econ. Entomol. 49(2):282-7.

Gerhardt, P. D., and D. L. Lindgren. 1955. Penetration of additional packaging films by common stored-product insects. J. Econ. Entomol. 48(1):108-9.

Heselev, M. 1978. Packaging material resistant to insect infestation. British Commonwealth Patent 1,568,936. Patent Office, London.

Highland, H. A. 1967. Resistance to insect penetration of carbaryl-coated kraft bags. J. Econ. Entomol. 60(5):451-2.

Highland, H. A. 1974. The use of chemicals in processing and packaging of stored-products to prevent infestation. Proc. First Int. Cong. on Stored-Product Entomol., Savannah, Ga. Oct. 7-11, 1974. pp. 254-60.

Highland, H. A. 1975. Insect resistance of composite cans. ARS-S-74. ARS, U.S. Department of Agriculture.

Highland, H. A. 1976. Materials, constructions and treatments for protecting packages from deterioration by insects. In. Proc. 3rd Internat. Biodegradation Symposium. p. 273-8.

Highland, H. A. 1977. Chemical treatments and construction features used for insect resistance. Package Development and Systems. pp. 36-8, May/June, 1977.

Highland, H. A. 1978a. Insects infesting foreign warehouses containing packaged food. J. Ga. Entomol. Soc. 13(3):251-6.

Highland, H. A. 1978b. Insect resistance of food packages - a review. J. Food Proc. and Preservation. 2:123-30.

Highland, H. A., R. V. Byrd, and M. Secreast. 1968b. Effect of ethylene vinyl acetate on the migration of piperonyl butoxide from coatings of synergized pyrethrins on kraft paper. ARS 51-28. ARS, U.S. Department of Agriculture.

Highland, H. A., L. D. Cline, and R. A. Simonaitis. 1977. Insect resistant food pouches made from laminates treated with synergized pyrethrins. J. Econ. Entomol. 70(4):483-5.

Highland, H. A., D. F. Davis, and F. O. Marzke. 1964. Insectproofing multiwall bags. Modern Packag. 37(12):133-8,195.

Highland, H. A., R. H. Guy, and H. Laudani. 1968. Polypropylene vs. insect infestation. Modern Packag. 41(2):113-5.

Highland, H. A., and E. G. Jay. 1965. An insect-resistant film. Modern Packag. 38(7):205-6,282.

Highland, H. A., E. G. Jay, M. Phillips, and D. F. Davis. 1966. The migration of piperonyl butoxide from treated multiwall bags into four commodities. J. Econ. Entomol. 59(3):543-5.

Highland, H. A., and P. H. Merritt. 1973. Synthetic pyrethroids as package treatments to prevent insect penetration. J. Econ. Entomol. 66(2):540-1.

Highland, H. A., M. Secreast, and P. H. Merritt. 1968a. Polyvinylidene-coated kraft paper as an insecticide barrier in insect-resistant packages for food. J. Econ. Entomol. 61(5):1459-60.

Highland, H. A., M. Secreast, and P. H. Merritt. 1970. Packaging materials as barriers to piperonyl butoxide migration. J. Econ. Entomol. 63(1):7-10.

Highland, H. A., and R. Wilson. 1981. Resistance of polymer films to penetration by lesser grain borer and description of a device for measuring resistance. J. Econ. Entomol. 74(1):67-70.

Joshi, H. C., and C. L. Kaul. 1965. Studies on the protectivity of jute bags impregnated with organic insecticides against red flour beetles and cigarette beetles. Indian J. Entomol. 27(4):491-3.

Lal, R., P. D. Srivastava, and P. Dhar. 1961. Efficacy of insecticide impregnated jute bags against Sitophilus oryzae Linnaeus (Curculionidae: Coleoptera). Indian J. Entomol. 22(3):204-10.

Langbridge, D. M. 1970. Treatment of paper to protect packaged food from insect attack. Appita. 24(1):45-51.

Laudani, H., H. A. Highland, and E. G. Jay. 1966. Treated bags keep cornmeal insect-free during overseas shipment. American Miller and Processor. 94(2):14-19,33.

Loschiavo, S. R. 1970. 4'(3,3-dimethyl-1-triazeno) acetanilide to protect packaged cereals against stored products insects. Food Technol. 23(4):181-5.

Mallis, A., W. E. Esterin, and A. C. Miller. 1961. Keeping German cockroaches out of beer cases. Pest Control. 29(6):32-5.

Rao, K. M., S. A. Jacob, and M. S. Mohan. 1972. Resistance of flexible packaging materials to some important pests of stored products. Indian J. Entomol. 34(2):94-101.

Watters, F. L. 1966. Protection of packaged food from insect infestation by the use of silica gel. J. Econ. Entomol. 59(1):146-9.

Wohlgemuth, R. 1979. Protection of stored foodstuff against insect infestation by packaging. Chemistry and Industry. 5:330-4.

Yerington, A. P. 1971. Insect resistance research. Good Packag. 32(9):12-13.

Yerington, A. P. 1975. Insect resistance of polypropylene pouches. Modern Packag. 48(5):41-2.

Yerington, A. P. 1978. Insects and package seal quality. Modern Packag. 51(6):41-2.

Yerington, A. P. 1979. Methods to increase the insect resistance of food shipping cases. USDA Advances in Agricultural Technology. AAT-W-7.

PHYSICAL AND CHEMICAL METHODS
FOR DETECTING INSECT FILTH IN FOODS

Russell G. Dent and Paris M. Brickey

Editor's Comments

This chapter describes the types of analytical methods
used to detect insects and insect filth in foods. The
methods can be and have been applied to other ingested
commodities such as drugs.

The Microanalytical Branch of the Division of
Microbiology, Office of Nutrition and Food Sciences, Center
for Food Safety and Applied Nutrition of the FDA (which the
authors represent) does most of the methods development in
this area.

The information on the theory of filth separation is
not new, but what is new is its presentation in a pragmatic,
easily understood, and useful way.

Note that little is said about the analysis for
possible toxic substances associated with insects. The
described chemical approach may have utility for this
possible need.

F. J. Baur

PHYSICAL AND CHEMICAL METHODS FOR DETECTING
INSECT FILTH IN FOODS

Russell G. Dent and Paris M. Brickey

Food and Drug Administration
200 C Street, S.W.
Washington, DC 20204

INTRODUCTION

The methods covered in this chapter are divided into three groups: separation of filth elements by visual examination, physical extraction of filth elements, and chemical detection of insect fecal metabolites. Usually with these conventional approaches the presence of rodent and avian filth can also be simultaneously determined. These methods are useful both as a quality assurance tool and as an index of sanitation effectiveness. In the section on the theory of light filth extraction, no one method is discussed because all the approaches are similar and based on the same physical concepts. This chapter outlines the types of analysis used in filth detection.

With the advent of the Food and Drug Act of 1906, methods of analysis began to be developed for filth in foods. Scientists in the food industry, the Food and Drug Administration, and the U.S. Department of Agriculture have jointly developed methodology for the detection of insect contamination of food and food ingredients. These methods allow industry to monitor the cleanliness of its products and regulatory governmental agencies to determine if effective sanitation is maintained and is in compliance with sections relating to filth and sanitation in the law. These methods are periodically reviewed and improved, revised, or dropped from the Official Methods of Analysis (7) published by the Association of Official Analytical Chemists (AOAC).

The AOAC is an independent international organization charged with obtaining, improving, developing, testing, and adopting uniform, precise, and accurate methods for the analysis of foods. Once such methods are collaboratively studied and adopted by the Association, they are recognized by the courts as a valid means for obtaining data for regulatory actions.

TYPES OF ANALYSES USED IN FILTH DETECTION

Macroscopic Methods

Macroscopic methods are primarily analyses for filth using the naked eye. These methods are often supplemented with a hand lens of 3X, 10X, or 60X for final confirmation. Some techniques used in this type of analysis are dry and/or wet sieving and/or sorting by hand of the filth elements from the product. By separating particles of different sizes, insect filth can frequently be segregated and verified easily. Some sorting techniques require a sequential sampling plan.

Example: Sampling plan for coffee beans to determine the numbers of defective beans per sample (insect infested or damaged, and moldy)

No. of Beans Examined in Sample	Stop Analysis Good	Continue Analysis Marginal	Stop Analysis Bad
500	37 or less	38 or more	
1000	85 or less	86 - 115	116 or more
1500	135 or less	136 - 165	166 or more
2000	185 or less	186 - 215	216 or more
2500	234 or less	235 - 264	265 or more

For example, if examining the first 500 beans, one finds 37 or fewer beans defective, the sample is considered acceptable and no further examination is required.

Another type of macroscopic examination is the use of x-rays. This type of examination is used for coffee beans and other seed products. X-raying detects insect damage in the bean and records the defects on film. Each seed is observed and treated in the same way as in the sequential sampling plan above. Some examinations require the analyst to cut open the food product, as in figs, and report the condition of each unit (e.g. moldy or insect-infested); results are reported as per category, number of units, weight, or size of an area. Analysis for heavy filth (sand, glass, or metal fragments) is considered a type of macroscopic examination. Such analysis depends upon differences in the specific gravity of the food product and the extraneous filth in solvents such as chloroform, carbon tetrachloride or, more recently, a safer substitute, trichlorotrifluoroethane. Macroscopic examinations may be performed on site; however, most are conducted in the laboratory.

Microscopic Examination

A microscopic examination is performed with a stereoscopic or compound microscope to observe elements which are so small as to be invisible or indistinct to the naked eye. This kind of examination is used to observe light filth (defined as insect fragments, rodent hairs, feather barbules, and other microscopic particles including organisms along with other extraneous materials [e.g., textile fibers] found in food). To promote good observation, light filth extraction methods have been developed which will concentrate the filth in an oil/aqueous trapping system at an interface. Then the interface is trapped off and filtered through ruled filter paper which allows the analyst to observe the elements conveniently. Most of the trapping systems are dependent on a standard set of apparatus combined with various reagents to arrive at the final result, the filter paper containing the filth elements. Once the filth elements are concentrated, they can then be identified by comparing the sample to authentic collections or standard text figures. Due to the nature and quantity of the equipment required for microscopic examinations these analyses are generally performed in the laboratory.

Theory of Light Filth Extraction Methodology

Light filth methodology is based on three very important principles:

1. Insect and rodent filth is generally lipophilic.
2. Insect and rodent filth is insoluble in most of the reagents used for filth extraction.
3. Plant and animal products can be treated with reagents to either make them hydrophilic or soluble.

Using the above principles, trapping systems have been developed to separate filth elements from a food product.

Before any method can be designed, the food product has to be examined for its characteristics. In the method design it is desirable to eliminate fine product particles and solubles which may interfere with the extraction. This can be done by acid digestion, wet sieving, or a detergent or solvent defatting pretreatment. Performing the above procedures alters the food product to make it more amenable to light filth extraction and at the same time removes possible interfering compounds or particles from the system. Some products do not require this pretreatment and can go directly into the trapping set up.

The kind of filth sought also has a bearing on the type of trapping system used. Presently, all trapping systems are designed to include an oil phase and an aqueous phase or a specific gravity separation, such as in AOAC Official Methods of Analysis (7) 44.029 Curculio Larvae in Pecan Pieces.

The object of an oil/aqueous phase system is to enhance the lifting power (contact angle) (18) of the oil globules and their ability to be dispersed throughout the immiscible aqueous phase. The oil globules then extract and concentrate the light filth. At the same time, the food product, which is not lipophilic, falls out of the system and the light filth, which is lipophilic, rises to the interface with the oil globules. The size and shape of the oil globules can be controlled by varying the composition of the oil and the aqueous phases. Once the type of filth to be retrieved and the product characteristics have been determined, the reagents can be selected to accomplish an optimal separation. A simple mineral oil/water trapping system usually separates rapidly. The oil globules are large and nearly spherical. The addition of ethanol or isopropanol makes the oil globule smaller and often prolongs the separation process. As the percentage of the alcohol increases, the oil globule shape begins to flatten and become disk-like when it adheres to the filth element. The size of these disk-like globules determines the type of filth that will rise in the greatest quantity, depending on the size, surface area, and weight of the fragments. The addition of Tween 80 - Versene causes further modifications of the oil globule and usually tends to make extraction times longer and the recoveries of smaller fragments greater.

If insects and other large filth elements can be observed macroscopically, then the system above would not be recommended. A sieving or pick out procedure would be more appropriate.

Similar changes can be made by altering the flotation oil characteristics. By thinning mineral oil with heptane, a desired globule shape usually may be obtained in any aqueous phase. Heptane may also be used by itself but it does not have the lifting power that heavier oils have.

Another type of filth separation is by specific gravity (as used for curculio larvae in pecans). Undiluted isopropanol is used initially, followed by dilution with water and agitation. This causes the larvae to float and the nut meats to sink. Although not used routinely in filth examinations, sugar and salt solutions of various concentrations can be used for filth separation where specific gravity differences can be utilized.

Extraction systems require specific types of equipment (glassware) to enhance their separation effects. By far, most filth extractions are accomplished in a Wildman trap flask. This consists of an Erlenmeyer flask containing a rod with a wafer or stopper attached to one end. This is a closed trapping system and it allows only the removal of the extracting oil from the top. The Corning percolator is funnel-like in shape and operates much the same as a separator does except that the top is open and the tubelation shutoff is accomplished using a rubber hose and clamp. This is an open system where the aqueous phase can be removed at will and replaced from the top by pouring the new phase through the trapped oil phase. This trapping technique is more dynamic but care must be taken in selecting its use with various reagent combinations because this may change the size of the oil globules and their ability to hold onto the particulate filth when the aqueous phase is removed. Simple decantation is also a technique where the filth can be removed either from the bottom or top of the solvent mixture.

There are two other techniques that are frequently used in conjunction with the above equipment. Deaeration is accomplished by boiling or drawing a vacuum on the aqueous phase. Cooling of the aqueous phase with the product sometimes improves separation of the two phases and a cleaner extraction. Both of these techniques are aimed at producing clean (minimal plant debris present) filter papers.

Chemical Examination

Because of the availability of sophisticated techniques and equipment, soluble filth can now be detected in food products much more efficiently than in the past. This section will trace the various examinations as they evolved for detecting the presence of insect contamination via the chemical identification of their metabolic products.

Chemical Indicators of Filth Contamination

Insects contain several identifiable metabolic compounds which can be detected by chemical techniques. Uric acid, a metabolite excreted in high concentrations in the fecal pellets of most insects, can be chemically determined and, therefore, serve as an indicator of insect contamination. Other excretory products, such as quinones, are also indicators of insect filth.

An Overview of Methods for Uric Acid Detection as an Index of Insect Infestation

An improved enzymatic-ultraviolet light extraction method for uric acid detection in flour was developed by Sen and Smith (13). Sen (14) collaboratively studied this method for the AOAC. Results of the study indicated very high yields averaging 103% recovery of uric acid. Later, Sen and Vazquez (15) showed a good correlation between the levels of uric acid in a food product and the numbers of insect fragments. Sengupta et al. (16) developed a method utilizing thin-layer chromatography (TLC) for detection of uric acid in spices and condiments. Mlodecki et al. (9) reported a method for determining the sanitary quality of dried mushrooms. This method separates and quantitates uric acid by extraction with an ammonium carbonate solution and purification with Sephadex G-10. A spectrophotometric determination is then made with Benedict's reagent. Laessig et al. (8) developed a method to separate uric acid from flour, other cereal products, and powdered milk by dialysis. This method can be automated to simultaneously analyze 30-40 samples and detects uric acid in the range of 0.04-0.32 mg/g sample. Pachla et al. (11) developed a high pressure liquid chromatographic (HPLC) technique which was combined with thin layer amperometric detection (TLAD) of uric acid. This method extracts uric acid from the sample with 3% sodium acetate and then concentrates the extract by centrifugation for analysis by HPLC and TLAD. Thrasher et al. (17) developed a thin layer chromatography method for uric acid in bird and insect excreta in foods. The method solubilizes the suspect excreta with lithium carbonate. The uric acid from the excreta is then separated and identified by TLC with a solvent system of butanol/methanol/water/acetic acid and a trisodium phosphate-phosphotungstic acid color developer. Detection results were 100% accurate at a 50 ng level and 75% accurate at a 20-25 ng level. Holmes (6) developed a fluorometric method for the detection of uric acid in flour. A 1/2 g sample is extracted with an acetate buffer at pH 11.8-12.0 with a 30 minute wait at 40°C. After centrifuging and filtering the supernatant, a fluorometer is used to determine the uric acid present. Roy and Kvenberg (12) developed a semi-automated colorimetric method for uric acid in multiple samples of several food products. The sample is pretreated with HCl, incubated for a time at 55-60°C, and then neutralized with NaOH. Uric acid is extracted with a sodium acetate solution and then treated with phosphotungstic acid for colorimetric determination of uric acid levels. The sample then proceeds through an immobilized uricase nylon coil where the uric acid is destroyed. A second determination of the uric acid is then made. The method sensitivity is ca. 5 micrograms uric acid/g sample. Galacci (3) improved on Roy and Kvenberg's instrumentation and recommended that his improved method be collaboratively studied. The method showed good correlation between uric acid, the levels of insect excreta pellets, and the levels of insect fragments in flour. The data further indicated that a uric acid value of 600 micrograms/50 g flour approximates the same degree of insect defilement as the FDA Defect Action Level of 50 insect fragments per 50 g sample. This suggested that flour samples exceeding 600 micrograms would contain violative amounts of insect filth. Sixteen samples can be done in 8 hours as opposed to eight per 8 hours with the official light filth method.

Brown et al. (1) developed a method of detecting uric acid levels in whole spices. This procedure utilizes a glucose analyzer. The enzyme uricase decomposes the uric acid and the reaction is monitored with an oxygen-sensitive electrode. Data obtained by this procedure showed good correlation with uric acid levels determined by AOAC methods.

DISCUSSION AND SUMMARY

In general, a method has its particular use in any given set of circumstances. However, there are times when it is more expeditious to choose a type of method for the time involved or its visual impact as evidence in court. Obviously, macroscopic and light filth examinations are more labor intensive, but the judicial impact is far greater because the results can be seen. Chemical examinations are very specific in the detection and quantitation of an undesirable filth element. The information gained by these examinations must be interpreted by peaks in a graph or changes in color through instrumentation. Both physical and chemical methods for detecting filth and uric acid have their place in quality control, sanitation monitoring, and regulatory enforcement.

This chapter describes the types of analyses used to detect insect filth: 1) by visual examination (macroscopic); 2) by aqueous/oil phase examination (microscopic); and 3) by chemical detection. The theory of light filth extraction (microscopic examination) is discussed.

LITERATURE CITED

1. Brown, S. M., S. Abbot, and P. A. Guarino. 1982. Screening procedure for uric acid as indicator of infestation in spices. J. Assoc. Off. Anal. Chem. 65(2):270-272.
2. Christensen, E. (ed.). 1983. Approved Methods of the American Association of Cereal Chemists, 8th ed. The Association, St. Paul, MN 55121.
3. Galacci, R. R. 1983. Automated analysis of flour extracts for uric acid and its correlation with degree of insect defilement. J. Assoc. Off. Anal. Chem. 66(3):625-631.
4. Gorham, J. R. (ed.). 1977. Training Manual for Analytical Entomology in the Food Industry. The Association of Official Analytical Chemists, Arlington, VA 22209.
5. Gorham, J. R. (ed.). 1981. Principles of Food Analysis for Filth, Decomposition and Foreign Matter. Food and Drug Administration, Washington, DC 20204.
6. Holmes, J. G. 1980. Note on fluorometric method for determination of uric acid in flour. Cereal Chem. 57(5):371-372.
7. Horwitz, W. E. (ed.). 1980. Official Methods of Analysis, 13th ed. The Association of Official Analytical Chemists, Arlington, VA 22209.
8. Laessig, R. H., W. E. Burkholder, and R. J. Badran. 1972. Routine and low-level determination of uric acid in dry milk, flours, and cereal grains. Cereal Sci. Today 17(10):328-330.

9. Mlodecki, H., W. Lasota, and T. Pustelnik. 1972. Uric acid content as an index of sanitary quality in dried mushrooms. Bromotologia I Chemia Toksykologiczna 5(4):487-489.

10. Muzzarelli, R. A. A. 1977. Chitin. Pergamon Press, Oxford, England, pp. 5-8.

11. Pachla, L. A., and P. T. Kissinger. 1977. Monitoring insect infestation in cereal products - determination of traces of uric acid by high pressure liquid chromatography. Anal. Chem. Acta 88(2):385-387.

12. Roy, R. B., and J. E. Kvenberg. 1981. Determination of insect infestation in food: a semiautomated colorimetric analysis for uric acid with immobilized uricase. J. Food Sci. 46(5):1439-1445.

13. Sen, N. P., and D. Smith. 1966. An improved enzymatic-ultraviolet method for determination of uric acid in flours. J. Assoc. Off. Anal. Chem. 49(5):899-902.

14. Sen, N. P. 1967. Collaborative study of the determination of uric acid in flour. J. Assoc. Off. Anal. Chem. 50(4):776-781.

15. Sen, N. P., and A. W. Vazquez. 1969. Correlation of uric acid content with fragment counts in insect-infested flours and wheat grains. J. Assoc. Off. Anal. Chem. 52(4):833-836.

16. Sengupta, P., A. Mandal, and B. R. Roy. 1972. Determination of uric acid in foodstuffs by thin-layer chromatography. J. Chromatogr. 72(2):408-409.

17. Thrasher, J. J., and A. Abadie. 1978. Thin layer chromatographic method for the detection of uric acid: collaborative study. J. Assoc. Off. Anal. Chem. 61(4):903-905.

18. Yakowitz, M. G., and W. V. Eisenberg. 1964. Contact angles as a measure of surface wetting properties in filth flotation analysis. J. Assoc. Off. Anal. Chem. 47(3):520-529.

DEFECT ACTION LEVELS IN FOODS

Ruth Bandler, Paris M. Brickey, and William V. Eisenberg

Editor's Comments

For those readers who are in the food industry, it is important that you be aware and stay aware of the concept of DALs and what the defined DAL values are for any of your products and raw materials. Remember, being within a DAL for a finished product is no defense if your facility is not in compliance with 402(a)(4) or the "may have become contaminated" section of the law (the GMPs section).

F. J. Baur

DEFECT ACTION LEVELS IN FOODS

Ruth Bandler, Paris M. Brickey, and
William V. Eisenberg[1]

U.S. Food and Drug Administration
Division of Microbiology
200 C Street, SW
Washington, DC 20204

INTRODUCTION

Defect action levels (DALs) in foods, as established by the Food and Drug Administration (FDA), are the levels of naturally occurring defects which are unavoidable and which do not constitute a health hazard to the consumer. These DALs are based on section 402(a)(3) of the Food, Drug, and Cosmetic Act, which defines a food as adulterated if it consists in whole or in part of any filthy, putrid, or decomposed substance or if it is otherwise unfit for food.

Defect action levels are set above zero because it is not possible, and never has been possible, to grow crops in open fields and to harvest and process them so that they are totally free of natural defects such as insects and mold, especially with the current goal to decrease the use of pesticides and related chemicals.

The action levels do not represent an average of the defects that occur in any of the food categories. The averages are actually much lower. The DALs represent the maximum levels for natural or unavoidable defects in foods produced under good manufacturing practices and are the limit at or above which FDA will take legal action against the product to remove it from the market. The fact that FDA has an established defect level does not mean that a manufacturer need only stay below that level. Defects resulting from poor manufacturing practices do not fall within the coverage of these limits and FDA emphasizes that compliance with defect levels will not prevent FDA from acting against a manufacturer who does not observe current good manufacturing practices. Insanitary plant conditions, for example, violate good manufacturing practices and render the food unlawful. Section 402(a)(4) of the Food, Drug, and Cosmetic Act defines an adulterated product in terms of insanitary conditions in the preparation, packing, or holding of food whereby it may have become contaminated with filth or whereby it may have been rendered injurious to health. Therefore, just being below the DALs is not sufficient for compliance. For example, regulatory action could be based on the finding of a few insect fragments, well below the DALs, in a product if suitable evidence of insect infestation is found in the stored raw material and/or storage area.

The U.S. Food and Drug Administration has had action levels for filth and decomposition in foods for many years. One of the first was established in 1911 for mold in tomato pulp. Limits were first set for insects in various fruits and vegetables in the 1920s. In 1938, the Food, Drug, and Cosmetic Act was passed and with it came the development of new and more sensitive analytical methods; this led to the establishment of limits for insect fragments and rodent

[1]Present address: 6408 Tone Drive, Bethesda, Maryland 20817

hair fragments in many foods. These action levels were known at various times as "tolerances," "Confidential Administrative Tolerances," "Field Legal Action Guides," and "Administrative Guidelines." They were held confidential to all except FDA regulatory officials until 1972 when they were first published as "Current Levels for Natural or Unavoidable Defects in Food for Human Use That Present No Health Hazard." They are presently provided for by part 21 of the Code of Federal Regulations, 21 CFR 110.99, Good Manufacturing Practice Regulations for Human Foods.

Defect action levels are established for products whenever it is necessary and feasible. The present list includes about 100 different products and approximately 200 specific DALs for these products. The list is being continually updated. New products are added to the list as time permits and as varieties of processed food change. Existing DALs are under study and review as current technology changes. The introduction of new agricultural strains, some of which are more resistant to biological or mechanical damage, and better processing techniques have caused a decrease in the levels of filth and decomposition in some food products. Combined with the development of newer and more sensitive microanalytical methods and the availability of new data bases for various products, these advances have resulted in changes or updates of many of the existing DALs. Most of the changes involve a lowering of the maximum permissible levels and thus require stricter controls for certain defects.

The establishment of DALs relies on data generated by a variety of sources such as Federal government agencies and industry. Usually the data are generated by FDA-sponsored in-house projects and outside contracts for the collection and analysis of various food products.

The Microanalytical Branch of FDA's Division of Microbiology is responsible for developing the new DALs and updating existing guidelines.

DEVELOPMENT OF A DAL

DAL development is a multi-stage process. The first stage is to identify a need for a DAL for a particular product. There are several ways in which this can occur.

1) Consumers bring certain problem areas to the attention of the agency. During investigation a problem may be discovered within a certain industry; therefore establishment of a DAL is initiated.
2) Industry often requests that DALs be established for a particular product or industry so that the various manufacturers have some standards to follow. This occurred with botanicals (herbs, e.g., peppermint). Here, the process is still at stage one. Because methods of analysis are presently being developed for the various botanicals, sample collection and analysis have not yet begun.
3) Suggestions for DAL development can also come from FDA regulators themselves. For example, many spices were being routinely sampled by our investigators but there were no DALs with which to compare them once they were analyzed. Therefore, we completed a collection and analysis program for 10 spices and are now in the midst of the data analysis.

Once a particular industry is identified for examination, the next step is to determine the scope of the industry or the extent of the problem. Are enough samples coming in that a complete DAL development program is necessary, or is the industry so small that each sample can be handled on a case-by-case basis?

If a decision is made to perform a DAL study, the types and forms of filth contamination must first be identified. Is the product subject to mold growth or insect or rodent contamination? Is the product marketed whole or unground (i.e., would one expect to find whole insects, rodent hairs, and fecal pellets) or is the product fairly well ground so that an analyst would have to search for filth fragments by microscopic analysis?

Once the filth variables of interest have been determined, a method of analysis for the product must be developed (if no suitable method exists). Often this may involve only adapting an existing method to the product at hand. Sometimes, as with the botanicals, completely new methods are needed. A new segment is therefore added to the total DAL development program.

Methods research is a long process, subject to extensive evaluation. After the initial research is completed, the method is exposed to in-house testing for ruggedness to ensure that the method is not overly sensitive to slight variations in performance. The method then receives interlaboratory testing (a collaborative study). A minimum of six laboratories from FDA district offices, state labs, and industry each performs the method to ensure that it can be used by other workers with other equipment and yield similar results. Usually, most methods are then granted official status by the Association of Official Analytical Chemists (AOAC) and are published in the AOAC book, Official Methods of Analysis (1).

SAMPLING PLAN

The next stage of DAL development is to set up the survey. First, the sampling plan must be developed.

1) The nature of the sampling sites is determined. Most sampling is done at the retail market; however, if the product is exclusively an import, sampling can be done directly at the port of entry.

2) Next, our statisticians determine the number of samples to collect. Under FDA policy, 1500 samples are collected to establish a new DAL. If the purpose of the survey is only to update an existing DAL, 500 samples are collected.

3) The number of sampling sites (i.e., for a retail survey, the number of stores) is based on the number of required samples. The Bureau of the Census has developed a list of "Standard Metropolitan Statistical Areas" (SMSAs), or "integrated economic and social units with a recognized urban population nucleus of substantial size" (2). Using a random number table, an appropriate number of geographic regions (or SMSAs) are selected from this list. Within each selected SMSA, five retail stores are visited, with the restriction that three are to be chain stores and two are to be independent stores. The collectors choose the individual stores. A specified number of retail units of each product are to be collected in each store; different brands are collected if possible, or else different production codes. In a recent survey of tomato products, conducted to update existing DALs, 500 samples of each tomato product were needed (e.g., paste, juice). The sampling plan called for visiting 23 randomly chosen SMSAs, yielding 23 SMSAs x 5 stores/SMSA x 5 units/store or 575 samples. The 15% surplus over the required minimum of 500 is a routine practice which allows for damage in shipment, lost samples, sample mishandling, and reserve samples. This type of sampling plan ensures adequate lot representation nationally.

4) Next, a suitable time frame is chosen for the study. Certain products are susceptible to seasonal variations which might affect insect populations and mold growth. Therefore, some of the surveys are conducted over a 2 or 3 year period (e.g., retail spices, wheat flour) to average out the climatic variations. In addition, sampling schedules are devised so that the survey is spread out over the total allotted time. Each SMSA is scheduled for sampling during a specific week so that each geographic region is visited during each season.

5) Finally, the personnel are chosen. The survey has two separate phases, sample collection and analysis, and these can be performed by different groups. Collections are made either by FDA investigators or by contractors. The analyses are performed by FDA district analysts,

in-house (FDA Division of Microbiology) or on contract, depending on resource availability. Outside assistance is used when available. For a recent wheat survey, U.S. Department of Agriculture (USDA) grain inspectors collected the samples and FDA analysts performed the analyses. In a wheat flour survey, collection was made by USDA agents as part of their routine quality check program of the flour mill production lines. USDA also performed the analyses.

After the sampling plan is developed, the sampling is carried out. Collection of samples is monitored by the Branch to ensure that the specified sites are being visited on schedule.

The samples are delivered to the laboratories and are analyzed. If the analyses are performed by a contractor, the performance is periodically audited to ensure that the analytical work is executed properly.

All data are sent to the Branch. Worksheets are reviewed for inconsistencies or problems which are corrected before data analysis begins.

DATA ANALYSIS

The statistical data analysis is done by the Divisions of Microbiology and Mathematics and is, of course, computer-assisted. First a statistical profile of the data is generated. This information includes frequency distributions of the filth variables and allows comparisons with levels in similar products (e.g., between spices) to show relative conditions of various products. Correlations between filth variables (e.g., between insect fragments and rodent hairs) are studied. Year-to-year variations are also examined to assess the effect of variations in climate. Finally, variables are chosen for which DALs will be developed. Data are collected for numerous types of filth, such as whole/equivalent insects of specified biological orders, bird contamination, and various types of mammalian contamination; however, DALs are usually set only for the more prevalent defects (e.g., insect fragments, rodent hairs, mold).

After the filth variables are chosen, the various defect limits must be established. Many factors are considered when developing these DAL limits. The frequency distribution of the individual filth variable is the starting point. Within-lot variability and analytical variability of the filth method are taken into account. The impact of a defect level on a particular food industry with regard to loss of food materials due to contamination is an important factor that is gauged before the limits are finally determined.

Once suitable levels for each filth variable are selected, the Branch makes its recommendation subject to extensive scientific, statistical, and legal review.

After approval, the new DAL is ready for release. A notice of availability to the industry is published in the Federal Register. The guidelines are published in an agency manual system, "Compliance Policy Guides." A list of action levels, "Current Levels for Natural or Unavoidable Defects in Food for Human Use That Present No Health Hazard," is available from

FDA Industry Programs Branch (HFF-326)
200 C Street, SW
Washington, DC 20204

The list is updated as needed; the most recent revision was in 1982. Finally, a summary of the statistical findings is submitted for publication in a food-related journal such as the Journal of Food Protection.

The status of the DAL program is constantly changing. Many establishments and updates have been completed (Table 1) and many others are in various stages of DAL development (Table 2).

Table 1. Recent DAL Establishments and Updates

Products	Defects
pineapple juice	mold
canned pineapple	mold
cranberry sauce	mold
unsweetened chocolate	insect fragments, rodent hairs
walnuts, pecans, brazil nuts	reject nuts
tomato paste, puree, juice, sauce, soup	mold, rot, fly eggs, maggots
macaroni, noodles	insect fragments, rodent hairs
canned peaches	mold, larvae
fruit nectars-apricot, peach, pear	mold
wheat	insect-damaged kernels, rodent excreta
wheat flour	insect fragments, rodent hairs

Table 2. DAL Projects in Progress

unpopped popcorn	infant cereal
crabmeat, tuna, sardines	green coffee beans
raisins	canned corn
canned greens	catsup, canned tomatoes
dried beans	dried fruits--apricots, peaches, pears, nectarines
figs, fig paste	
cocoa powder	tropical fruit nectars and pastes-- guava, papaya, mango
pickle relish	
pimientos	whole spices (11)-import
frozen berries	processed spices (10)-retail

SUMMARY

Defect action levels (DALs) are the maximum levels for natural or unavoidable defects in foods produced under good manufacturing practices. DAL development is a multi-stage process. Specific food products and filth defects are chosen, methods of analysis are developed, sampling plans are devised for representative coverage of an industry, samples are collected and analyzed, and suitable filth levels are selected. DALs have been developed for about 100 different products and new products are added whenever it is necessary and feasible.

LITERATURE CITED

1. Association of Official Analytical Chemists. 1984. Official Methods of Analysis, 14th ed. Association of Official Analytical Chemists, Arlington, VA.
2. U.S. Office of Management and Budget. 1975. Standard Metropolitan Statistical Areas, revised ed. U.S. Government Printing Office, Washington, DC.

EPA RESPONSIBILITIES AND CONCERNS

Fred J. Baur and Phil Hutton

Editor's Comments

It was the intent to have this chapter written by EPA. This proved not to be feasible. Mr. Hutton was of considerable assistance in providing background information, suggestions for the text, and most importantly, clearing the manuscript through EPA management. It therefore reflects official EPA attitudes and concerns.

F. J. Baur

EPA RESPONSIBILITIES AND CONCERNS

Fred J. Baur and Phil Hutton

1545 Larry Avenue
Cincinnati, Ohio 45224
and
Technical Support Section
Insecticide/Rodenticide Branch (TS-767C)
Office of Pesticide Programs
United States Environmental Protection Agency
Washington, D.C. 20460

INTRODUCTION

The Environmental Protection Agency (EPA) regulates pesticides and their use under the authority of the Federal Insecticide, Fungicide, and Rodenticide Act (FIFRA) as amended (December 17, 1980), and the Federal Food, Drug, and Cosmetic Act (FD&CA). EPA enforces FD&CA tolerances for pesticides but not the entire FD&CA which is, of course, primarily under the jurisdiction of the Food and Drug Administration. FIFRA is the legislative origin for most federal regulations concerning the manufacture, sale, and use of insecticides in the United States.

The aim of this book is to inform those industries or companies that use insecticides, not those that manufacture and sell them. Hence this chapter will concentrate primarily on the proper use of insecticides. The key points are: (1) use only insecticides which are registered for the application of interest; (2) use them only in accordance with their labels; and (3) if the insecticide is classified for restricted use it must only be applied by or under the supervision of a certified commercial applicator. It is essential that the label is read prior to use, the information on the label including the limitations and restrictions is fully understood, and that the label directions are followed.

REGISTRATION OF INSECTICIDES

The processing of applications for registration through the Office of Pesticide Programs (OPP) consists of two steps: (1) application and approval of registration; and (2) if the use is on food or in food areas, the establishment of tolerances.

Application and Approval

The Office of Pesticide Programs carries out the registration of proposed pesticide products by means of a product manager system located in the

Registration Division. Determination of the product manager within the Registration Division is a function of the active chemical ingredient(s) in the pesticide product. The following chart lists the current product manager (PM) assignments at EPA for insecticides and rodenticides:

EPA PRODUCT MANAGERS, INSECTICIDE/RODENTICIDE BRANCH

INSECTICIDES/RODENTICIDES	NAME	TELEPHONE (703)-
Branch Chief	Herbert Harrison	557-2200
Technical Support Section Head	Alexandre Tarsey	557-2783
Carbamates, Chlorpyrifos, EPN, Miticides, Some Organophosphates	Jay Ellenberger, PM 12	557-2386
Chlorinated Hydrocarbons, DDVP, Some Organophosphates	George LaRocca, PM 15	557-2400
Rodenticides, Fumigants, Some Organophosphates	William Miller, PM 16	557-2600
Biologicals, Repellents, Pyrethroids	Timothy Gardner, PM 17	557-2690

Aspects of the product studied, as appropriate, include chemistry, toxicity, efficacy, possible persistence of residues in the environment, and effects on non-target species. In these efforts the product manager is assisted by a team of individuals and selected scientific personnel drawn from the Registration Division, the Hazard Evaluation Division, and the Benefits and Use Division. The team conducts a detailed administrative review of the application, the proposed label, and the accompanying or referenced data. When all scientific reviews are completed, the product manager has the decision whether to approve or reject the application based upon the risks and benefits of the proposed use and then so informs EPA management and the prospective registrant. If a new active ingredient or significant new use is involved, a notice of the application and/or the registration will be published in the Federal Register.

No insecticide should be used unless it is registered. In checking the label, your applicators need to ensure the presence of an EPA registration number which indicates the product has been evaluated and approved for those sites of application appearing on the label.

In fiscal year 1982, the Agency processed the following registration actions; 193 new chemicals, 28 new biological chemicals, 11,993 "old" chemicals, and 8,834 amendments to previous registrations.

More details on the registration process can be found in Section 3 of FIFRA, which is available from the Registration Division, OPP/EPA.

Setting of Tolerances

The second part of the registration process is tolerance setting. A tolerance is the residue of the pesticide legally permitted in food or feed and it represents the maximum residue allowable in the commodity. It is important to note that the tolerance level must also not be exceeded in any finished product whether for human or animal use, by any processing of the agricultural product or commodity.

The FD&CA, Sections 408 and 409, is the authority for the establishment of tolerances. The responsibility for this passed from FDA to EPA on December 2, 1970, when EPA was established. While EPA now establishes the tolerances, FDA has retained the responsibility for their enforcement, except for meat, poultry, and eggs, which are enforced by the Meat and Poultry Inspection Service (MPIS) of the USDA. No pesticide for which the proposed use presents a likelihood of residues in or on food or feed can be registered without first having established a tolerance or having been exempted from the requirement of a tolerance because of its nature, such as very low toxicity or an extremely short residual life.

The tolerance setting process requires the registrant to submit a petition for tolerance along with the data necessary to establish safe levels covering the residues resulting from the proposed use. Scientific reviews are conducted by the Residue Chemistry and Toxicology Branches of the Hazard Evaluation Division. The toxicologists calculate a "no-effect level" (NOEL) in the diet which is defined as the highest level of pesticide that elicits no signs or other toxic reactions (such as cholinesterase depression or neurotoxicity) in selected test animals (two species). The no-effect level is then generally multiplied by a suitable number, usually 100, to provide an appropriate safety factor. This safe level is termed acceptable daily intake. This is then compared to the actual residue levels found in or on the raw agricultural commodity or food item as a result of the proposed use, and an appropriate tolerance level for the pesticide is established. After registration, the acceptable daily intake levels are monitored by the FDA through market basket studies to ensure that amounts actually appearing in food do not exceed those projected by the submitted data.

It is EPA policy that tolerance levels not be established at levels higher than necessary. Current tolerances may be found in the Code of Federal Regulations (CFR) 40, 150-189, 1983 and information on food additives related to pesticides can be found in CFR 21, 170-199. Useful lists in these regards appear at the end of the chapters.

CLASSIFICATION OF PESTICIDES

Pesticides (active ingredients) may be registered for general use, registered as classified for restricted, or both. A specific product, however, may be registered for general use or classified as a restricted use product but not both. A pesticide product will be registered for general use if EPA concludes that its proper use will not generally cause unreasonable adverse effects on man or the enviroment. If the pesticide, even though properly used, may generally result in adverse effects on the environment or possible injury to the applicator, it will be classified for restricted use. As of this writing, formulations based upon some 70 active ingredients were classified for restricted use. Those actives are:

acrolein	bromodiolone	chloropicrin
aldicarb	calcium cyanide	chlorophacinone
allyl alcohol	carbofuran	clonitralid
aluminum phosphide	chlordimeform	cycloheximide
amitraz	chlorfenvinphos	DBCP
azinphos-methyl	chlorobenzilate	demeton
brodifacoum	chlorpyrifos	diallate

338

dichlorvos	hydrocyanic acid	phorate
dicrotophos	isofenphos	phosacetim
diflubenzuron	lindane	phosphamidon
dioxathion	magnesium phosphide	picloram
disulfoton	mesurol	pronamide
endrin	methamidophos	propetamphos
EPN	methadithion	sodium cyanide
ethoprop	methomyl	sodium fluoroacetate
ethyl parathion	methyl bromide	starlicide
fenamiphos	methyl parathion	strychnine
fensulfothion	mevinphos	sulfotepp
fenvalerate	milban	sulprofos
flucythrinate	monocrotophos	TEPP
fluoroacetamide	nicotine	toxaphene
fonofos	nitrofen	zinc phosphide
heptachlor	paraquat	
hoelon	permethrin	

Current information on which products are classified for restricted use can be obtained from EPA, OPP, Registration Division, Process Coordination Branch, 401 M Street, S.W. Washington, DC 20460.

The most important aspect of classification is that restricted use products can only be purchased and applied by certified applicators or individuals under the direct supervision of a certified applicator. However, it is desirable that anyone applying insecticides in a food processing plant be a certified applicator even if no restricted use products are used there. The educational experience necessary to become certified is beneficial in learning safer and more efficient techniques of application for anyone involved in routine pesticide application.

CERTIFICATION

Certification is obtained through the appropriate agency of the state in which the applicator seeks to use pesticides. Usually this is the State Department of Agriculture. The local county extension service is the contact for definitive information on certification in your state. The extension services provide training under cooperative arrangements with EPA. There are two kinds of certified applicators; private applicators (which are generally farmers), and commercial operators. Commercial operators apply pesticides for hire snd are certified by category (such as field crops, fruit and orchards, nursery and ornamentals, etc.). States vary in their category descriptions for applications in food handling plants but most are at least similar to the EPA federal category #7: Industrial, Institutional, Structural, and Health Related Pest Control.

Certification usually requires passing a test on basic application techniques, pesticide safety, label comprehension, and precautionary measures. Many states require that certified applicators attend pertinent seminars or presentations to update their knowledge on new developments in pesticide application.

All states now have certification plans which have been approved by EPA as required. Forty eight of the states, plus Puerto Rico, the Virgin Islands, the District of Columbia, and Guam have their own plans. The two remaining states, Colorado and Nebraska, rely on the federal program. In these two instances, contact of both the state offices and the regional EPA offices (in

Denver and Kansas City, respectively) is recommended.

APPLICATION OF INSECTICIDES IN FOOD HANDLING ESTABLISHMENTS

The admonition to read, understand, and follow the label cannot be stressed too often. It is unlawful "to use any registered pesticide in a manner inconsistent with its labelling." Such misuse is a violation of federal law and most state laws and this statement appears on most federally registered pesticide products. Applicators should read the product label carefully prior to purchase in order to determine if the material can be applied in those areas of the plant where pest control is needed. Likewise, all precautionary labelling should be carefully studied to make sure that some of the desired uses are not prohibited.

It should be noted that there are some situations where the label does not have to be followed directly. Examples are: 1) pests not appearing on the label can be the intended target of the application if the site(s) appear on the label; 2) applying the pesticide at less than the labelled dosage; 3) employing methods of application not prohibited on the label; and 4) mixing a pesticide with a fertilizer when not prohibited on the label. When employing any of the above exceptions, the applicator should be assured that the application will be safe and effective, by consulting with recognized experts such as the local Cooperative Extension Service.

USE OF INSECTICIDES WITHIN FOOD HANDLING ESTABLISHMENTS

A food is defined in Section 201 of the FD&CA as "(1) articles used for food or drink for man and other animals, (2) chewing gum, and (3) articles used for components of any such article."

A food handling establishment is an area or place other than a private residence in which food is held, processed, prepared and/or served -- a very inclusive definition.

Food areas of a food handling establishment include places for receiving, storage, serving, preparing, packaging, and enclosed processing equipment.

Insecticides are generally described as non-residual or residual. The non-residual insecticides are those products applied to obtain insecticidal effectiveness only during the time of application. The two types of application most common for non-residual products are space applications, such as by fogger, and contact, such as with a wet spray. Most food plants include both treatments in their insect control program. The residual insecticides are those products which are applied to obtain insecticidal effects for several days or longer. Residuals are applied as general, spot, or crack and crevice treatments. General treatment is application to broad expanses such as walls, floors, and ceilings or as an outdoor treatment. Spot treatment is application to non-contiguous level areas not to exceed two (2) square feet per spot. These areas must not be in contact with food or food handling equipment and will not ordinarily be contacted by employees. These areas may occur on floors, walls, and the bases or undersides of equipment.

In food areas, residual pesticides may be applied by crack and crevice treatment only. Crack and crevice treatment is the application of small amounts of insecticides into cracks and crevices in which insects hide or through which they may enter the building. Such openings may lead to voids such as hollow walls, equipment legs and bases, motor housings, and switch boxes. (Note: <u>Never</u> apply water-based sprays to switch boxes or other electrical circuits due to the danger of electrical shock!) Normally crack and crevice application instructions appear on any label for which this technique is mandatory. Be alert to special instructions because residual insecticides used in food processing plants have a definite long-range potential for the contamination of the final product. For this reason, residual insecticides can only be applied in food areas if the active ingredient has a tolerance for this use and/or the material is labelled for crack and crevice application. Note that special equipment such as bulbous dusters, insertion tubes, pin-stream nozzles, etc. is necessary for proper application. One further note, if the food processing plant is under the jurisdiction of the USDA Meat, Poultry, and Egg Inspection Service (MPIS), all chemicals used in the facility must be on the MPIS approved list and be cleared for use with the MPIS Inspector in Charge prior to application.

STORAGE AND DISPOSAL

The requirements for storage and disposal are listed on the label. These requirements must be followed. Good general comments include: 1) control pesticides under "lock and key"; 2) never store pesticides near food or feed products; 3) never transfer pesticides to another container except for application equipment; 4) avoid storage under extreme conditions of heat or cold; 5) storage areas should have good ventilation; 6) have a fire extinguisher and gas mask easily available; 7) never dispose of used containers in areas where drainage is poor or leaching is likely to occur; and 8) keep a complete inventory of all pesticides in a separate office from the storage area. This can be critical to emergency personnel in the event of fire, flood, etc.

Some states and local jurisdictions have special storage and disposal restrictions, so make sure that all personnel handling pesticides are aware of any local laws.

SAFETY

The safe use of pesticides from a personal standpoint is covered in another chapter (22). A reminder is in order, however, to have on had the proper equipment such as goggles, gloves, respirators, etc. and the means of promptly treating a possible accidental exposure including the appropriate antidotes. The telephone number and location of the nearest poison control center, hospital, and fire department should also be immediately available. In fact, it is a good idea to place such numbers on the application equipment in order to facilitate help in an emergency situation.

ENFORCEMENT

One of EPA's main responsibilities is to ensure that pesticides are registered and then used according to label directions-- the right material used for the right need in the correct way. FIFRA has delegated the primary responsiblity for pesticide-use violations to the states. Their main efforts are: 1) to conduct marketplace sampling and testing of pesticides to make sure the concentration(s) of active ingredient(s) is (are) in accordance with the registration and the label declaration; 2) to make sure that products do not contain illegal impurities; 3) to monitor experimental permits and temporary tolerances; 4) to check on licensed dealers to ensure that sales of restricted products are not being made to uncertified users; 5) to monitor complaints; and 6) to conduct use investigations to assure that products are being used correctly. Note that the food industry can and does get involved in experimental uses through cooperative programs designed to obtain the data necessary to register new uses of pesticides in food handling plants.

Generally the state inspector will check for proper records and may take samples of pesticides on the premises for chemical analysis, although this is usually more likely to occur at pesticide wholesalers, retailers, and distributors than at food processing plants. Make sure good up-to-date records are kept in easily understandable form. Results of any chemical analysis and a copy of the collected sample label are then forwarded to EPA to ascertain any suspected violations. Be cooperative when and if a state inspector visits your plant. The enforcement program is a major factor in ensuring that the pesticide products you purchase are up to strength and are suitable for the purposes claimed on the label.

COMPLIANCE CONCERNS

The amended FIFRA makes it unlawful to use any registered pesticide in a manner inconsistent with the labelling. FIFRA confers on the states, and ultimately EPA, the responsibility to enforce the law against the possible misuse of insecticides. Use and misuse inspections are indispensable elements of enforcement. FIFRA and rule 41 of the Federal Rules of Criminal Procedure are the sources of statutory authority to procure warrants authorizing entry for the purposes of inspection and sampling.

State agencies give precedence to reported cases of possible misuse over use observations and inspection activities. It does, however, have a routine inspection program on a cross section of pesticide uses, during and following actual application. The purpose of use inspections is to develop data on the common practices of applying pesticides to: 1) encourage the proper use of pesticides, and 2) determine whether or not pesticides are being used in accordance with their labelling. The data of concern are: were the labels read and understood; were the labels followed; is the application and protective equipment maintained in clean and good working order; are the pesticides properly stored; and are excess pesticides disposed of in a way which minimizes impact on the environment.

Use investigations are conducted by Regional and State Pesticide Investigators (following the guidelines as listed in sections 15 and 16 of the EPA Pesticides and Toxic Substances Enforcement Division Inspectors Manual).

The states are the primary enforcement authority in any cases of misuse. EPA must, therefore, defer to the states for enforcement of any documented instances. If the state does not act within 30 days or it does not plan to, EPA may begin proceedings.

Many documented cases of misuse involve agricultural uses, including aerial application. Not surprisingly, the food industry has been occasionally involved, most commonly by the use of residuals in edible food areas of plants. Two misuse examples are: 1) the use of diazinon by a pest control operator in a meat handling facility--the label called for crack and crevice treatment, the avoiding of exposed surfaces and no introduction into the air. The application, observed by a county agricultural employee, was a liberal spray applied up the vertical surfaces of walls and equipment in such a manner that the spray was airborne; and 2) the use of dichloron in a warehouse, containing bagged corn kibble and bagged texturized vegetable protein, by the warehouse manager-- the label instructions called for a coarse spray, paint brush, or spot application.

The firm, on the recommendation of the insecticide manufacturer, was using a fogger. As one would anticipate, the stored food products contained pesticide residues. These examples are illustrative of the key use point-- was it in accordance or consistent with the labelling? That is EPA's main compliance concern to the user industry. Secondary concerns are: 1) the improper storage and disposal of pesticides and pesticide containers; and 2) ineffective (electronic) devices for which extensive claims are made to control pests but which are safe for humans and pets.

SOURCES OF INFORMATION

The adage, an ounce of prevention is worth a pound of cure, is very true about the proper use of pesticides. Where can one go for help? Contacts include the local agricultural extension service, the state agency involved with environmental protection, the local health services, the watts line of the EPA Office of Pesticide Programs (1-800-531-7790), or the Regional EPA Offices. Following is a list of regional EPA offices:

EPA REGION 1
JFK FEDERAL BLDG.
BOSTON, MA 02203
(Connecticut, Maine, Massachusetts, New Hampshire,
Rhode Island, Vermont)
617-223-7210

EPA REGION 2
26 FEDERAL PLAZA
NEW YORK, NY 10007
(New Jersey, New York, Puerto Rico, Virgin Islands)
212-264-2525

EPA REGION 3
6TH AND WALNUT STREETS
PHILADELPHIA, PA 19106
(Delaware, Maryland, Pennsylvania, Virginia,
West Virginia, District of Columbia)
215-597-9800

EPA REGION 4
345 CORTLAND STREET, N.E.
ATLANTA, GA 30308
(Alabama, Georgia, Florida, Mississippi, North Carolina,
South Carolina, Tennessee, Kentucky)
404-881-4727

EPA REGION 5
230 S. DEARBORN
CHICAGO, IL 60604
(Illinois, Indiana, Ohio, Michigan, Wisconsin, Minnesota)
312-353-2000

EPA REGION 6
1201 ELM STREET
DALLAS, TX 75270
(Arkansas, Louisiana, Oklahoma, Texas, New Mexico)
214-767-2600

EPA REGION 7
324 EAST 11TH STREET
KANSAS CITY, MO 64108
(Iowa, Kansas, Missouri, Nebraska)
816-374-5493

EPA REGION 8
1860 LINCOLN STREET
DENVER, CO 80203
(Colorado, Utah, Wyoming, Montana, North Dakota,
South Dakota)
303-837-3895

EPA REGION 9
215 FREMONT STREET
SAN FRANCISCO CA 94105
(Arizona, California, Nevada, Hawaii, Guam,
American Samoa, Trust Territories of the Pacific)
415-556-2320

EPA REGION 10
1200 SIXTH AVENUE
SEATTLE, WA 98101
(Alaska, Idaho, Oregon, Washington)
206-442-1220

SUMMARY

The Environmental Protection Agency (EPA) is charged with the
responsibility for the implementation and enforcement of FIFRA, the Federal
Insecticide, Fungicide, and Rodenticide Act. Its main thrust is to control
the manufacture and use of pesticides via the pesticide registration
process and the setting of tolerances.

EPA delegates to the states the primary responsibility for ensuring the
proper use of insecticides. Investigators from the states and the regional
EPA offices visit food manufacturing plants and other using facilities to
determine whether the use of pesticides is in accordance with the registered
label.

Labels on pesticides must be approved and up-to-date; they must be read
and understood; and they must be followed completely.

FDA'S REGULATORY ACTIONS/ATTITUDES

Elaine R. Crosby

Editor's Comments

Mrs. Crosby in her chapter emphasizes that she is speaking of attitudes of the current administration. Her basic premises were valid in the past and are judged to be so for the future.

Voluntary compliance needs stressing. Just as pure foods (absence of filth) cannot be justified economically, neither can our economy pay for an FDA large enough to give complete oversight of the industries it has the responsibility to regulate.

F. J. Baur

FDA'S REGULATORY ACTIONS/ATTITUDES

Elaine R. Crosby
Small Business Representative

Food and Drug Administration
3032 Bryan Street
Dallas, Texas 75204

A discussion of FDA's regulatory actions/attitudes should begin by mentioning the Agency's continuing effort to carry out the Administration's regulatory reform program. What does regulatory reform mean to FDA? It is worth emphasizing at the outset that regulatory reform is not "deregulation." The current Administration is not committed to the root-and-branch eradication of Government. What the Administration intends to achieve is not the elimination of regulation but the limitation of regulation to those areas in which it is genuinely necessary and, within those areas, to assure that regulation reflects an appropriate balance between the attainment of the regulation's purpose and the costs the regulation imposes.

Thus, when a problem is identified that seems to require Government intervention, an analysis begins to determine whether the proposed form of intervention is overly intrusive, burdensome, disproportionately felt by small business, or simply too expensive in relation to the objective sought. Note, however, that the decision to undertake this analysis presupposes a more fundamental decision. Namely, the Government has a legitimate interest in dealing with the problem in question whether it concerns public health, safety, or food adulteration by rodents or insects.

One of the basic principles of regulatory reform is trust in the integrity of the private sector and confidence that in most instances it can be relied on to safeguard the interests of consumers.* Another is confidence that consumers can usually make appropriate decisions concerning their own interests provided they are informed about the available choices. A third principle of regulatory reform that must be kept in mind is when private sectors do not adequately deal with a problem, the Government will take appropriate regulatory action. While the Government stands ready to relinquish some of its responsibility by granting a more important role to industry initiatives and consumer education, nothing in the philosophy of regulatory reform requires it to abandon the fundamental public health mission embodied in laws such as the Federal Food, Drug, and Cosmetic Act.

That law, which became effective 47 years ago and which replaced and amplified a statute that was itself 30 years old, represents a broad consensus that Government has a proper and necessary role in safeguarding the integrity of the most important products on which the American public depends. Over the years, FDA has attempted to translate the mandate of the law into an effective and workable system of public health protection. Some elements of that system may need adjustment. FDA's goal is to adjust the system to give effect to the principles of regulatory reform without compromising those achievements that everyone agrees were necessary and should be continued.

The subject of regulatory reform should not be left without emphasizing a final point. In discussing the ways in which FDA is trying to reduce Government regulation, do not lose sight of the private sector's responsibility in assuring the success of this enterprise. Remember, the goal of regulatory reform is not to reduce overall public health protection. Rather, it is to reallocate responsibility for maintaining that protection by scaling back the Government's presence to allow more room for the private sector to address society's problems. It follows that regulatory reform will succeed only if members of the private sector recognize a heightened responsibility to market safe, effective, clean, and informatively labeled products.

*EDITORIAL NOTE:

Mrs. Crosby states above "One of the basic principles of regulatory reform is trust in the integrity of the private sector and ..." This is appropriate and well stated. There is no question in my mind that regulatory agencies such as FDA, EPA, etc. have to rely greatly upon industry for voluntary compliance and most informed consumers would agree. Note the emphasis on informed. I vividly recall the considerable flak that FDA received when they first released their DALs in 1972 (see Chapter 25) which recognized the well known fact among the food industry that there is no such thing as a pure food. To put this point into the proper perspective for this book, as long as foods are made from agricultural products, they will inevitably contain some filth caused by insects. Economically it can be no other way. But back to Mrs. Crosby's statement. Yes, there must be trust and FDA has to trust industry. Since industry is made up of individuals none of whom are perfect and since insects are not considered health risks relative to food, there are numerous instances where it is difficult for industry to decide whether to scrap or not - is the dollar loss worth it? Individuals and people in industry will take chances, variably so. I am glad Mrs. Crosby said "in most intances" industry can be relied upon. I am confident industry will continue to merit this trust, as it is in their best self interest to do so.

The next point that should be covered in a discussion of FDA's regulatory actions and attitudes is the broad and very fundamental issue of "enforcement." Enforcement is what happens when something goes wrong with the social agreement between FDA and the private sector. The graphs on the next two pages show a breakout of the types of enforcement as well as the number of legal actions which have occurred from October 1, 1981, to June 30, 1982. These are broken out into the various products which FDA regulates.

There are two items on the graphs which require additional discussion. The first is that the food area received the largest number of legal actions with a total of 115 seizures, 189 recalls, 5 citations, 7 prosecutions, and 3 injunctions in Fiscal Year 1982. The second item which is apparent is that the total number of legal actions for all products decreased from FY81 to FY82. This decrease in enforcement numbers has caused considerable controversy.

According to Ralph Nader's Health Research Group (HRG), FDA's enforcement actions for Fiscal Year 1981 decreased 45 percent from the average level of the past four years. A more recent analysis by HRG showed a further decrease to 66.4 percent for the first half of Fiscal Year 1982 when compared with the same period for the years 1977 through 1980. The Health Research Group is of the opinion that the decline in enforcement is the product of an evil partnership between FDA and regulated industry as a result of the Reagan Administration.

As a word of caution, FDA has not gone soft on enforcement. FDA has, since its inception, had three distinct but related roles. It is a science agency; it is a consumer education agency; and it is a law enforcement agency.

Everyone at FDA agrees that the Agency has important scientific and consumer education obligations. Without downplaying the significance of those obligations, however, industry should be aware that most FDA employees view our most vital function as ensuring that the requirements of the Federal, Food, Drug, and Cosmetic Act are honored by those who fall under our jurisdiction.

This is not a partisan view. There is no constituency of which FDA is aware that opposes enforcement of laws designed to protect the safety of the American public and the integrity of the market place. Industry has as strong an interest as consumers in compliance with standards that assure safe, clean, effective, and forthrightly labeled products. Such compliance promotes confidence in the products of American industry. It also allows the majority of businesses to market their goods with the security that their interest in selling honest products will not be threatened by the unscrupulous minority. Indeed, FDA is often told by businesses that it should enforce the law even more vigorously than it does in those areas that have been assigned a lower regulatory priority.

Seizures, Recalls, and Regulatory Letters

FY 81 vs. FY 82 (October 1 to June 30)

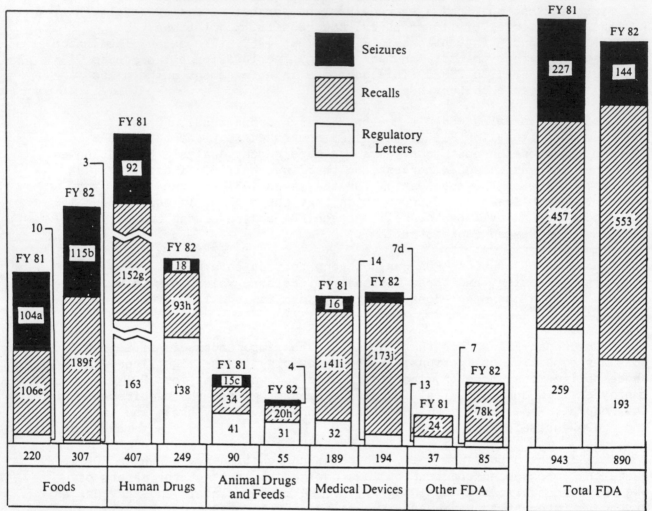

a. Includes 8 mass seizures.
b. Includes 10 mass seizures.
c. Includes 3 mass seizures.
d. Includes 1 mass seizure.
e. Includes 3 Class I recalls.
f. Includes 20 Class I recalls.

g. Includes 7 Class I recalls.
h. Includes 1 Class I recall.
i. Includes 11 Class I recalls.
j. Includes 40 Class I recalls.
k. Includes 41 Class I recalls.

NOTE: Beginning the first quarter of FY82, enforcement action counts reflect Agency determinations rather than counts at some intermediate level (e.g., bureau recommendations). Sources are as follows:

Seizures — Enforcement Policy Staff, ACRA. Counted when approved by General Counsel.

Recalls — Emergency and Epidemiological Operations Branch, EDRO. Counted when approved by Bureau.

Regulatory Letters — Freedom of Information Staff, ACPA. Counted when filed with FOI Staff.

Citations, Prosecutions, and Injunctions

FY 81 vs. FY 82 (October 1 to June 30)

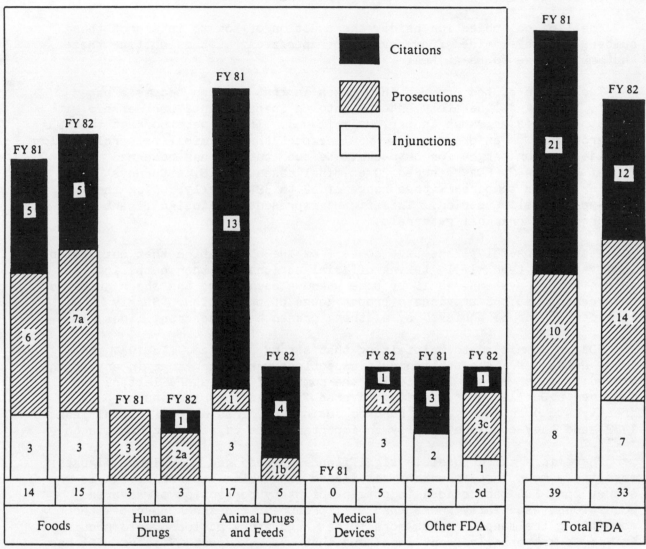

a. Includes 1 prosecution for contempt of court.
b. Includes 1 request for a grand jury investigation.
c. Includes 1 request for a grand jury investigation and 2 prosecutions for violation of Title 18, U.S.C.
d. Does not include 1 civil penalty.

NOTE: Beginning the first quarter of FY82, enforcement action counts reflect Agency determinations rather than counts at some intermediate level (e.g., bureau recommendations). Sources are as follows:

Prosecutions, Injunctions, and Civil Penalties — Enforcement Policy Staff, ACRA. Counted when approved by General Counsel.

Citations — Individual bureaus. Counted when approved by bureaus.

And yet, the fact remains that enforcement statistics are down. The decreased figures are striking and deserve close scrutiny. There is plenty of room for legitimate debate about what the numbers themselves mean. In the first place, to argue that these enforcement figures necessarily mean decreased consumer safety presents what FDA views as a serious conceptual problem. To do so is analogous to observing that a decrease in traffic tickets automatically means that more traffic violations are going undetected. It is equally plausible, and perhaps more compelling, to conclude that a decrease in traffic tickets means that compliance with the law is increasing.

That logical objection aside, the question of how to interpret these numbers can be examined from several perspectives. Let me outline those that make sense to us at FDA.

The burden of the argument HRG makes is that FDA has become a paper tiger, that it deliberately entered into an inappropriate partnership with industry. This argument is seriously flawed. If HRG were correct that FDA has grown lax on enforcement, we would expect that criminal referrals would provide a major target for cutbacks. No such cutback has occurred. In Fiscal Year 1982, FDA referred 20 criminal cases to U.S. Attorneys' Offices. This compares with a range of 22 to 32 annually during the four-year baseline period. This figure represents an insignificant difference in criminal referrals.

A second useful perspective comes from the observation that enforcement involves more than simply taking official actions and then compiling them in monitoring systems--as if by some unknown mechanism, the sheer number of enforcement actions provides a proper gauge of protection. Surely enforcement can be achieved by a strong presence on the front lines.

FDA is committed to maintaining that strong presence. In 1981 the total number of FDA establishment inspections was higher than in any of the preceding five years. Admittedly, the number of regulatory letters has declined. But letters reporting adverse findings have increased greatly--to over thirty-five hundred during the first half of Fiscal Year 1982--and they continue to play an important role in securing compliance.

There is a third possible significance. These new compliance figures may simply mean that voluntary compliance is working very well. It is no secret that FDA has decided to obtain voluntary correction of adverse conditions, especially where that will produce the desired result without committing the manpower that more formal regulatory actions would require. And it is not exactly front page news that over the past 25 years, FDA has relied on a wider variety of compliance measures than just the traditional court actions. FDA has traditionally shown its willingness to use innovative and frequently more efficient enforcement approaches to replace traditional ones. FDA's objective is to achieve compliance, not to spike the enforcement indicators.

Fourthly, other factors may well contribute to the decline in the more traditional measures of enforcement activity. In the past decade, FDA has increasingly relied on broad regulatory standards as a means of ensuring better and more systematic compliance--Good Manufacturing Practices immediately come to mind. It is entirely possible that because of these initiatives, the major industries within FDA's jurisdiction have achieved compliance rates high enough to diminish the need for ad hoc regulatory actions.

Finally, the nature and sophistication of industry--and its problems--have changed over the past quarter century. In the late 1950's, for example, a single inspector might check several breweries owned by different companies. Now, in contrast, a much smaller number of national firms exert a greater influence over the market. These firms maintain full-time quality control staffs with computerized practices that easily outstrip what smaller firms could once do. Is it any surprise that compliance has improved?

All of these remarks about compliance numbers are merely observations. What we are still talking about is how to interpret the numbers. FDA is not convinced that anyone will be able to provide an explanation of what, exactly, the figures mean. The issues are complicated, they involve many variables, and we simply do not adequately understand all the factors that may influence their movement.

But we can tell you what these declining enforcement figures do not mean! They do not mean that FDA has gone soft! It may be that at least some of the decline reflects decisions FDA has made about how best to encourage compliance. FDA is not wedded to any particular enforcement. The Agency certainly is not trying to lower the formal enforcement indicators to win popularity among those it regulates. Therefore, should it appear that our current approach is jeopardizing consumer protection, we will change that approach with absolutely no qualms. Should higher enforcement statistics be the result, so be it; although in that case, our interest will remain in overall compliance, not in one set of figures of limited significance.

The mission of the Food and Drug Administration is consumer protection. To the maximum extent possible, the agency carries out this mission by encouraging voluntary compliance on the part of responsible industry officials. However, we are a law enforcement agency, and we will not hesitate pursuing our legal sanctions which include: seizures, injunctions, publicity, civil penalties, revocation of licenses, withdrawal of product approval, or criminal prosecution when we deem that such actions are necessary.

In summary, insects and subsequent insect problems have existed since the beginning of time. Even though the food industry is continually making technological advances, the presence of insects will never be eliminated. The most that can be hoped for is to control their activities and to prevent their adulteration of food. To accomplish this end, the agency relies to a large extent on industry's integrity and their willingness to achieve voluntary compliance.

NOTE: Portions of this chapter were taken from Dr. Arthur Hull Hays' 9/15/82 Speech to Health Industry Manufacturers Association and Chief Counsel Scarlett's 9/16/82 remarks to the Regulatory Affairs Professional Society

ILLUSTRATIONS OF INSECT PROBLEMS AND SOLUTIONS
(Case Histories)

Editor's Comments

One should always try to learn from the experiences of others. The previous chapters provided information on control. This chapter shares how some individuals have used their knowledge to practical advantage. It contains some real down-to-earth information.

F. J. Baur

ILLUSTRATIONS OF INSECT PROBLEMS AND SOLUTIONS
(Case Histories)

Contributors:

Fred J. Baur, Wendell E. Burkholder, Elaine R. Crosby,
Thomas E. Evers, James W. Gentry, Don Gilbert, Donald S. Hanson,
James W. Hartman, Harold Leyse, Ernest A. R. Liscombe,
Raymond V. Mlecko, George T. Okumura, John O'Reilly, John V. Osmun,
J. H. Rutledge, Kenneth O. Sheppard, Robert C. Yeager

INTRODUCTION
by Fred J. Baur

Experience is the best teacher. This adage is also very true with insects and their control. The purpose of this, the last chapter, is to enable the reader to learn from the experience of others. The illustrations of problems speak to different species, how the problem was discovered, how it was handled, what was learned, etc. As insect behavior patterns remain unchanged, the lessons from the illustrations should stand the test of time. They could, however, be tempered by changes in the amount of concern which industry may have with insects and their control as a result of possible changes in consumer and regulatory attitudes.

Most of the contributors were authors of this book. Not all are listed.

In addition to the illustrations which will follow, a list of "insect problems" submitted by one of the authors is worth sharing. The entire list is included since the source is FDA, hence is an indication of what that agency sees and reacts to. The contributor also stated, quite accurately, "a little sanitation sense is all it takes to correct them." The list is:

1. Insects breeding in encrusted food ingredients inside balance pan.
2. Insects breeding in encrusted food on processing equipment.
3. Insects breeding in "boots" of closed processing systems.
4. Poor rotation of stock which results in the establishment of an insect population.
5. "Morgue" areas not isolated thus causing the spread of insect infestation.
6. Insects in rodent bait boxes.
7. Insects breeding under tarps or other types of covers for food vats.
8. Insects (cadelles) hidden in wooden bins.

AN FDA CASE HISTORY--INADEQUATE PEST CONTROL PROGRAM

June 1971* An FDA factory inspection of a donut manufacturer revealed
           building defects which could serve as vermin entryways. Live

*Under section 704 of the Food, Drug, and Cosmetic Act, designated employees of the Agency, upon presenting credentials and a written notice, are authorized to enter, at reasonable times, and to inspect, within reasonable limits and reasonable manners, any firm falling under the Agency's jurisdiction. The federal courts have interpreted "at reasonable times" as any time a firm is operating or open for business. Inspectional frequencies are based, to a large degree, on conditions discovered during previous inspection.

and dead insects were observed in the food storage areas as well as in the manufacturing equipment. Additionally, rodent excreta pellets were noted in the raw material storage area, in manufacturing equipment, and in waste material.

Analyses of in-plant samples of finished product were negative.

Follow-up inspection in January 1972 revealed sparsely distributed live insects on the exterior of unopened bags of raw material. Inspection was classified as not "actionable" at this time.

March 1973   Reinspection disclosed no insect or rodent activity. Minor building defects were noted; no contaminated products were found. Classified inspection as Voluntary Correction Following Inspection.

March 1974   Inspection revealed rodent excreta pellets in warehouse section and one section of manufacturing area. Building defects which could permit insect and rodent entry were noted. Some evidence of fruit flies was noted in one corner of the manufacturing area and numerous poor employee practices were noted. No product contamination found. Firm scheduled for routine reinspection.

October 1974   Examination of three lots of food which had moved in interstate commerce revealed evidence of adulteration. Live _Lasioderma serricorne_ (cigarette beetles) were found on and in bags of soy flour and a lot of donut mix was found contaminated with rodent excreta and rodent gnawings. A second lot of donut mix was discovered with live _Tribolium castaneum_ (red flour beetles) and cockroaches on and in the bags and the outsides of the bags were littered with rodent excreta. Other lots of food including salt, donut pyro, and whey were found adulterated with insects including _Tribolium castaneum_ (red flour beetles), _Lasioderma serricorne_ (cigarette beetles), and cockroaches as well as rodent evidence.

December 1974   At a section 305** hearing held with the local FDA district the firm provided written proposals for improved sanitation; they agreed to obtain the services of a professional exterminator and to increase in-house rodent and insect control. They felt personnel were adequately trained to inspect and reject insanitary raw materials, thus ensuring that adulterated raw materials would never be used in production of donuts and/or donut premixes. All adulterated products were voluntarily destroyed.

January 1975   The local district office recommended prosecution of both the firm and the firm's president on the basis of the October 1974 inspection. Recommendation was disapproved by

**Under section 305 of the Food, Drug, and Cosmetic Act, any person against whom criminal proceedings are contemplated must be given the opportunity to present his or her views regarding the proceedings before the violations are reported by the Departmental Secretary to the U.S. Attorney's Office.

headquarters on the basis that all documented violations were from the October 1974 inspection.***

August 1975   Reinspection of the firm revealed live and dead insects, insect larvae, cockroaches, and live flies throughout the firm, including the donut mix and donut/sweet roll manufacturing area. Live and dead insects were found in the firm's processing equipment including blended mixes of cinnamon/sugar and a hushpuppy mix. Additionally, live and dead insects and live larvae were found inside cases of parsley and on the exterior surface of other lots of raw materials.

October 1975   A section 305 hearing was held with the firm informing the local district office that all contaminated products had been destroyed and admitting that in-plant management was not experienced enough to handle potential insect/rodent problems. Expert assistance had been obtained by securing the services of a sanitation consulting firm. This firm promised to make FDA-type inspections at least six times during the next year.

Additionally, an outside exterminator had been hired to perform pest control functions on a twice-a-month basis and to inspect the firm, submitting written reports of their findings to management.

As a preventive measure, the firm purchased automatic closing overhead doors and air curtains. They also acquired insect electrocuting units and installed these in strategic locations throughout the plant. The firm purchased commercial vacuum sweepers for use during their clean-up procedure and prepared and submitted to each employee written instructions for plant sanitation. The firm established in-house inspections to be conducted twice a week and now requires all raw material suppliers to provide them with a Food and Drug Guarantee.

November 1975   Local district office recommended prosecution of firm and firm's president and vice-president. FDA Headquarters concurred with the recommendation. Request for prosecution was submitted to U.S. Attorney's Office.

November 10, 1976   Prosecution proceedings filed by U.S. Attorney's Office citing three adulterated products: donut mix, sugar, and parsley flakes. All products were adulterated within the meaning of section 402(a)(3) (containing insects and insect larvae) and section 402(a)(4) (products held under insanitary conditions).

***Legal actions pursued by the Agency follow established procedures prior to submission to the U.S. Attorney. A District's recommendation for prosecution must first receive the Regional Director's approval. The recommendation is then submitted to the appropriate Bureau, to the Agency's Associate Commissioner for Regulatory Affairs, and lastly, to the Agency's General Counsel. Each level can propose changes, recommend alternative legal actions, approve or disapprove of the recommended action. Although the district can appeal, normally a disapproved action is forsaken.

<u>November 29, 1976</u> Firm was arraigned before federal magistrate. On guilty
                pleas, the individual defendants were fined a total of
                $3,000. Case closed. Firm relocated donut mix manufacturing
                operation to another facility. Follow-up routine inspection
                by Agency reveals no sanitation or bacteriological problems.

VOLUNTARY COMPLIANCE IS CHEAPER THAN REGULATORY COMPLIANCE

Facts

Firm is a large bakery which supplies products to a three-state area.
Recent inspections by state officials had revealed continuing evidence of rodent
and/or insect activity (<u>Gnathocerus cornutus</u>-broad horned flour beetle and
<u>Tribolium confusum</u>-confused flour beetle). An inspection by FDA found extensive
evidence of insect activity, some minor evidence of rodent activity, and a
number of poor employee practices. The firm had begun to combat the infestation
problem and management stated that an attempt would be made to correct the poor
employee practices.

A follow-up inspection four months later found the insect problem remained
as evidenced by the finding of live insects in ingredients, dead flies in
greased troughs awaiting batter, and a dead fly in unbaked muffin dough. The
firm stated that $3,500 had been spent since the last inspection for materials
involved in the corrective actions taken.

Solution of the Problem

After a meeting with the firm's management in which various regulatory
options versus voluntary actions were discussed, the firm entered into a
voluntary agreement with FDA. The agreement provided for many of the same
provisions as would a Consent Decree of Permanent Injunction had injunctive
action been taken through the courts. These included, among other things,
destruction of adulterated materials and products, correction of construction
defects, and establishment of a sanitation control program with the
responsibility for its operation assigned to a person competent to maintain the
facility and equipment in a sanitary manner.

Since signing of the agreement, the firm has developed a number of internal
procedures to improve sanitation, has hired an outside sanitation consultant,
has developed a sanitation control program, and is placing more emphasis on
employee education. Management is now more aware that it has a responsibility
to introduce and reinforce to employees the proper procedures to prevent
contamination of food products and to improve the overall sanitary conditions in
the plant.

Although subsequent inspections have revealed some minor insect problems,
the overall condition of the facility is much improved. More importantly, when
problems are encountered, management is quick to resolve them.

Lessons Learned from the Problem

Management may, at times, recognize that a problem exists in a facility, but
may be hesitant or unwilling to expend the capital and/or effort to achieve
lasting compliance. When faced with the alternative of a costly legal action,
responsible officials often change their attitude and recognize that cooperation
with FDA to attain compliance through voluntary actions is the best way to go.

Management is often lax in its responsibility to provide basic education to
employees in the area of good sanitary practices. It often fails to realize
that even after the basics are provided, repeated reinforcement and repetition
of basic rules are required. Also, in many situations, management fails to
understand (or chooses to ignore, because of added expense) the need for the
hiring of an expert in a certain field of sanitation, i.e., pest control,
correction of structural deficiencies, etc., to bring the facility into
compliance.

VOLUNTARY ACTIONS GET RESULTS

Facts

The firm is a relatively small bakery which supplies breads, rolls, cakes, doughnuts, and other pastries to retail and institutional accounts within the city in which it is located. FDA's first inspection of the firm after it came under new management revealed scattered insect infestations (<u>Tribolium confusum</u> - confused flour beetle) in equipment and various areas of the bakery. A detailed letter to the president outlined the FDA inspectional findings to alert him to a problem which could rapidly escalate if not promptly corrected.

The follow-up inspection three months later revealed a continuing insect problem in both the production area and raw materials storage room. Some cleaning had been done and was still in progress. The local state sanitarian was called and arrangements made to conduct a joint FDA/state inspection that same week. This inspection confirmed an extensive insect problem in the firm's equipment and in its raw materials. A second letter was sent by FDA to the firm outlining the inspectional findings. The firm's response indicated that a number of corrective actions were underway to correct the existing problems.

A comprehensive follow-up inspection four months later revealed that, despite efforts of management, the insect problem remained, and that some kind of action would be necessary to bring the firm into compliance.

Solution of the Problem

To bring a regulatory action against the small firm would result in a possible shut-down, causing a shortage of bread and other bakery products to consumers in a comparatively isolated area. Responsible officials of the firm were making an effort to eliminate the insects, but appeared to lack the necessary expertise to achieve success. After discussions among the firm's management, state officials, and FDA, a voluntary agreement was entered into by all three parties. This agreement set forth the conditions which must be met by the firm to assure compliance with regulations--both state and federal. It provided for the destruction of all adulterated products within the bakery, correction of structural defects, elimination of all insects and filth contributed by them from the facility, and the establishment of a sanitation control program and internal inspection program within the facility.

The firm met the conditions of the agreement and is currently monitored by the state sanitarian as well as FDA. The most recent report from the state indicated that improvements made within the firm are "phenomenal." The president is quoted as saying that "this is becoming like a disease, as I can see improvement, I want to do more."

Lessons Learned from the Problem

Although management of a facility is often willing to make the effort to correct violative conditions, there sometimes is a lack of knowledge, expertise, or capital to achieve compliance of a lasting nature. Direction from outside the firm may be necessary to achieve the desired results. At times, cooperation in the form of a voluntary agreement can accomplish the end result with far less cost to the firm, the state, and/or the federal government, while affording the consumer the same degree of protection.

SICK BABIES

The problem was discovered in a UV electrocutor in a baby food cereal plant. There was nothing wrong with the unit, per se; it was simply never serviced and dead insects had accumulated for well over a year. When finally detected, it contained thousands of <u>Trogoderma</u> larvae and from there the infestation had spread throughout the plant. A thorough inspection revealed larvae in such places as tight crevices under the rubber gaskets of the observation port in the flour sifter and in several electrical conduits leading to switches on food product conveyors and the package fillers.

It is no wonder that larvae were being found in baby cereal by distraught
mothers as far apart as California and New Jersey. Several cases of
gastroenteritis were reported.

The solution to the problem was not easy. In fact, there may well have been
Trogoderma present long before they were caught in the electrocutor.

A pest management program was devised. The first step was to locate all
possible points of infestation. Next, immediate corrective measures were taken
to the extent possible before complete fumigation could be scheduled. This
involved thorough vacuuming, installation of a new type of door gasket,
construction of several cleaning ports in machinery, caulking of many cracks and
crevices, installation of additional UV light/electrocutors, purging of ducts
and conveyors, etc. One ULV application of dichlorvos was made, and a number of
aluminum phosphide tablets were inserted into entrances to infested electrical
conduits. As quickly as possible, a fumigation using methyl bromide was
performed. Immediately thereafter, all conduit openings and switch boxes were
sealed.

INSECT SIGNIFICANCE BASED ON IDENTIFICATION

There are times when the significance of an insect species must be
considered when evaluating a situation in order to avoid the unnecessary
destruction of valuable food. In some instances it should be reasonable to
simply remove insects from food products if it can be done safely and
effectively. Proper identification of the insect and knowledge of its habits
are very important.

An example occurred a few years ago involving a bulk railcar of corn meal
which had been shipped from the midwest to a food processor on the west coast.
The corn meal was destined for use as an ingredient in a human food product and
would thus undergo further processing. The customer reported that inspection of
the load upon arrival showed the presence of several live confused flour beetles
on top of the product and the shipment was being rejected. The time of year was
early April when average temperatures in the midwest were not consistently warm
enough to begin in-transit fumigation of shipments. However, there had been
brief periods of warm weather and the shipment in question had traveled into the
warmer climate of California.

Investigations were begun immediately in the milling operation even though
there were no known problems with confused flour beetles. At the same time, a
request was made for the customer to forward specimens of the insects via air
express. Upon arrival of the insects, it was determined that they were not
confused flour beetles. They were identified as hairy fungus beetles. This
same misidentification has occurred several times in the past. If the two
species of insects are side by side for comparison, even an inexperienced
technician could readily distinguish the difference. However, when the hairy
fungus beetle is present alone, it is possible for someone not trained in
entomology to misidentify it as a flour beetle.

Important elements to consider when assessing the significance of an insect
invasion would be the species involved, number present, habits, feeding
preferences, capabilities, etc. The presence of several live confused flour
beetles in a bulk rail car shipment of milled cereal product would suggest
infested product. That insect tends to shy away from light so the obvious
presence of several live ones would lead one to suspect even greater numbers
down in the lading. Also, they would feed on sound product and reproduce in
that environment. That represents an infestation and may require a rather
stringent approach such as rejection. The presence of several live hairy fungus
beetles that have somehow managed to enter a bulk shipment of milled cereal
product may not be cause for great alarm. The insect derives part of its common
name because it feeds on fungus. It is not going to feed and reproduce in
product that is in good condition. They are strong flyers and are attracted to
light. In warm weather they may be present in very large numbers, especially in
places like rail yards where there may be decayed food or grain spillage. Due

to their small size and flying ability, they may find a means of entering a bulk rail car. Usually the insect is there merely as an incidental invader and will remain on top of the load.

In the case cited, the misidentification and information about the hairy fungus beetle were discussed with the customer's corporate quality control personnel. The decision was made to fumigate the shipment with phosphine gas. After aeration, the top few inches of the load were vacuumed off and discarded. It was felt that those measures would render the remaining product quite acceptable. Close monitoring of the scalper in the unloading system did not show further insects. Fortunately in this case reason prevailed and an expensive rejection was averted.

Certainly, the author would not advocate the use of an insect infested ingredient in the manufacture of a human food product. However, the term "infested" should not be used to describe a few incidental insects that happen to get inside a shipment. The key is to identify the pest and then determine what it truly represents in the way of a hazard.

BE SURE OF THE SPECIES

A pest control company recommended that a food processing plant be fumigated because it was infested with thousands of cigarette beetles. After critical examination, the insects turned out to be hairy fungus beetles which were attracted to the wet grain next to the building. After the wet grain was eliminated, the pest disappeared and the fumigation was not necessary.

THE IMPORTANCE OF INSECT IDENTIFICATION

One of my first lessons in pest control involving insect identification came in a costly way. We had a warehouse that was infested with a strange orange colored insect with black tips at the end of each wing cover. We assumed they came from some commodity in the warehouse and our cure-all at that time was to fumigate with hydrocyanic acid gas. (This was before the discovery of DDT.) We did fumigate but it didn't seem to phase these insects. Later, when we finally had the insect identified and found out our pest was a wharf borer, we were able to understand why we failed. The insects were emerging from some wet wood beneath the warehouse floor and our gas would never penetrate down there.

This was a valuable lesson and I have never treated an unknown pest since.

WINGED PSOCIDS IN THE SOUTHWEST

A problem developed that involved bulk corn grits being shipped in railroad hopper cars from a mill in the midwest to a brewery in the southwestern part of the country. The report received from the customer indicated that six hopper car shipments were to be rejected because there were numerous tiny flying insects and webbing inside the cars. Because of the magnitude of the complaint, someone immediately traveled to the location to investigate.

The six cars were being held in a rail yard and inspection confirmed the presence of numerous tiny insects plus webbing inside all of the cars. Specimens of the insects were taken to the brewery laboratory where they were examined microscopically and identified as one of the winged psocids, an insect that does produce webbing. It was assumed that the insects had gained entry at some point while the cars were in transit.

Early the following morning, a return trip was made to the rail yard. From a position on top of one of the hopper cars, the investigator made a very important observation. The morning dew glistened on a thin layer of silken webbing on virtually every weed, bush, and tree in the general vicinity. Closer examination revealed that the entire area was heavily populated with the same winged psocids. The condition was documented by photographing the webbing on vegetation and collecting insect specimens for comparison to those taken from the shipments. All such information was presented to the railroad involved.

Not only did the railroad cover the cost of claims in this case; they further initiated an intensified program of clean-up, vegetation control, and insecticide spraying. No further difficulties were encountered thanks to good observation by the investigator and cooperation by all parties involved in initiating the necessary corrective measures.

The author would like to add to the above that railroad hopper cars are certainly a vast improvement over boxcars for transporting bulk food items. However, it is still difficult at times to seal a hopper car to the extent of totally preventing entry by minute insects. Also, it is not unusual for bulk shipments of food to be held in rail yards for periods ranging from several hours to perhaps a few days. Such rail yards may have adverse conditions present such as junk or debris, grain spillage, heavy weed growth, stagnant water, etc. Especially during warm weather, those types of conditions may enhance the development of large populations of a variety of insects. The insects may be attracted by the odor of food commodities or the warmth of the shipment during cool nights, and some may gain entry into the railcars. For those reasons, many shippers elect to fumigate bulk food shipments in-transit with one of the phosphine-generating products in order to protect the lading until it reaches destination.

HOUSEFLIES

Problem
Houseflies entering through enclosed forklift corridor.

Description
Plant and grounds neat and clean. Rear yard not paved. Product waste disposed of via conveyor to large bin in rear yard. Disposal vehicles filled at bin and waste hauled away. Conveyor and bin hosed out daily.

Findings

Minimal waste spilled from conveyor each day had accumulated in soil to extent that rotting material provided development site for larvae.

Solution

Paving yard and modifying conveyor to prevent spillage provided permanent solution.

Lesson

Know life history and requirements for development.

GOOD SANITATION SAVES BIG BUCKS

Facts
Firm is a manufacturer of various dried pasta products. Firm had established a sanitation control program a number of years ago. Food and Drug Administration inspections revealed that the sanitation program was being followed and that the firm was in an acceptable state of compliance. However, an FDA inspection in February 1983 disclosed extensive insect activity (Tribolium castaneum - red flour beetle) in the regrind room, where scrap pasta is reground into flour for reintroduction into the manufacturing process. Live insects were found in, on, and near processing equipment.

When management was advised of the insect problem, over 6,000 pounds of reground flour were diverted to animal feed use. In addition, management placed a hold on all pasta in inventory and recalled all production, which had been made using the reground flour during the previous 30 days. Over 200,000 pounds of this pasta were later destroyed under FDA supervision. Management took these voluntary steps to preclude FDA regulatory action, as Section 402(a)(4) of the Federal Food, Drug, and Cosmetic Act states that a food is adulterated if it is prepared, packed, or held under insanitary conditions whereby it may have become contaminated with filth. Processing a food product or its ingredients in insect infested equipment thus constitutes a violative condition.

Solution of the Problem

Management made an immediate investigation to determine the cause of the insect problem in the regrind room. It was determined that about six months prior to the inspection, three holes had been cut through the ceiling of the regrind room located on the first floor and the floor of the second story to install pipes to convey, via gravity, scrap pasta from a second-floor hopper to the regrind mill on the first floor. The section where the pipes transversed through the ceiling and floor formed a shelf-like structure which had not been sealed after the pipe installation. Flour dust soon accumulated in this shelf-like area and eventually became a breeding place for insects.

Although management was able to locate the insect breeding area, it could not explain why plant workers had not reported the presence of insects in the regrind room to management.

Lessons Learned from Problem

A responsible member of management should inspect areas subjected to construction or renovation. All potential insect/rodent entryways/harborages should be sealed.

Employee education as to good sanitation is a necessity. Employee turn-over and the need to constantly make employees aware of the need for following good manufacturing practices dictate that such education must be a continuing process.

If management had followed both of the recommendations above, the firm could have saved over $100,000, which represented the cost of conducting the recall and the destroyed production.

INFESTED UNUSED EQUIPMENT--HEAT STERILIZATION CORRECTED PROBLEM

Described here is a problem that has occurred at least once in most mills.

On Thursday, a member of a food plant's sanitation staff was making a full sanitation inspection of a milling department that was once a separate building from the rest of the plant but has since been connected by walls that were used to form a connecting enclosure. Now connected to the mill is a modern packaging and batch mixing department.

What was observed by the sanitation inspector at a floor wall juncture, just under some overhead pneumatic lines, were numerous confused flour beetles. Earlier there had been various pieces of milling equipment scheduled for removal, but some time had passed and that activity had not taken place. Once the findings had been identified, the extent of the insect activity was determined to be confined in and around various pieces of the unused equipment.

All visible insect activity was treated with an approved contact pesticide and all carcasses removed. The next day, insects were once again discovered crawling about on the floor around the machinery. It was determined the insects were in hidden areas within the equipment and spraying pesticide only would not correct the problem. Since a weekend downtime was approaching, the room was cleaned where the equipment was stored, doors and screened windows were shut, and the area prepared for heat sterilization. Friday, first shift, the room was cleaned. By second shift, unit heaters in the room were turned on. After 3 hours, temperatures in the room were consistently reading 130°F. At that time, the thermostats were set and the temperature recorded every 2 hours for the next 30 hours straight.

At the conclusion of the heat sterilization exercise, the heaters were turned off and the room was opened up. Upon entering the area where the infested equipment was located, it was noticed that numerous dead insects were lying on the floor. Several were noticed on ledges within the equipment. These were again cleaned up. The equipment was removed that day and taken to a separate warehouse location for permanent storage. The next day, a follow-up inspection showed no sign of insect activity. Later that month, 2 of the previously affected pieces of equipment were observed on the receiving dock of the warehouse. It was obvious that the equipment was being prepared for transportation back up into the mills for reinstallation. A quick but thorough inspection of the interiors revealed no insect activity which would have certainly jeopardized the quality of the food and the sanitation of the systems that were to be involved.

DERMESTID BEETLES

Problem
Beetles found at frequent intervals in light trap in cereal packaging room.

Description
Room and equipment cleaned daily. No obvious source of infestation.

Findings
Inch by inch search of area initiated. Larvae found in small depressions in block wall where cereal dust had accumulated.

Solution
Vacuum cleaning of wall and filling of depressions provided permanent solution.

Lesson
Painstaking searches of less than obvious sites are often required.

A BOUT WITH FUNGUS GNATS

About three years ago, I received a call from a homeowner saying that his home was infested with small flies. Upon inspection, I found literally millions of tiny flies all over this house. They were so numerous that it would be impossible to put a plate of food on any table without having flies in the food before it could be eaten.

This particular home was new, in fact some inside construction was still going on. It was a geodesic dome home built in a series of triangular sections. The inside of the framework was covered with a 4 mil polyethylene vapor barrier and over this the inside was covered with the traditional sheet-rock. The outside of each triangle was covered with a plywood sheathing. Three holes were placed in the outside sheathing of each triangle and foam insulation was introduced into the void between the sheathing and the inside vapor barrier. Following the introduction of the foam insulation, the outside was covered by black tar-paper and then roofed with wooden cedar shingles. In several areas on the inside where there were some small gaps in the sheet-rock, flies were found alive behind the vapor barrier. By placing tape around several electrical boxes that protruded through the vapor barrier, many flies were captured on the sticky inside of the tape. It was definitely then determined that the flies were indeed coming from inside the walls. The flies were identified as Phorid's or fungus gnats, and they were breeding and feeding on fungi within the walls because the foam insulation did not dry properly. The insulation people were called in and they did acknowledge the presence of the flies and also agreed to pay to have the home fumigated. I felt that if we could cover the entire structure with 6 mil poly and left the windows and doors of the home open, we could introduce the gas under this outer poly tarp and get sufficient penetration of the walls to kill the gnats.

We proceeded as planned, covered the entire structure, introduced methyl bromide at 1/2 lb. per 1000 cubic feet of space within the structure, used several large fans to circulate the gas, and maintained this exposure for 24 hours. It did the job, but I did note some things during the exposure that led me to believe control would have been achieved without using any methyl bromide at all.

It was a warm sunshiny day and as the sun heated the air under the covering, I noticed moisture was building up under the tarp until it actually ran down the inside. This resulted in drying out the walls to such an extent, that neither the gnats or the fungi could survive, providing a lasting solution to the problem.

INSECTS IN ANIMAL FEED TUBS

Facts

An internal food plant audit turned up an insect infestation on the formulation mezzanine. Sawtoothed grain beetles were found in some of the covered containers used for storage of the minor raw materials and also in several containers used for materials that had been degraded to animal feed. Nothing was found in the process system.

Explanation

Less than a month before this audit, the plant had found a new, more expedient outlet for materials degraded to animal feed. The new program involved the buyer supplying large white plastic tubs in which all animal feed product was to be placed. These tubs were put on specially designed 4-wheel dollies at the plant and located at strategic spots throughout processing-packing (including the formulation mezzanine). After some investigation, which included an examination of recently delivered "clean" tubs, it was determined that the beetles were being brought into the plant randomly on these clean feed containers. The feed buyer washed all tubs before redelivering them to one of several food plants with which they had contracts. However, the feed buyer's plant was infested (sawtoothed grain beetles, no less) and the cleaned tubs in storage were picking up many of these tiny insects and carrying them into the unsuspecting food plant.

Moral

Everything that comes into a food plant can be a potential problem relative to insects and needs to be closely scrutinized. This includes not only those sensitive raw materials which quite naturally will be suspect but also materials and equipment that would seem to present no attraction to insects.

THE IMPORTANCE OF PROPER PLANT AND EQUIPMENT DESIGN

Pictured on the next page is a "problem" that occurred a number of years ago at a prepared mix plant. What was learned was basic then and now.

Facts
1) The operation was a batch process.
2) The finished mix was stored in troughs (see picture) prior to feeding the packing line via a dump station.
3) An alert dump operator, upon folding back the canvas cloth protective covering (see picture), noted several dozen insects clustered mostly in an oval/circular area about 10-12 inches in diameter on the surface of the mix in a central location.
4) Both plant and staff management responsible for quality assurance/GMPs (present terminology) attempted unsuccessfully to locate the source of the insects, identified as cigarette beetles.

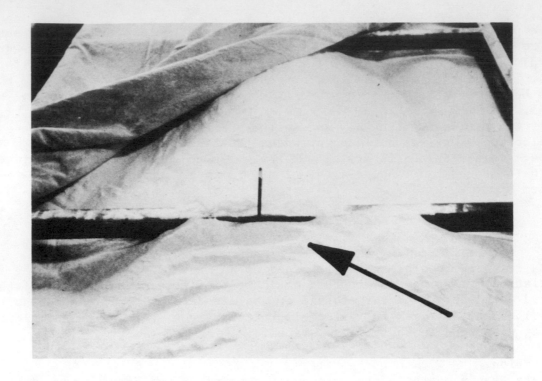

5) The "problem" reoccurred, unchanged, 2-3 more times in the next 3-4 months.
6) The dedicated, knowledgeable, responsible plant individual located the source.
7) It was the hollow support bar seen above with a pencil sticking in a hole.

Explanation
1) As a test item, several of the troughs had the canvas covers replaced with see-through plexiglass hinged covers.
2) When the hinges attached to the hollow support bars were removed, not all holes were sealed shut.
3) Both insects and mix could and did enter the hollow bars (more than one trough was involved). A good harborage was provided.
4) Periodically, as the population dictated, the beetles would "swarm" out to seek other worlds to conquer, but were confined by the canvas, fortunately.

Moral
1) Always construct a plant with pest control in mind using proper sanitary design commensurate with the raw materials, finished product, process involved, and available dollars.
2) Be particularly mindful of modifications made in plant or equipment (this illustration) as these may adversely impinge on GMPs concerns. This is an all too common omission.

Solution
    In the design of plants and equiment, always involve several professional disciplines to obtain multiple inputs on "best" sanitary handling. Always include selected manufacturing and QA/GMPs personnel. Selection of other invitees is based on scope of the project including nature of process and raw materials used. As examples, give consideration to environmental health and microbiology experts.

GERMAN COCKROACHES

Problem
Cockroaches found in stainless mixing vats each morning, especially those filled with clean water.

Description
Several small rooms connected by open doorways to accommodate forklift delivery of raw materials. Floors and walls tiled. No cracks, holes or seams. Room and equipment spotless.

Findings
Decorative wood frames around forklift doorways provided diurnal resting space for cockroaches.

Solution
Removal of wooden panels.

Lesson
Know habits of pests and utilize this knowledge in searching out pest.

INVASION OF ENVIRONMENTAL BEETLES

Facts
A second-floor food processing area had only one outside wall with windows and these windows were sealed shut. Still, environmental night-flying small beetles had invaded this area. These insects were not a food-infesting type and did not appear to live long inside. However, the results were similar to an infestation inasmuch as some of the insects (mostly dead) were found in the food equipment and in product.

Explanation
This happened during late summer, in an old building located alongside a river. These small beetles were plentiful in the outside environment at night. The plant operated 24 hours a day, and the light and warmth of the processing room attracted these night-fliers to the outside of the windows. Even though the windows themselves had been carefully sealed against their entry, age had opened up tiny cracks between the building wall and the window casings. These cracks were not necessarily straight through and, therefore, were quite hard to detect, but multitudes of beetles were able to find their way through the maze and into the room. Once inside they were not particularly concerned about where they landed and stayed.

Moral
1) Most insects are attracted to light and warmth and can find passage to these attractants through entries that are almost undetectable to an inspector. Older buildings are more vulnerable to this type of invasion and require close scrutiny.
2) While it is evident that the food-infesting insects must be of greatest concern to a food processor, environmental (or casual) insects can become problems. After all, consumers do not differentiate between a food-infesting beetle and an environmental night-flying beetle; to them a bug is a bug.

CLOVER MITES

Although clover mites are seldom if ever mentioned in the same breath with food processing or stored-product pests, a very real potential source of contamination in foods does exist and frequently surfaces.

Clover mites are not insects, but arthropods belonging to the class Arachnida (spiders, ticks, and scorpions), order Acarina (mites and ticks). They normally feed on plants outdoors, but sometimes enter structures in such numbers that they become an important pest. Geographically, they are found in general throughout the United States. Our sources indicated that the greatest number of reported cases come from east of the Mississippi. As more and more plants locate in the "wide open spaces" we can look for an increase in clover mite problems.

Initially such reported problems were related to structures surrounded by well fertilized and groomed lawns. We are rapidly learning, however, that clover mites are not only capable of traveling considerable distances but are readily found migrating from shrubs and trees.

In this case the clover mites were not only found next to and on the building at ground level but on the roof as well and in considerable numbers. Such a condition prompts the question - Are they air borne? The answer is - they can be and no doubt are. In spite of the improbability of large numbers being deposited in one area - it is apparently possible if the source is sufficiently close and at a similar height. This plant also has large air in-take fans on the roof. We definitely established, however, that they were also traveling across an adjacent railroad track bed, the non-vegetated area between tracks and building, and then definitely crawling up the building. From one or more of the upper levels clover mites gained access to a mixing vat prompting the need to destroy a product of considerable monetary value.

Control has been complete through the use of a miticide applied to the trees, railroad bed, around and on the building, and on the roof. This outside treatment is being performed monthly until a heavy frost. The one-time inside treatment consisted of pyrethrum/piperonyl butoxide aerosol. The outside service will also be started again next year prior to clover mite season to avoid any such problem from recurring.

Hopefully you won't have the experience but if you begin to see "red moving dust" on window sills or related areas, on the building itself, or on the grounds around it - don't depend on a "dust" mop. Get your PCO or his counterpart involved.

PERSISTENT GERMAN COCKROACH INFESTATION

Facts
The making area of a food plant continued to have German cockroach problems even after a thorough clean-up plus several months of diazinon crack and crevice treatments coupled with monthly pyrethrins space treatments.

Explanation
Older insulation on steam lines, on cold water lines, and especially covering a large oven centrally located in the making area had become broken, frayed, and open at many places. An earlier infestation had entrenched the cockroaches within the insulation and a somewhat dusty making process supplied an adequate amount of food. Evidently the cockroaches had found a very compatible habitat somewhere between the insulation and the hot/cold surfaces. The space treatments affected this population very little and the residual managed to control, to some extent, only those cockroaches that left their cozy home because of overpopulation or, maybe, just plain curiosity. Anyway, the community within the insulation continued to thrive.

Moral
There are several lessons to be learned here. First, even what appears to be a very clean plant can have cockroach problems. Appearance cleaning alone will not always eliminate cockroach harborages. Second, indiscriminate use of insecticides (even when label directions are followed) is not the most effective

use. Cockroach harborages need to be found and the insecticide carefully directed into these harborages, in this case, directly into the voids in the insulation. Third, poorly maintained process-related equipment can readily become a pest harborage. The only permanent solution is to eliminate potential harborages. In this case, unnecessary insulation was removed, and the necessary insulation was repaired and sealed.

TROPICAL FISH PET STORE COCKROACHES

Sometime ago we received a call from a tropical fish pet store. They had a severe problem with an explosive population of German cockroaches. The owner of the fish shop knew the dangers of spraying his shop with regular insecticides. He had called three or four pest control firms before he called us. Generally speaking, I think pet stores avoid any pesticides or pesticide applicators. He was very cautious as to what products would be used and the manner that they would be administered. We made an inspection of the building and discovered that the flower plant store next door and the apartments above would be additional problems. After a lengthy question and answer session, he hired us. We agreed to do a cockroach control program without using any sprays that would normally be dispensed from aerosols or air compressed sprayers. We also agreed on a time table when the owner of the pet shop could expect some results and final eradication. Many fish tanks, like those in homes or apartments, can have conventional residual insecticides sprayed in the general vicinity. Precautions in these situations usually require an old newspaper over the top of the tank to keep mist from drifting into the tank, and disconnecting the air pump supply for a short period of time. This aquarium shop had dozens of fish tanks full of guppies, oscars, kissing fish, and many other exotic, beautiful, and expensive prize fish. Fish names were unfamiliar to this novice and it didn't matter. Any misapplication would be costly. We accomplished the control feat with the use of a vacuum cleaner (to suck up cockroaches), insecticidal dusts, sticky type roach traps, and roach paste and baits. The brand names are numerous. The results were great and faster than anticipated. The tedious way we placed the control products and the locations in which we placed them was what got results. The reason we were able to take on this challenge, where others walked away or were shown to the door, was knowledge! That knowledge comes from several sources. One has to understand the pest, its habits, etc. One also has to understand the arsenal of products that are registered for use on that particular pest problem. The label of that product is very important. How to use it, where to use it, cautions, and so on. The pest knowledge comes from belonging to the national and state associations, knowing your university entomologists. The knowledge also comes from trade magazines and books such as the one you are now reading. It's up to the individual to consider all the facets and put the right program into motion.

CIGARETTE BEETLES

Problem
Adult beetles in storage area. No known source of infestation.

Description
Area neat and clean with no obvious source for insects. Area had been recently fumigated and problem apparently solved but there was concern about a recurrence of the problem. One piece of used equipment stored here. In discussions plant personnel were asked if equipment had been checked. Assurances were given that the equipment had been dismantled and thoroughly cleaned prior to storage therefore could not be the source.

    Large numbers of dead beetles in unused equipment.

Solution
    Periodic checking and cleaning of equipment scheduled.

Lesson
    Unused equipment often collects residue in sufficient quantity to support insects.

IS YOUR PRESSURE POSITIVE?

Location
    A round-the-clock dry grain processing operation.

Problem
    Night-flying insects presented potential contamination.

    Although the windows were screened, the need for air circulation prevented the use of a fine mesh, thus providing easy entrance for numerous night-flying insects.

    In spite of correcting the direct light source to avoid its being visible to the outside, the problem continued.

    Since there was no way to apply a pesticide without almost certain product contamination, ultraviolet light traps were tried. Some minor relief resulted, but trapping was not the answer.

    The solution for the problem was not apparent until it was realized that a negative pressure existed within the structure. The insects were literally being "sucked" into the processing area.

Control
    Change to at least equal or preferably a positive pressure. It worked in this instance and it may help you.

VALUE OF LIGHT TRAPS (ELECTROCUTORS)

    A plant had leased a number of reconditioned boxcars for the sole transfer of finished products from the plant to distribution outlets. The plant has an inside rail loading platform where the cars sometimes await loading. The dock area is equipped with insect light traps. During normal maintenance of the traps, cigarette beetles were noted in the tray (prior to any detection in processing or storage). A search was made for the source and the beetles were found in a boxcar. The car was fumigated but the problem recurred. A further search located the real source in accumulations behind the wooden walls.

    This illustration serves to show the value of light traps and confirms the admonition of "watch those boxcars!" Make sure they are clean and check what might be behind false walls and what is in channels. In this instance, new wood was placed on top of old wood without any inspection or cleaning of the wall voids.

    Some key points on the use of traps are:

1) Insect light traps are valuable diagnostic tools--the main value in many food plants.
2) The traps require periodic maintenance--including identification of species and removal of tray contents.
3) They will detect any flying insect attracted to lights, not just flies.
4) The periodic use of residual in the trays can help prevent the trap from being a source of possible plant infestation.

AN INSECT PROBLEM OF A METAL CAN STORAGE AREA

Facts

An insect problem existed in a can storage warehouse of a food processing company. The insects were _Trogoderma variabile_, the (dermestid) warehouse beetle. Insects were falling or somehow getting into the empty cans. Since there was little food available for the insects the question of concern was where were they breeding.

Explanation

The solution was derived via telephone after the insect identification was made. One of the questions asked the warehouse owner was, "do you have insect light traps?" The answer was yes. It was determined that more than a month had passed since they were emptied and cleaned. I asked them to check the light trap for insects. A few days later the warehouseman walked into my office and handed me a large plastic bag with dead moths that had been caught in the light trap and hundreds of live and fat _Trogoderma variabile_ larvae and adults.

Moral

Always clean light traps at least once a month and where permissible use an insecticide or mineral oil treatment to kill the insects and to prevent secondary infestations by dermestid beetles.

PREFERABLY AVOID BUT AT LEAST MONITOR LONG-TERM STORAGE

Facts

A warehouse was more than 35 years old and floor cracks had developed. Canned food products in corrugated containers had been placed in long-term storage. The clamp truck driver lifted some product and began to drive away as cans began to fall from the bottom of the load. Examination revealed that termites had eaten away most of the bottom of each corrugated container.

Explanation

The warehouse was located in a major termite zone. There had been no underslab treatment with termiticide at the time of construction or subsequently. The evidence of termite attack on pieces of wood lying on the ground alongside the warehouse had been ignored.

Moral

1. Pretreat ground under slab for termites. 2. Be alert and treat later if necessary. 3. Always rotate stocks and look for potential insect problems.

INDIANMEAL MOTH IN A PET FOOD INGREDIENT

Several years ago a call was received from a customer who was a pet food manufacturer. The customer reported that a boxcar shipment of toasted corn flakes in burlap bags was infested with Indianmeal moths. The product in question was a dog food ingredient. The customer further claimed that the problem had not been detected until after the product had been unloaded into their warehouse. As a result, the entire warehouse was now infested and they felt that the shipper should pay to have it fumigated. That was a pretty serious charge which naturally triggered an investigation.

The first step was thorough inspection of the manufacturing process involved. As expected, no problems were indicated. In fact, in this particular case, the product underwent a cooking process that virtually rendered it sterile. Also, the product moved directly from processing to a finished product holding bin to packaging to the railcar. Time in the holding bin was a matter of hours. There was minimal opportunity for infestation prior to loading. New burlap bags were used for packaging and thorough inspection of the bags showed no insects. During packaging each bag of product was code-dated. A check of the cleaning/coopering sheet for the boxcar involved did not indicate anything unusual. It appeared that the earliest potential for insect invasion would have been while the shipment was in transit.

A request was made to be allowed to visit the customer facility to follow up on the shipment in question. Due to good past relations with the customer, the request was granted. A tour of the warehouse did reveal a general infestation with Indianmeal moth. Further, back in a corner of the warehouse one pallet of the same ingredient was infested to the extent that webbing was so thick it appeared similar to polyethylene sheeting. Obviously this was not a condition that developed overnight. A check of the code-dating of that particular pallet revealed a simple failure to rotate the stock. The product was six months old. Key elements were an understanding of the life cycle of Indianmeal moth, knowledge of the webbing laid down by the larvae, and the time required for such a condition to develop.

A meeting was requested with management of the customer company at which all pertinent facts were reviewed, beginning with the investigation conducted back at the manufacturing plant. Further, technical assistance was provided in outlining the steps necessary to bring the immediate problem under control. The customer personnel readily recognized the weaknesses of their inbound ingredient inspections and stock rotation. The end result was a customer who was grateful for the assistance, and the shipper did not pay for fumigation of the warehouse.

AN INSECT PROBLEM OF A FOOD PROCESSING PLANT

Facts

A Trogoderma variabile (warehouse beetle) problem was suspected in a food processing plant. Pheromone traps were placed throughout the plant. In one area near several tall storage tanks the traps, located on the floor, contained a number of male warehouse beetles. A visual search resulted in locating an infestation of the beetles in the accumulated dust on the top surface of the storage tanks. Many pupae were present, especially large (female) pupae. The males usually have one less larval growth stage (instar) and emerge earlier than females. Appropriate clean-up and other control measures were taken. Subsequent pheromone trapping indicated the infestation was under control.

Explanation

The pheromone traps indicated a probable site of infestation. The prompt follow-up inspection located the source of the beetles. Prompt attention to the problem prevented future problems since the clean-up procedures and treatment effectively eliminated the females before they emerged as adults from the pupae.

Moral

Timing is critical in phermone trapping for Trogoderma. Prompt attention to sudden increases in male trap catch may enable locating and elimination of females before they have a chance to mate and reproduce. Trapping of males will also reduce the males available for mating when the females emerge as adults.

THE UNCOOPERATIVE WAREHOUSE MANAGER

Occasionally in the course of investigating an insect complaint, reason may not prevail and a lack of cooperation leads to unnecessary expense and strained relations between a shipper and customer. Such was the case several years ago when a food warehousing company phoned to say that a boxcar shipment of dry dog food was being rejected because of insect infestation. A request was made of the customer to visit the warehouse and discuss the problem. Permission was grudgingly granted. The customer was located in the midwest and the time of year was midsummer.

The first indication of a difficult situation was a very cool reception upon arrival at the warehouse. Most companies are very cordial and helpful when a supplier is concerned enough to follow up on a complaint in person. The warehouse manager was asked to describe the nature of the problem. He simply stated that when his people opened the boxcar they saw live insects and rejected the shipment. When asked what type of insects were involved, he said he did not know and did not care. He further stated that the local railroad agent had been called in on the problem and he had a sample of the insects. The boxcar had been moved from the warehouse.

Next, the railroad agent was contacted and a request made to jointly inspect the shipment. He was most cooperative and agreed to do so immediately even though the temperature that day was nearly 100°F. A very methodical examination was made of the entire lading. To summarize the inspection, the boxcar was in good condition, the bales of dog food were in good condition, and no insects, live or dead, were found. The railroad agent admitted he could find nothing wrong when he had previously checked the car. In fact, he said the loading personnel should be complimented on the appearance of the shipment. When asked about the sample of insects reported to have been found by the customer, he produced one very small fungus beetle which had been found in the doorway. He said there had been another but it got away. This was all very disturbing.

A return trip was made to the warehouse to again try to discuss the matter with the manager. The suggestion was made that he reconsider acceptance of the shipment based on the fact that all available information and observations did not provide reasonable grounds for rejection. He responded that their policy was whenever they found any live insects the shipment was to be rejected with no exceptions. By this time, diplomacy was becoming difficult. A hypothetical situation was posed. What would he do if his people opened a boxcar and a couple of insects flew in while they were unloading it? Would he have them simply close the car and reject it? He said that was being argumentative and refused to discuss the matter further. The shipment remained rejected. As follow-up, the recommendation was made that shipments no longer be made to that warehouse.

The above incident occurred during a period when the Food and Drug Administration was focusing inspection activity on the food warehousing industry. Fear of regulatory action had apparently led this company to adopt policies so rigid that a number of food manufacturers were refusing to ship product to them. It was truly tragic that reason could not prevail because their extreme caution created a very expensive problem for several companies, including the warehouse.

INSECT-FREE ENVIRONMENT

Western Kansas is in the middle of the bi-annual flight of the army worm moth. Wintering in east Kansas, the army moth spends the summer along the eastern slope of the Rocky Mountain range. All points in between are plagued with the massive flights twice each year; in May, the moth heads for the Rockies, and in September, they return to winter on the eastern edge of the plains.

A pharmaceutical plant, in this zone, whose product destination is hospitals, nursing homes, and the like, could expect to encounter some difficulty in quality assurance. Except for a good engineering department and wide-awake management, perhaps they might be in trouble at least twice each year.

(1) Design lighting with the insect in mind; i.e., sodium vapor (outside) in all areas within 250 feet of the plant; install mercury vapor - in a perimeter fashion - at 250 feet distance from the plant. This provides an

intercept point safely away from the potential insect concentration at plant entry areas.

(2) Triple doors - 2 entrance foyers - at each frequently used plant entry point. Include light traps inside each foyer.

(3) Trash collection areas - compactor and dumpster - are at least 150 feet away from any plant entry point.

(4) Lunch room is in the center of the plant area; i.e., not next to an outside wall where odors can attract insects.

(5) Positive air pressure (make-up air) properly filtered and conditioned to room temperature.

(6) Regular employee seminars wherein everyone knows the values of good housekeeping, no food carried outside the lunch room, keeping doors shut, and related good manufacturing practices.

(7) Construct a 3 foot washed rock barrier around the entire plant perimeter.

(8) Install light traps in all entrance areas, including dock areas, and more traps at 30 to 50 foot intervals throughout the entire facility.

Sounds pretty good, but this isn't exactly the sequence of events. The plant in question installed the program "after" receiving a plant shut down for 6 weeks and destroying the contaminated finished product, plus making front page news in several major newspapers. The total cost is unknown but an educated guess placed the cost to be in excess of $5 million.

LIGHTING AS IT AFFECTS INSECTS

The company required a complete new facility to produce a new detergent. The committee for new plant site locations selected a nice level tract about 5 miles from a small town in north central Ohio. The farmer normally raised corn, beans, and alfalfa crops and since he still owned several more acres joining this tract, he would continue.

The new 100,000 square foot plant, upon completion, would operate 3 shifts in order to meet sales demands for the new product. All went well during the winter start-up period, and until early June when the crops began to take shape. Almost overnight, at least 50% of the finished product was contaminated with insects.

The time of arrival at the plant site is late in the afternoon, but it is obvious we will not have to wait until sundown to observe the problem! The principal reason for the insect problem is a poor choice of the kind of lighting. Mercury vapor contains an abundance of ultraviolet energy which can be considered the insects' daylight. Mercury vapor to bugs is like a "whistle" to a hound dog at supper-time. He wants it bad!

The parking lot, the dock areas, especially the plant interior, were super bright in ultra-violet energy. Even the make-up air intake on the roof was lit up perfectly for the corn borer, alfalfa insects (several species), midges, mosquitos, and several 100 other kinds of field insects. The design of industrial lighting with insects in mind is a relatively new concept. High pressure sodium vapor light produces an insignificant amount of ultraviolet energy and with twice the lumen output per watt, you can use 50% less lamp wattage and still maintain the same level of effective light.

All lighting at this plant should be high pressure sodium vapor, except for one thought: if mercury vapor is effective to attract and concentrate night-flying insects, then leave the existing fixtures (1000 watts each) on the perimeter (250 feet) to intercept and concentrate the incoming insects at a safe distance away from the plant. It works because all of the insects commonly found in this part of the country can not see more than a 100 feet; and, the greatest percentage can not see more than 25 feet. With the lighting changes and a few more physical methods such as light traps, tighter security of open doors at night, filtering make-up air, and keeping a better control on weed height near the plant site...the plant has been operating without insects ever since. Of course the first mistake was made by the plant site acquisition committee when they passed up good industrial sites on their way to see the farmer about lower cost land!

AFTER 100 YEARS...INSECT PROBLEM!

A large (1 million square foot) plant in the midwest has been operating successfully, without any significant problems with flying insects, for about 100 years. But, when sundown comes in the spring of the year and all of the insects off the nearby river suddenly change their habits and want in your plant by the millions, you're apt to holler "calf-rope!"

On-site, the quality control manager is about ready for retirement, but his insect problem has to be solved first and foremost, like right now! He is quite appreciative of corporate engineering providing the new 1/4 million dollar lighting installation. For all these past years, he's had to make out with the original incandescent lights. "It sure is nice," he says, "to see what you're doing for a change." The only thing wrong is that the new lighting is mercury vapor and the ultra-violet radiant energy from the new lamps were concentrating large volumes of insects at every plant entry point, as well as to air intakes.

Corporate engineering was pround of their new 1/4 million dollar capital expenditure. As a result, it took almost a month for engineers to agree to setting up a test to compare mercury vapor with other kinds of light. The test site and procedures were designed and the comparison between mercury vapor and high pressure sodium took place. Results! For every insect concentrated at the sodium vapor lamp...there were 112 insects at the mercury vapor.

Sodium vapor lighting contains a very small amount of the ultraviolet energy which insects find attractive. Mercury vapor lighting, on the other hand, contains a great amount of it. By using sodium vapor lighting around buildings and placing large, mercury vapor lamps 150 to 200 feet away from buildings, a potential pest problem can be diverted away from high-traffic areas.

In addition, be sure not to locate any lighting directly over entrances. Instead, place sodium vapor lights away from entrances and direct the light back towards the building you're treating. This simple procedure will dramatically enhance your ability to prevent a flying insect problem from developing on the premises.

LIGHT TRAPS CEASE TO FUNCTION

Upper New Jersey terrain can hatch some monster insect volumes to plague food industries. This 125,000 square foot plant had made a rather successful effort to control insect presence inside the production areas. Preventive measures, including proper lighting and a liberal installation of light traps, worked very satisfactorily for them in the start-up year. But, by the middle of the second year, things were not working too well. The thought occurred to management that more traps were needed in several areas.

New blacklight lamp replacements had been received in April and it had been assumed that all light traps had received the required annual lamp replacement as scheduled. The quality assurance director had called the trap manufacturer to send someone as soon as possible for re-evaluation and placement of additional traps. The lamp replacement question came up in the brief discussion about more traps being required. Quality control stated the lamps had been changed as scheduled wherein the question was again restated: "Are you sure?"

The answer, of course, the maintenance chief replaced only one lamp because only one lamp was not operating. Replacement of the lamps brought immediate relief of the fly population build-up.

Blacklight lamps have a maximum useful life of 7000 burning hours as an insect attractant. If light traps are operated as designed - 24 hours per day, year 'round - then 10 months is about all you can expect the lamp to provide adequate energy levels for effective insect response. This is particularly true for the housefly response.

FLY PROPAGATION ON PREMISES

The panhandle section of Texas is desolate to the house fly and stable fly. Nothing to eat, no water to quench a thirsty habit, no place to lay eggs for the reproduction. You wouldn't expect to find a fly problem in such a location. Especially, enough to cause USDA meat inspectors to shut down a beef cattle process area 3 or 4 times per week. What is the cost of an hour shut-down? How much does it cost to lose the production of 400 beef cattle per hour...plus over 800 employees standing still? This doesn't include the overhead of a 1 million square foot facility!

Walk down the railroad spur with me. There's an engine coming along about now, and we step back to let the big dude have the right-of-way. As the engine goes by us, the weight presses the rails down, and out spurt fly maggots (larvae) by the 100's. Well, here is the source for a monster fly problem in the middle of the panhandle area of Texas.

We go into a conference room for a briefing of our visit, after making a complete tour of the plant, inside and outside. What is being done about the rail spur, we query! "Oh, we're on top of that situation. We lime the rails daily!", the general manager replies.

We advise lime is not a deterrent to fly larvae and everyone looks at each other in profound amazement. Now, we're not talking about uneducated folks. The size of this processing plant merits some of the best management in the business. The entire area is concrete - including between the rails. The sanitation above ground is near perfect at all times. Quality assurance people, staff engineers, maintenance people, on-staff pest control director...no one knew the habits of the fly. Not sure they are supposed to know...until today.

Solution: Saturate - not just surface spray - beneath the rail spur on a weekly basis, for about 3 weeks, and monthly, thereafter. Use a residual chemical approved for outside application and...alternate the chemical formulations otherwise the target insect will likely become resistant.

HIDE BEETLE HABITS

A shipment of powdered milk in a freight car was infested with hide beetles when it reached its destination. When the shipment was loaded it was insect-free. Where did the pest come from? The manufacturer was not familiar with the habits of the hide beetle. The larva of this species gnaws into wood or similar items and builds a shelter before becoming an adult. The wooden pallets under the powdered milk bags were infested and when the adults emerged, the eggs were laid on the products.

RAILCAR INFESTATION

Facts

In early summer when temperatures as high as 90-95°F were just beginning to be recorded in the southern states, two shipments of food products from a southern plant were found to be infested (sawtoothed grain and confused flour beetles) when they arrived at their destination (distribution warehouse in the south). There had been no sign of infestation in railcar or product when the cars were loaded. A check of the plant's finished product warehouse, where some of these same products were still in storage, turned up no problems. However, at just about the same time, plant inspectors found a grain beetle infestation in a third railcar being readied for loading.

Explanation

The previous fall the railroad put new floors in six food-grade boxcars. The new floors had been constructed directly over the old floors without cleaning out food accumulations that had become imbedded in the cracks. These cars were then, more or less, consigned to hauling finished product from this particular food plant to various distribution warehouses. Throughout the winter and spring months (cold-cool temperatures) the cars were used without problems. However, with the return of warm weather, eggs hatched and surviving adults became more active and soon found passageways through the new floors and infested the inside of the cars.

Moral

1) Potential insect harborages must be eliminated; covering them up may only lead to bigger problems later on.
2) While cold weather may kill some insects, survivors in all stages will become more active when favorable conditions return.
3) Railcars and trailers with "covered over" damage are prime insect-problem suspects.

DON'T TRUST ANYBODY

Facts

Grain beetles and flour beetles were seen crawling on individual units as a rail car of a food ingredient was unloaded. Most surprising in light of the excellent condition of the recently rebuilt railcar!

Explanation

A new floor had been installed in the car _over_ the old dirty and cracked floor. Flour residues under the new floor provided a perfect breeding ground for the insects.

Moral

Don't trust anybody! Examine all receipts. Be specific on any rebuilding of conveyances that you contract for.

STILL ANOTHER RAILCAR STORY

This is not an insect success story but it has a good message.

A corrugated tote (such as are used to distribute nuts and other raw materials) dump operator noted some insects on top of the outside of the plastic bag containing a raw material. The insects were identified as the dreaded warehouse beetle. As is customary with the involved company, a search was immediately undertaken to locate the source. It was never found and the problem never arose again; but, there was an interesting finding, hence this reporting.

Flutes of corrugated are a favorite hiding place of insects. Therefore, the totes were suspect and a trip was made to the supplier. No warehouse beetles were found. However, a sealed car containing brand new totes was opened, stock removed, the clean corrugated on the floor folded back, and, to mixed surprise and chagrin, it was noted that the floor had a generous coating of corn starch (see picture on next page). Could the railcar have survived a trip of at least several days in warm weather without having become infested? And would the totes not have become infested? This is doubtful.

This illustration is an example of the need for proper cleaning and inspecting of conveyances or contaminated products or materials may be brought into a plant or warehouse.

INSPECT CONVEYANCES, EVEN TRUCKS

Facts

A truckload of an insect susceptible packaged food product was half unloaded when a sanitation inspector examined the truck. Broken lining along one side was pulled out and the pelletized material found there was stirred with a spatula. It immediately became "alive" with a heavy infestation of cigarette beetles. Close examination of floor turned up more beetles in cracks. The product was examined and beetles were found in the outer case corrugation and between the individual packages and the outer case.

Explanation

The pelletized material was a high-protein fish food which had been shipped in bulk as a previous load. The shipping plant had a policy of inspecting conveyances before loading but had failed to do so in this case.

Action

The material was reloaded and the entire truck contents fumigated. Fortunately, no insect penetration of individual retail units had occurred.

IMPROPER PLACEMENT OF WARNING PLACARDS FOR PHOSPHINE

Facts

1) Hopper cars of flour were routinely fumigated with prepackaged pellets of aluminum phosphide before leaving the mill.
2) In inclement weather, difficulty was experienced in getting masking tape used to hold the warning placards on the exterior of the hatch covers to stick to the wet metal.
3) Workers knew that placards were mandatory and decided to place one end of the placard under the rim of the hatch before it was latched. In this way, the placard could not blow off.

4) When a car on which the placards had been so clamped arrived at destination, the placards were wet and some of the prepackaged pellet strips were found to be charred when the hatch covers were opened.

Explanation
1) The placards had acted as a wick during a rainstorm and water contacted the prepackaged pellets.
2) The water caused a rapid breakdown of the aluminum phosphide, the flash point of phosphine was reached, and a small flame resulted which charred the moisture-permeable paper on the package of fumigant.

Moral
When fumigating hopper cars with aluminum phosphide do not clamp the warning placards under the rim of the hatch covers. One never knows when inclement weather will be experienced while the railcar is in transit.

INFESTED RODENTICIDE

Pictured below is a comparatively unusual example of infested rodent cake found in a packaging supplier's plant. Loose rodent bait and even packaged bait are notorious for attracting insects and becoming infested. In this example, the drug store beetle was the culprit. The moral is--monitor your rodent bait for possible evidence of insects as well as for a possible source.

AMERICAN COCKROACH PROBLEM IN A COLD PROCESSING ROOM

Facts
A food product manufactured at a plant required a cold processing step that was carried out in a room kept at about 50°F. One would not expect to have insect problems in this environment and it is true no such problems were ever evident during operations. However, at each shutdown period, when this room was allowed to warm up, American cockroaches appeared.

Explanation

The plant laboratory was located right next to the cold processing room and, in fact, the two shared a common wall. In order to keep the lab side of the wall from "sweating," the cold room side of this wall was insulated. In the past, the lab had seen an occasional American cockroach but the problem appeared to be solved through regular clean-ups and residual insecticide treatments. However, the cockroaches had gotten into the space between the lab-cold room wall and were able to breed in this space protected from the cold. From here it was a rather easy matter for the insects to find their way to the food attractant in the processing room whenever the environment became more compatible.

Moral

While the insects may not be able to live in the immediate vicinity of a food attractant, should they find more compatible conditions on the fringe areas, they will become established and continually attracted to the food whenever the environment is favorable.

THE ABSENCE OF FOOD DOES NOT MEAN NO INSECT PROBLEMS

Do plants, or areas of plants, that do not use or handle foods or food materials intended for human consumption ever have problems with insects? The answer is a resounding yes! There are probably two main reasons; namely, (1) insects are ubiquitous and numerous, and (2) their behaviors differ from humans' including what they use for food.

The essence of this unusual problem was a non-food plant with extensive roof leaks over an area of tens of thousands of square feet. The cause was holes made by insects. Roof construction (felt/asphalt) contained no materials to serve as food to the identified ground and scarab beetle species. What is the probable explanation?

1) The beetles were probably living in a pile of wood chips for a waste fuel boiler. Wood chips can serve as food for ground beetles and the decayed wood present as food for scarab beetles.

2) The beetles were attracted to the mercury vapor lights mounted on the roof and illuminated for safety/security reasons.

3) Beetles landed on the roof at night and with the arrival of daytime and its heat (this plant is located in the southern United States) the beetles would burrow into the roof to avoid desiccation (see picture on next page).

A possible solution to such a problem is to use low-pressure sodium vapor lights.

The message is always to be alert to possible insect problems even in areas which contain no food, for humans or insects.

MISCELLANEOUS INPUTS

1. A shipment of paper products packed in 0.5 mil polyethylene bags was contaminated with confused flour beetles. Shipment was in corrugated boxes via trailer from the Caribbean to the U.S. At least one bag was penetrated (entry hole and beetle inside observed). Corrugated could be reused, an average of 2-1/2 trips. Trailers were routinely inspected and fumigated. Problem disappeared prior to location of source.

2. Packaging is frequently a source of insect contamination:
   - grain beetles in flutes
   - cockroaches in fiber drums
   - sawtoothed grain beetles, ground beetles in polyethylene caps
   - crickets, cockroaches, scavenger beetles, carpenter ants, mites, psocids in glass containers.

3. Psocids are considered a lower consumer and regulatory risk because of ubiquitous nature and small size. Eradication can be most expensive.

4. Corrugated in addition to offering hiding places in the flutes offers food in the starch used to seal the flutes to the flat sheets. At this writing starch is always used. The adhesives and glues used for the flaps are not attractive to insects.

5. Do you have a dropped ceiling in areas where insect infestation is a concern? Be aware of it as a potential problem area. Fogging and/or ULV applications have a tendency to drive insects up and such a ceiling can provide an insect haven. Accumulated "edible dust" can and does provide sufficient food as well as a breeding environment. Therefore, be sure to remove sufficient ceiling tiles to permit the insecticide to reach the space above the ceiling. If the ceiling is not a "push-out" but a dead space does exist above it, consider an access panel unless you know it to be securely sealed.
   Similar problems can develop in hollow tile walls; in, on, and behind switch boxes; behind door and window framing, baseboards, moulding, etc. If such areas are subject to effective sealing, well and good, but if not take advantage of the space and treat with approved pesticides in an approved manner.

6. About 25 years ago the western United States were actively fighting the quarantined insect called the khapra beetle. Positive identification is very important in initiating an eradication program and in preventing the spread of

this beetle which looks similar to other related species in the adult and in the larval stages. After many years of toil this insect was eliminated at a cost of over $10,000,000, not including the man-hours spent by the governmental officials. About 800 food plants and warehouses were shut down and fumigated.

Recently (1981) the khapra beetle reared its ugly head again in the eastern United States. There was at least one misidentification of this insect and the building was unnecessarily fumigated.

In possible instances of infestations by the khapra beetle, only a specialized expert should identify this insect.

Dr. Dennis S. Hill
'Haydn House'
20 Saxby Avenue
Skegness
Lincs.